Low Temperature Biology of Insects

Low temperature is a major environmental constraint impacting the geographic distribution and seasonal activity patterns of insects. Written for students and academic researchers in environmental physiology and entomology, this book explores the physiological and molecular mechanisms that enable insects to cope with a cold environment and places these findings into an evolutionary and ecological context.

An introductory chapter provides a primer on insect cold-tolerance and subsequent chapters in the first section discuss the organismal, cellular and molecular responses that allow insects to survive in the cold, despite their, at best, limited ability to regulate their own body temperature. The second section, highlighting the evolutionary and macrophysiological responses to low temperature, is especially relevant for understanding the impact of global climate change on insect systems. A final section translates the knowledge gained from the rest of the book into practical applications, including cryopreservation and the augmentation of pest management strategies.

DAVID L. DENLINGER is Distinguished University Professor at Ohio State University. He is a recipient of the Recognition Award in Insect Physiology, Biochemistry and Toxicology from the Entomological Society of America and a member of the National Academy of Sciences. He is a Fellow of AAAS, the Entomological Society of America, and the Royal Entomological Society.

RICHARD E. LEE, JR. is Distinguished Professor of Zoology at Miami University, Oxford, Ohio. His honors include several teaching awards, the Benjamin Harrison Medallion from Miami University, and election as a Fellow of AAAS, the Entomological Society of America, and the Royal Entomological Society.

Low Temperature Biology of Insects

Edited by

David L. Denlinger

Department of Entomology
Ohio State University
Columbus, Ohio

Richard E. Lee, Jr.

Department of Zoology
Miami University
Oxford, Ohio

Cambridge
University Press

CAMBRIDGE UNIVERSITY PRESS
Cambridge, New York, Melbourne, Madrid, Cape Town, Singapore,
São Paulo, Delhi, Dubai, Tokyo

Cambridge University Press
The Edinburgh Building, Cambridge CB2 8RU, UK

Published in the United States of America by Cambridge University Press, New York

www.cambridge.org
Information on this title: www.cambridge.org/9780521886352

First published 2010

Printed in the United Kingdom at the University Press, Cambridge

A catalog record for this publication is available from the British Library

ISBN 978-0-521-88635-2 Hardback

Contents

Contributors

Jeffrey S. Bale
School of Biosciences, University of Birmingham, Birmingham, UK

Brian M. Barnes
Institute of Arctic Biology, University of Alaska, Fairbanks, USA

Mark Bayley
Department of Biological Sciences, Zoophysiology, Aarhus University, Aarhus, Denmark

William E. Bradshaw
Center for Ecology and Evolutionary Biology, University of Oregon, Eugene, Oregon, USA

Martin A. Carrasco
Department of Biological Sciences, University of Notre Dame, Notre Dame, Indiana, USA

Steven L. Chown
Centre for Invasion Biology, Department of Botany and Zoology, Stellenbosch University, Matieland, South Africa

Peter Convey
British Antarctic Survey, Cambridge, UK

David L. Denlinger
Department of Entomology, Ohio State University, Columbus, Ohio, USA

John G. Duman
Department of Biological Sciences, University of Notre Dame, Notre Dame, Indiana, USA

Martin Holmstrup
National Environmental Research Institute, Department of Terrestrial Ecology, Aarhus University, Silkeborg, Denmark

Christina M. Holzapfel
Center for Ecology and Evolutionary Biology, University of Oregon, Eugene, Oregon, USA

Raymond B. Huey
Department of Biology, University of Washington, Seattle, Washington, USA

Vladimír Košťál
Biology Centre of the Academy of Sciences of the Czech Republic, Institute of Entomology, Ceske Budejovice, Czech Republic

Richard E. Lee, Jr.
Department of Zoology, Miami University, Oxford, Ohio, USA

Roger A. Leopold
USDA Agricultural Research Service, Biosciences Research Laboratory, Insect Genetics and Biochemistry Research Unit, Fargo, North Dakota, USA

Xia Lin
Department of Biological Sciences, University of Notre Dame, Notre Dame, Indiana, USA

Volker Loeschcke
Department of Biological Sciences, Ecology and Genetics, Aarhus University, Aarhus, Denmark

M. Robert Michaud
Department of Entomology, Ohio State University, Columbus, Ohio, USA

Philip K. Nickell
Department of Biological Sciences, University of Notre Dame, Notre Dame, Indiana, USA

Johannes Overgaard
Department of Biological Sciences, Zoophysiology, Aarhus University, Aarhus, Denmark

Sindre A. Pedersen
Department of Biology, Norwegian University of Science and Technology, Trondheim, Norway

Joseph P. Rinehart
USDA Agricultural Research Service, Biosciences Research Laboratory, Insect Genetics and Biochemistry Research Unit, Fargo, North Dakota, USA

Todd Sformo
Institute of Arctic Biology, University of Alaska, Fairbanks, Alaska, USA

Brent J. Sinclair
Department of Biology, University of Western Ontario, London, Ontario, Canada

Jesper G. Sørensen
National Environmental Research Institute, Department of Terrestrial Ecology, Aarhus University, Silkeborg, Denmark

Janet M. Storey
Institute of Biochemistry, Department of Biology, Carleton University, Ottawa, Ontario, Canada

Kenneth B. Storey
Institute of Biochemistry, Department of Biology, Carleton University, Ottawa, Ontario, Canada

Kent R. Walters
Department of Biological Sciences, University of Notre Dame, Notre Dame, Indiana, USA

Karl Erik Zachariassen
Department of Biology, Norwegian University of Science and Technology, Trondheim, Norway

Preface

Tiny, wingless snow flea freezes
overnight as stiff as tin.
Does he mind or even notice
just what sort of state he's in?
Things warm up and – leaping lizards! –
Snow flea's jumping as before,
none the worse for being frozen
on the icy Arctic shore.
(Eileen Spinelli, 2007 *Snow Flea*)

Although the tilt of the earth's axis is only a little over 23°, this is enough to provide the dramatic contrast between the heat of summer and the cold of winter as our planet rotates around the sun. These pronounced seasonal changes in temperature dictate the seasonal patterns that dominate the temporal patterns of life on earth. On a much shorter timescale, the daily rotation of the earth on its axis, yielding the day/night cycle, adds the important dimension of thermoperiodism that generates daily variation in the number of heat units that reach the earth's surface.

Certainly all animals are influenced by daily and seasonal cycles of temperature, but it is perhaps ectotherms, such as insects, that are most severely affected. With, at best, limited abilities to regulate their own body temperature, insects are at the mercy of the prevailing temperatures. Insects commonly use seasonal shortening in day length (photoperiod) as an environmental token to reliably signal the advent of inimical temperatures. For many insects these photoperiodic cues are used to program diapause, a period of developmental arrest used to wait out the winter months. And others, such as the Monarch butterfly, couple long-distance migration with entry into a reproductive diapause at a protected hibernaculum. Though entry into a dormant state is a key component for winter survival, dormancy by itself does not assure low-temperature survival. To this must be added a suite of physiological adaptations that enable the insect to survive at temperatures that are often well below zero. Adaptations for low-temperature survival are not restricted to overwintering species. Such adaptations are common, even among species that lack the capacity for diapause, as well as in non-diapausing stages of species that possess the capacity for diapause. Nearly all species are also confronted with daily changes in temperature that demand finely tuned physiological adaptations that enable the insect to function effectively over a wide temperature range. Fluctuations in temperature are particularly important during warm seasons when insects are especially active and thus need to make fine adjustments in their physiology to respond to daily temperature swings that frequently exceed 10–20 °C. And, even in tropical regions of the world, temperatures during the night can drop precipitously.

It is the goal of this book to outline the mechanisms used by insects to survive and respond to low temperature, then to reflect upon how these adaptations are expressed in a broader ecological and evolutionary context, particularly with respect to global climate change, and finally to examine practical applications that can be derived from low-temperature studies of insects.

In many ways this volume has its origin in our earlier book, *Insects at Low Temperature*, published in 1991. But, it's not exactly a revision, because the earlier book covered a number of topics not discussed here. In this book we focus on topics that have emerged or significantly grown since 1991. In many respects the field has blossomed, matured and diversified in the nearly two decades that have elapsed since the original book was published. The tools of molecular biology enormously enhance our understanding of mechanisms of cold-hardiness, and such updates are emphasized herein. We have not included chapters here that were well covered in the 1991 book and remain timely, thus readers will still find the earlier book to be a good source for literature on water relationships, the biochemistry of cryoprotectants, the effects of cold on morphogenesis, the relationship between diapause and cold-hardiness, thermoperiodism and numerous accounts of species' (e.g. fresh-water invertebrates, honey bees, silk moths) adaptations to low temperature.

Readers are introduced to the fundamentals of low-temperature insect biology in Chapter 1 (Lee) and the remainder of the first section of the book focuses on physiological and molecular responses to low temperature, including a discussion of rapid cold-hardening (Lee and Denlinger), antifreeze proteins and ice nucleators (Duman *et al.*), lessons from genomics, proteomics and metabolomics (Michaud and Denlinger), cell structural modifications (Kostal), oxidative stress (Storey and Storey) and cross-tolerance between cold, desiccation and environmental toxins (Holmstrup *et al.*). The second section of the book examines ecological and evolutionary responses to low temperature and includes chapters on a macroscale perspective of cold-hardiness (Chown and Sinclair), fitness consequences of low temperature (Huey), consequences of global climate change (Bradshaw and Holzapfel), genetic variability in the evolution of cold-tolerance (Overgaard *et al.*), and life-history adaptations to polar and alpine environments (Convey). A brief final section captures two important practical aspects of low-temperature biology. The one chapter documents advances in insect cryopreservation (Leopold and Rinehart), and the final chapter discusses the implications of cold-tolerance for insect pest management (Bale).

As editors of this volume, we are especially grateful for the participation of the contributing authors, who enthusiastically joined in this effort. Thanks also to Chris Hudson, Dominic Lewis and Caroline Brown from Cambridge University Press for their assistance with this project. We gratefully acknowledge the National Science Foundation, the National Institutes of Health and the United States Department of Agriculture for their support of research in our laboratories.

David L. Denlinger
Columbus, Ohio

Richard E. Lee, Jr.
Oxford, Ohio
October 2009

PART I PHYSIOLOGICAL AND MOLECULAR RESPONSES

Low Temperature Biology of Insects, ed. David L. Denlinger and Richard E. Lee Jr. Published by Cambridge University Press.

1

A primer on insect cold-tolerance

RICHARD E. LEE, JR.

1.1 Introduction

Low temperature affects insects differently based on the severity of the cold and the duration of exposure. Life stage and acclimation state also have a major impact on an insect's response to low temperature. Many temperate and polar insects enhance their cold-tolerance seasonally in preparation for winter, as short, cool days in autumn trigger cold acclimatization, as well as entry into the metabolic depression of diapause. However, insects also have the capacity to make significant and rapid adjustments to even slight changes in environmental temperature, as would occur on a summer's day.

This introductory chapter seeks to provide a short primer on the physiology of insect cold-tolerance that will be useful to students and others new to the area of study. This overview of basic concepts in insect cold-tolerance intends to provide a context for later chapters providing in-depth reviews of specific areas. Specifically, this primer focuses on regulation of supercooling and ice nucleation, and basic adaptations promoting cold-tolerance. Suggestions for conducting and clearly reporting experimental results on insect cold-tolerance are also included. Since this volume is intended to update and complement our previous book, *Insects at Low Temperature* (Lee and Denlinger, 1991), this synoptic chapter will emphasize articles published during the past 20 years and topics not covered elsewhere in this volume.

1.2 Types of insect cold-tolerance

Chilling and cold are relative terms; consequently, the temperature ranges they represent vary depending on the species in question. For a tropical species,

Low Temperature Biology of Insects, ed. David L. Denlinger and Richard E. Lee Jr. Published by Cambridge University Press.

10 °C may be sufficiently low to cause chill coma (a loss of locomotory capacity due to impaired neuromuscular function), while a polar or alpine species may be able to walk normally at −5 °C or below (e.g. a Himalayan midge can walk at −16 °C (Kohshima, 1984)). Cold-hardening refers to the capacity of an insect to increase its level of cold-tolerance. It may result in a lowering of the chill-coma temperature, an increase in the ability to survive extreme cold, or the acquisition of the capacity to survive internal ice formation.

Numerous systems for categorizing types of insect cold-tolerance have been proposed, which have generated spirited discussion at numerous conferences and in the literature (Ramlov, 1998; Sinclair, 1999; Nedved, 2000; Chown and Terblanche, 2007; Hawes and Bale, 2007; Chown *et al.*, 2008; see Chown and Sinclair, Chapter 8). Though particular categories and names vary, it is generally agreed that there are three basic types of cold injury corresponding to the primary types of insect cold-tolerance (Fig. 1.1).

Chilling intolerant refers to species that succumb to the direct effects of low temperature without internal ice formation. Depending on the species, such chilling injury may occur at temperatures above or below 0 °C. This category may be usefully subdivided into direct and indirect chilling injury. Direct chilling injury, or cold-shock injury, results from brief exposures to cold (on the order of minutes to hours) that damage cell membranes by causing a phase transition from the liquid-crystalline to the gel state, and the lateral separation of membrane proteins (Levitt, 1980; Larcher, 2001). Cold-shock injury is a major problem for the cryopreservation of various mammalian cells and tissues, particularly spermatozoa, and recent evidence indicates similar injury occurs in insects, as Lacoume *et al.* (2007) reported that cold-shocked (1 h at −18 °C) males of the parasitoid wasp, *Dinarmus basalis,* had markedly reduced sperm stores and fertilized fewer females than control wasps. (See Lee and Denlinger, Chapter 2, for a review of cold-shock injury and the rapid cold-hardening response that protects against it.)

Indirect chilling injury occurs over extended periods of days to weeks and is found in a wide range of plants and animals. This type of injury occurs at higher temperatures, from a few degrees below zero to 10–15 °C. Indirect chilling injury is important commercially because it determines the shelf life of fruits and vegetables, particularly those of sub-tropical and tropical origin. It is a major problem for storage of chilled human tissues and organs for transplantation (Taylor *et al.*, 2007). Understanding long-term chilling injury is also important for predicting the establishment of non-native pests, storing biological control agents for mass release, and the use of cold for quarantine treatment of imported produce and other commodities (Hallman and Denlinger, 1998; Bale, Chapter 14).

Like direct chilling injury, indirect chilling injury appears to be caused, at least in part, by thermotropic damage to cell membranes causing metabolic imbalance

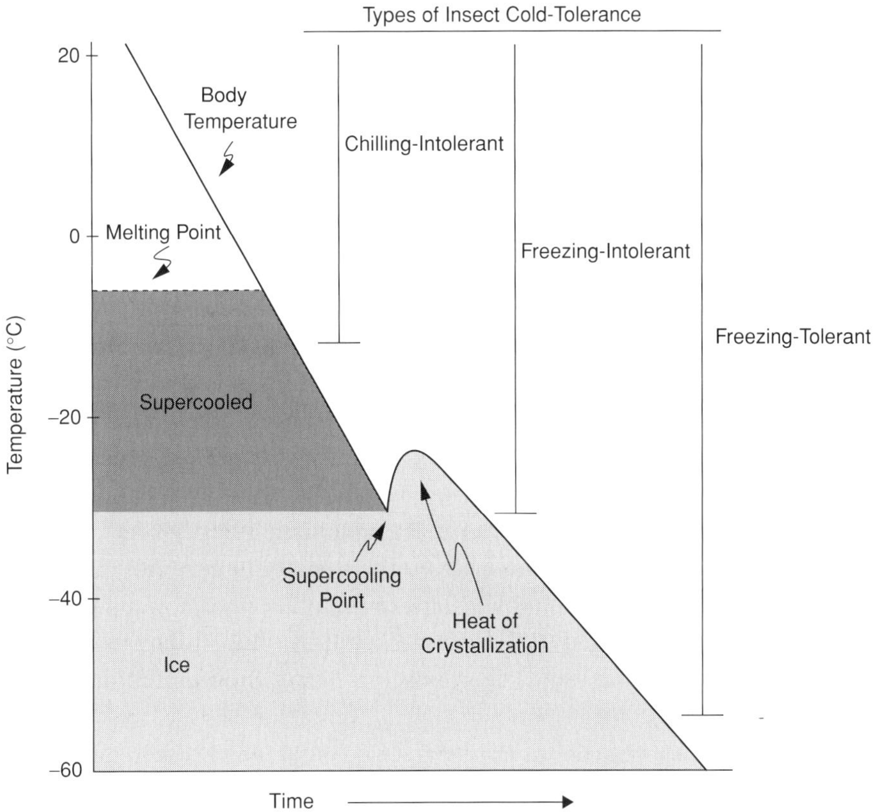

Figure 1.1 Insect responses to low temperature. Body temperature (bold line) in relation to hemolymph melting point, supercooling point (or temperature of crystallization), and release of the heat of crystallization as body water freezes. Bars on the right indicate representative ranges for different types of insect cold-tolerance. (Adapted from Lee, 1989.)

and loss of selective membrane permeability. The magnitude of chilling injury is closely associated with a loss of ion homeostasis, particularly increases in potassium ions and decreases in sodium and magnesium ions in the hemolymph, and loss of electrochemical potentials across cell membranes (Kostal *et al.*, 2004; Kostal *et al.*, 2006; Kostal *et al.*, 2007). Cryoinjury to the cell membrane is consistent with the observed decreases in membrane potential of nerves and muscles, and impaired coordination within the neuromuscular system (Denlinger and Lee, 1998). Chilling injury may also result from oxidative stress during cold storage (Rojas and Leopold, 1996; Storey and Storey, Chapter 6).

Interestingly, mortality caused by indirect chilling may be prevented, or damage caused reversed or repaired, by a brief warming episode in the midst of weeks of cold exposure (Chen and Denlinger, 1992) or by brief periods of daily warming

(Turnock and Bodnaryk, 1993; Renault *et al.*, 2004; Colinet *et al.*, 2006). During these brief intervals of warming some insects can re-establish ion gradients (Kostal *et al.*, 2007) or up-regulate proteins involved in energy metabolism, protein chaperones (Hsp70/Hsp90), or cytoskeletal components (Colinet *et al.*, 2007; Michaud and Denlinger, Chapter 4). Other reports suggest these warming periods may allow for recharging depleted energy reserves (Chen and Denlinger, 1992) or for removal of accumulated toxic metabolites (Leopold *et al.*, 1998; Colinet *et al.*, 2007).

Freezing intolerant or *avoidant* species can survive cold as long as ice does not form within their bodies. These insects are not injured by cold-shock and can tolerate chilling to sub-zero temperatures. By far, most insects exposed to sub-zero cold in nature are freezing intolerant. Although *freezing susceptible* is sometimes used to describe this type of cold-tolerance, this name is somewhat ambiguous because it can be interpreted to mean the organism has the potential to freeze internally (which, of course, all organisms do), instead of indicating that internal ice formation is lethal to it. In winter, freezing-intolerant species rely on mechanisms by which they seasonally increase their cold-tolerance and capacity to remain unfrozen by supercooling – in some cases, to extremely low temperatures of −40 °C or even below −60 °C (Ring and Tesar, 1981). The capacity to supercool increases as body size and water content decrease, and with the accumulation of low-molecular-mass solutes.

Freezing-tolerant species can survive the freezing of their body water. Insects that survive freezing typically do so only over a specific temperature range from the initiation of ice formation to some lower temperature (Fig. 1.1). As ice forms in the extracellular space, only water molecules join the growing ice lattice and dissolved solutes are excluded. Consequently, as freezing continues, solutes become increasingly concentrated (termed freeze concentration) in the hemolymph, causing an osmotic outflow of water from surrounding cells. As cells are dehydrated, intracellular solutes are freeze concentrated until osmotic equilibrium is re-established.

1.2.1 Cryoinjury due to freeze concentration

There is general agreement that the freeze concentration of solutes results in freezing injury; however the actual mechanism remains unclear (Mazur, 2004; Muldrew *et al.*, 2004). One hypothesis, sometimes referred to as solution-effects injury, posits that extracellular freeze concentration causes an excessive concentration of extracellular and intracellular electrolytes that damages the cell membrane and leads to cytolysis. Alternative explanations focus on the specific effects of cellular dehydration. Meryman (1968) suggested that excessive osmotic shrinkage may exceed a minimum critical cell volume, from which the cells cannot recover. Membrane destabilization or loss of membrane materials during dehydration may also occur (Steponkus and Lynch, 1989). These explanations for freezing injury are primarily based on studies with red blood cells, plant protoplasts and other isolated cell suspensions from humans or other mammalian models that do not naturally experience severe cold. Since few studies have been done with freeze-tolerant insects or other ectotherms, some caution should be used in extending explanations to naturally cold-hardy animals.

1.3 Mechanical injury

The growing ice lattice can also damage tissues and organs as ice forms between adjacent cell layers. As the ice mass grows, it pulls tissue layers apart, causing mechanical injury that remains after thawing. Mechanical injury may also result from recrystallization within frozen tissues. In this scenario, small ice crystals shrink as they lose water molecules to growing crystals. Recrystallization occurs especially rapidly at high sub-zero temperatures – the range in which many freeze-tolerant ectotherms winter. Ice-binding proteins can inhibit recrystallization and, thereby, prevent its damaging effects within body tissues (Duman *et al.*, Chapter 3).

1.4 Supercooling

During cooling, the body temperature of most insects closely tracks that of their immediate environment due to their small size and limited ability to generate heat (Fig. 1.1). A few notable exceptions are some winter-active moths and a few other endothermic insects that can generate sufficient heat to remain active, and can even fly at low winter temperatures (Heinrich, 1993). As insects are cooled to temperatures below 0 °C, ice does not form immediately in their body tissues for two reasons. First, ions, sugars, amino acids, proteins and other dissolved solutes in an insect's hemolymph colligatively depress the melting point (MP) by 1.86 °C per osmole of solute. For example, if an overwintering insect had a hemolymph osmolality of 1.5 osmoles, its MP would be −2.79 °C and no ice could form until the temperature dropped below this value.

Second, once cooled to temperatures below its hemolymph MP, an insect usually enters a supercooled state, in which its body water remains unfrozen. The capacity to supercool is inversely related to water volume (Angell, 1982). A few microliters of water readily supercool to −15 to −20 °C, while under special conditions supercooling continues to its limit, near −40 °C. Most insects behave as though they are small vessels of water and, whether collected in summer or winter, supercool at least a few – if not, many – degrees before ice forms spontaneously in their tissues. The temperature at which ice forms is termed the supercooling point (SCP) because it denotes the limit of supercooling (Fig. 1.1). The SCP is readily determined using thermocouples to detect the heat of crystallization, which is released as body water freezes. For this reason, the SCP is sometimes referred to as the temperature of crystallization (T_C).

Supercooling capacity is the difference between the hemolymph MP and the SCP. So an insect with an MP = −1.5 °C and an SCP = −22.5 °C would have a supercooling capacity of 21 °C. Note that an insect with a low SCP would have a high supercooling capacity, and vice versa. Freezing-intolerant insects enhance their supercooling capacity in winter, while freezing-tolerant ones generally diminish this capacity. Extensive lists of SCP values for diverse arthropods are available (Sømme, 1982; Lee, 1991; Sinclair, 1999; Turnock and Fields, 2005).

For both freezing-intolerant and freezing-tolerant insects, the SCP represents a major transition point. For freeze-intolerant species it represents the lower limit of their potential survival, while for species that tolerate freezing it marks the beginning of a radical change in their physiological state, characterized by anoxia, cellular desiccation, build-up of metabolic wastes and depressed metabolic activity (Storey and Storey, 1988). However, even in the frozen state some complex physiological processes continue, including cryoprotectant synthesis (Storey *et al.*, 1981; Walters *et al.*, 2009) and diapause development (Irwin *et al.*, 2001). When assigning an insect to one of these categories, it is critical to keep in mind the conditions and temperatures the insect would experience in nature.

Using ecologically relevant cooling (often $< 0.5\ ^\circ\text{C min}^{-1}$) rates and appropriate durations of exposure are especially important in laboratory experiments (Lee, 1991). For example, if a wintering insect is naturally exposed to severe cold for weeks at a time, then laboratory exposures lasting only minutes to hours may reveal little about tolerance in the field. Freeze-tolerance should not be determined by thawing insects shortly after the SCP is reached and then assessing survival. In laboratory determinations of the SCP, the freezing exotherm is often very brief – lasting only seconds to minutes – due to relatively high rates of cooling and the rapid removal of the heat of crystallization by the surrounding heat sink. Even though the freezing exotherm has dissipated, an equilibrium amount of ice in the insect has not necessarily formed. In freeze-tolerant larvae of the gall fly (*Eurosta solidaginis*), only 47% of body water froze during the first 6 h of exposure to −23 °C, while at least 48 h were required to reach an equilibrium level of ice formation under these conditions (Lee and Lewis, 1985). Furthermore, slow rates of freezing may be critical to allow time for the mobilization of cryoprotective responses as ice forms within the body (Holmstrup *et al.*, 1999). Consequently, it is often necessary to hold an insect for extended periods at temperatures near the SCP before it can be judged freezing-tolerant.

In describing experimental treatments or reporting results, care should be taken to clearly distinguish between whether an insect was simply exposed to a sub-zero temperature or whether actual freezing of its body water occurred. For example, if it was stated that an insect was exposed to freezing, it is unclear whether the author meant the insect was cooled to a sub-zero temperature or whether ice formed internally. Since many insects readily supercool for extended periods, especially at high sub-zero temperatures, it is important to report whether the insects actually froze internally by determining the SCP. This distinction may also become confusing when an insect has a low hemolymph freezing point (FP); for example, in an insect with an FP of −4 °C, exposure to −3 °C could not induce internal ice formation. Clearly distinguishing whether an insect remains supercooled or freezes is essential for correctly categorizing the type of cold-tolerance.

1.5 Initiation of freezing and ice nucleation

Despite the innate tendency of insects to supercool, ice eventually forms. In biological systems it is generally believed that ice forms by a heterogeneous mechanism in which a non-water substrate provides a template that stabilizes aggregations of water molecules, leading to ice formation at higher temperatures than would occur otherwise (see reviews by Vali 1995; Wilson *et al.*, 2003). However, Zachariassen and colleagues (2004) recently challenged this view in freeze-intolerant insects. They argue that supercooling capacity is closely correlated with an insect's water volume, and the pattern of freezing is more similar to homogeneous mechanisms of nucleation.

The most efficient ice-nucleating agent in an organism will determine the temperature at which ice begins to form. As the ice lattice grows, the heat of crystallization often increases the body temperature by several degrees or more (Fig. 1.1). Due to this increase in body temperature and because freeze concentration of solutes further depresses the hemolymph MP, it is unlikely that other less active, endogenous nucleators will induce freezing at other locations in the body.

Several distinct classes of ice-nucleating agents have been identified in insects: (1) ice-nucleating proteins, (2) crystalloid compounds and (3) ice-nucleating microorganisms (Table 1.1). The ice-nucleating activity of these endogenous nucleators ranges from highly efficient, in the range of −2 to −5 °C, to ones with little activity (<−18 °C).

1.5.1 Ice-nucleating proteins

Zachariassen and Hammel (1976) first reported the presence of an ice-nucleating agent in the hemolymph of freeze-tolerant beetles, which led to the discovery of ice-nucleating active proteins. Since then, ice-nucleating proteins with activity in the range of −6 to −10 °C have been reported from the hemolymph

Table 1.1 *Classes and efficacy of endogenous ice-nucleating agents compared to inoculative freezing by environmental ice*

Type	Ice-Nucleating Activity (°C)	Reference
Ice-nucleating proteins	−6 to −9	(Zachariassen and Hammel, 1976)
Crystalloid compounds	−7 to −10	(Mugnano *et al.*, 1996)
Ice-nucleating bacteria	−2 to −10	(Kaneko *et al.*, 1991a,b; Lee *et al.*, 1991, Lee *et al.*, 1993a)
Ice-nucleating fungi	−5	(Tsumuki *et al.*, 1992)
Inoculation by environmental ice	−1 or <MP of hemolymph, little or no supercooling	(Tursman *et al.*, 1994)

of various freeze-tolerant insects. (See Chapter 3 by Duman *et al.* for additional information on ice-nucleating proteins.)

In freezing-tolerant species these unusual proteins function to insure extracellular ice formation and prevent extensive supercooling and the resulting rapid propagation of ice that would be more likely to cause cryoinjury (Zachariassen, 1985; Zachariassen, 1992; Duman, 2001). By insuring freezing at a high sub-zero temperature, the ice lattice forms more slowly as environmental temperatures decrease, thereby lessening the osmotic stress caused by the freeze concentration of solutes in the extracellular space. Cellular dehydration also increases the capacity of cells to supercool and prevents lethal intracellular freezing. Also, by freezing at a high sub-zero temperature, ice formation and growth occur more slowly allowing insects and other freezing-tolerant animals to make physiological adjustments, such as mobilization of cryoprotectant, as ice is formed in their bodies (Storey and Storey, 1988; Lee and Costanzo, 1998; Holmstrup *et al.*, 1999). Note, however, that ice nucleation at temperatures near 0 °C is not required for freezing-tolerance; some insects supercool extensively and still survive freezing (Ring and Tesar, 1980).

These nucleators may also function to conserve energy and water during winter, since frozen insects have lower metabolic rates (Scholander *et al.*, 1953) and lose less water than ones that remain supercooled (Zachariassen, 1992). Frozen larvae of the gall fly, *E. solidaginis,* have a metabolic rate 47% lower and a water loss rate 35% lower, than supercooled larvae (Fig. 1.2; Irwin and Lee, 2002).

Crystalloid inorganic compounds comprise another category of endogenous ice-nucleating agents. Larvae of *E. solidaginis* often contain more than 30 calcium phosphate spherules in their Malpighian tubules (Mugnano *et al.*, 1996). Many of these spherules exhibit ice-nucleating activity between −8 to −10 °C that closely matches the larval T_c of −9.4 °C. During the larval-to-pupal transition these spherules

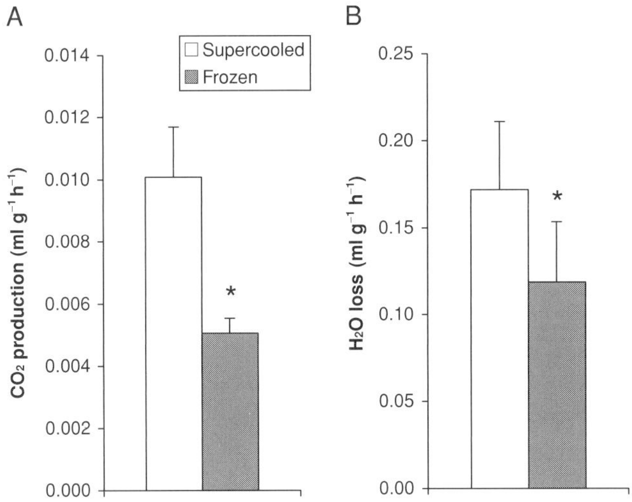

Figure 1.2 Metabolic rate (CO_2 production) (A) and water loss (B) of frozen and supercooled goldenrod gall fly larvae (*Eurosta solidaginis*) at −5 °C. Frozen larvae had significantly lower metabolic rates and water loss. (From Irwin and Lee, 2002)

disappear concomitant with a decrease in T_c to −18.3 °C. Since the larval hemolymph and most other tissues lack efficient nucleators, as evidenced by their capacity to supercool extensively, these spherules appear to determine the larval T_c. Uric acid, calcium carbonate, potassium phosphate and other crystalloids are known from various overwintering insects; commercial preparations of these compounds have ice-nucleating activity (Mugnano *et al.*, 1996). Consequently, it seems likely additional freeze-tolerant species will be discovered that use these nucleators to insure freezing at high temperatures.

Ice-nucleating active (INA) micro-organisms possess the unique trait of catalysing ice nucleation at temperatures as high as −1 °C (see reviews in Lee *et al.*, 1995c). In INA bacteria, aggregations of proteins in the outer membrane function as templates on which embryonic seed crystals can form and induce freezing at very high sub-zero temperatures. These epiphytic, and frequently plant pathogenic, bacteria are well known for their role in promoting frost damage to crops. By promoting freezing damage, the ice-nucleating phenotype may function to facilitate bacterial invasion of the plant tissues and access to nutrients (Lindow, 1983).

In the early 1990s, two research groups independently isolated INA bacteria from the gut of insects (Kaneko *et al.*, 1991a; Lee *et al.*, 1991). Some of the bacterial isolates exhibited maximal INA thresholds near −2 °C. In 1992, Tsumuki and colleagues isolated an ice-nucleating active fungus (*Fusarium* sp.) from freeze-tolerant larvae of the rice stem borer. When sterile larvae were fed a suspension of fungal mycelium, SCPs increased from −20 to −5 °C. This notable discovery was the first time an INA microbe was isolated from a freeze-tolerant insect, suggesting a mutualistic relationship between the fungus and host. By insuring freezing at a high sub-zero temperature, the likelihood of the freeze-tolerant host surviving the winter increases, while the fungus receives shelter and food. Three species of INA bacteria have been isolated also, from a freeze-tolerant frog, *Rana sylvatica*, which suggests that these bacteria may also promote freezing in this species as well (Lee *et al.*, 1995a).

Exploration related to using INA microbes for control of insect pests also began in the late 1980s. Since most pest insects are freezing-intolerant and rely on enhanced supercooling for winter survival, this strategy was based on using INA bacteria to diminish the insect's supercooling capacity and increase the chance they would die from freezing during the winter. Strong-Gunderson *et al.* (1990) were the first to demonstrate that ingestion of INA bacteria, *Pseudomonas syringae* and *Erwinia herbicola,* elevated the supercooling point of a beetle from −16 °C to −3.5 and −4.4 °C, respectively. Intensive study was then directed toward using INA microbes against stored grain pests (Fields, 1990; Fields, 1993; Lee *et al.*, 1992; Lee *et al.*, 1993a; Lee *et al.*, 1995b). These studies confirmed that the cold-tolerance of various beetles and larvae of the Indian meal moth could be reduced using INA bacteria and fungi.

Our laboratory focused on the Colorado potato beetle as a candidate for biological control using INA bacteria (Lee *et al.*, 2001). This species, a devastating pest of potatoes in North America and Europe, has developed resistance to a range of insecticides. As they prepare for winter, freeze-intolerant adults depress their SCP to about −7.6 °C and burrow a few centimeters into the soil (Lee *et al.*, 1994). Since beetles survive cooling to temperatures immediately above their SCP, this temperature can be used as an estimate of their lower lethal temperature. Application of a freeze-killed preparation of *P. syringae* elevated their SCP to −3.7 °C (Lee *et al.*, 1994). After several studies that evaluated the efficacy of various bacterial strains for elevating the SCP and for long-term retention in the gut, we selected two strains for a long-term field test (Costanzo *et al.*, 1998a; Castrillo *et al.*, 2000, 2001). Ingestion of either *Pseudomonas fluorescens* or *P. putida,* significantly elevated the beetle SCP to about −4.4 °C. However, after seven months of overwintering under field conditions, only beetles fed *P. flourescens* had SCPs (−4.2 °C) that were significantly elevated compared to controls (Fig. 1.3). Based on these SCP values, environmental cooling to −5 °C would cause no mortality in untreated control beetles, while >95% of the INA-treated beetles would freeze and die. This study demonstrated that beetles retain ingested INA bacteria and that SCPs remain elevated over extended periods under winter conditions.

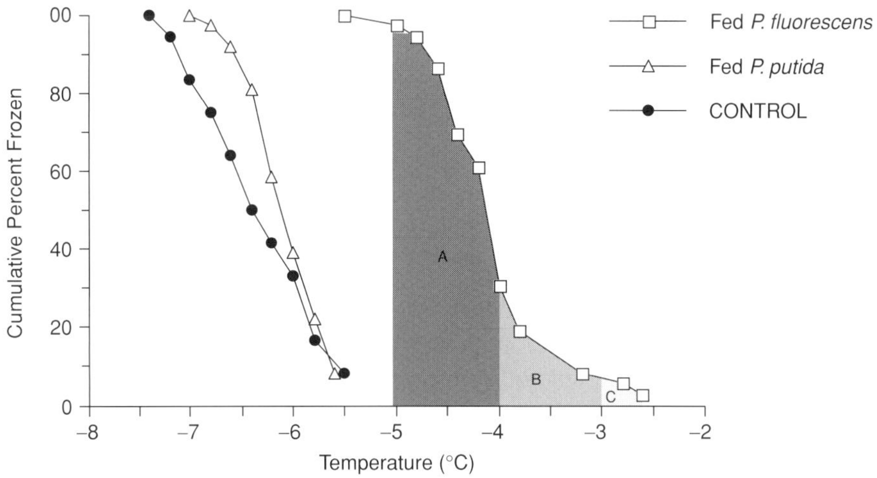

Figure 1.3 Cumulative freezing profile based on supercooling points of individual Colorado potato beetles treated with ice-nucleating *Pseudomonas fluorescens* F26–4C or *P. putida* Hr6–1, and measured after wintering in the field for 7 months. In beetles fed *P. fluorescens* F26–4C, hatched areas represent expected mortality at sub-zero temperatures based on the range of supercooling values exhibited by recovered beetles in May: (A) >95 to 50% mortality at −5 to −4 °C; (B) ~50 to 10% at −4 to 3 °C; and (C) <10% at −3 °C. (From Castrillo *et al.*, 2001)

A potential drawback of using INA micro-organisms for biological control is reduction of cold-tolerance in non-target species. Application of suspensions or dry preparations of INA bacteria readily reduce the supercooling capacity of a wide range of insects, including natural predators (e.g. lady beetles and spiders) that are important agents for biological control of insect pests (Lee *et al.*, 1993a; Tanaka and Watanabe, 2003). When spiders ingested prey that had fed on INA bacteria, their supercooling point increased by 10 degrees to −7.2 °C (Tanaka and Watanabe, 2003). Since spiders are intolerant of freezing, this increase indicates a marked reduction in their cold-tolerance. Consequently, broad application of INA agents might directly reduce the cold-hardiness of potential biological control agents, but also indirectly by transfer through the food chain.

1.5.2 Inoculative freezing

Despite the various classes of ice nucleators described above, the best agent for inducing insect freezing at the highest possible temperature is contact with environmental ice. This process, commonly termed inoculative freezing (or sometimes secondary nucleation) causes water to freeze at temperatures immediately below the FP of body fluids. Here environmental ice seeds the freezing of body water through natural body orifices or directly across the cuticle. When freezing occurs under these conditions, the insect experiences little or no supercooling, so temperature of crystallization is a more appropriate term to describe the temperature at which freezing began (Lee, 1991). Since insects frequently occupy hibernacula in which their bodies are in close contact with snow or frozen substrate, SCP determinations done on isolated, dry insects in the laboratory may not reflect the temperature at which they would freeze in nature.

For some terrestrial arthropods the capacity to survive freezing requires ice inoculation. Larvae of the drosophilid, *Chymonyza costata,* were initially classified as freezing-intolerant based on their inability to continue development after supercooling and freezing at −20 °C (Shimada and Riihimaa, 1988). However, when freezing was initiated by inoculation at −2°C, diapausing larvae survived to −80 °C. Other species that require ice inoculation to survive freezing include a tipulid larvae (Gehrken and Southon, 1992), a tenebrionid adult beetle (Gehrken *et al.*, 1991) and the centipede, *Lithobious forficatus* (Tursman *et al.*, 1994).

For other insects, survival depends on avoiding inoculative freezing. With the exception of some larval dipterans (Danks, 1971), aquatic stages of insects and other freshwater invertebrates are freezing-intolerant and have little capacity to supercool (Moore and Lee, 1991; Oswood *et al.*, 1991); these species are particularly susceptible to inoculative freezing at temperatures slightly below the MP (Frisbie and Lee, 1997). Low temperature survival in these aquatic species hinges on thermal buffering by their aquatic environs and heat liberation as ice forms within water bodies, and their capacity to avoid contact with external ice (Danks, 2007).

During migration to and at their mountain wintering sites in Mexico, monarch butterflies (*Danaus plexippus*) are especially susceptible to inoculative freezing. All misted adults froze and died during a 24-h exposure to −4 °C, while all dry individuals survived this treatment; furthermore, the elevation of the SCP was directly correlated with the water content of these individuals (Fig. 1.4; Larsen and Lee, 1994). At their wintering sites, exposure to rain and snowstorms can wet monarch bodies, making them susceptible to lethal freezing, and sometimes causing devastating losses of millions of monarchs (Anderson and Brower, 1996). Oberhauser and Peterson (2003) modeled the potential effects of global climate change in the monarch's winter range and predicted that expected increases in precipitation and decreases in temperature would make current habitats inadequate for winter survival within 50 years.

Susceptibility to inoculative freezing varies depending on intrinsic factors. The efficacy of inoculative nucleation differs depending on the anatomic site of application and the ice-nucleating agents used (Steigerwald *et al.*, 1995; Lee *et al.*, 1996). Diapausing pupae of the flesh fly, *Sarcophaga crassipalpis*, are more resistant to

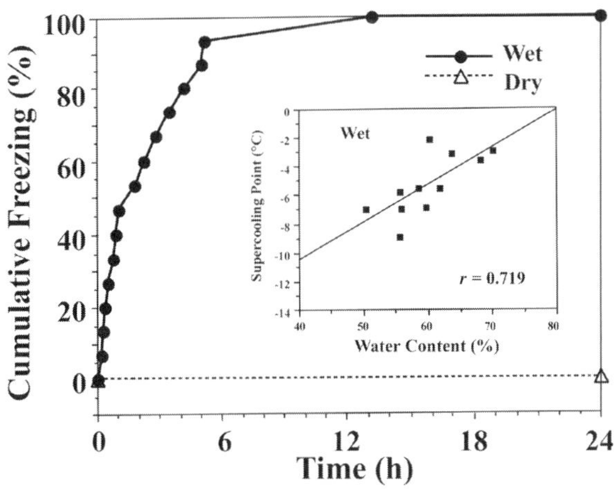

Figure 1.4 Effect of wetting on freezing and supercooling capacity of monarch butterflies (*Danaus plexippus*) during 24 h of exposure to −4 °C. Insert: Correlation between supercooling points and water content. (Adapted from Larsen and Lee, 1994.)

inoculative freezing than non-diapausing ones (Kelty and Lee, 2000). This difference is apparently related to increases in the amount of lipid in the winter puparium, as well as non-lipid components. Seasonal changes in antifreeze proteins associated with the subcuticular epidermal layer can also increase resistance to transcuticular inoculation by ice (Chapter 3 by Duman *et al.*).

External environmental factors, particularly hydric conditions and ice-nucleating agents present in the soil, can also influence susceptibility to inoculation. Although isolated larvae of the gall fly, *E. solidaginis*, have SCPs of −9 °C, galls still green and fully hydrated in early autumn freeze at −4.5 °C, causing the larvae within to freeze inoculatively (Layne *et al.*, 1990). Freeze-intolerant eggs of the migratory locust, *Locusta migratoria*, that are held in moist soil for extended periods have higher mortality rates (Qi *et al.*, 2007). Soil composition and moisture levels and environmental ice nuclei significantly affect supercooling capacity and cold-tolerance of Colorado potato beetles and other ectotherms (Costanzo *et al.*, 1997; Costanzo *et al.*, 1998b).

1.6 Cryoprotectants

The discovery of glycerol in cold-hardy insects in the late 1950s (Salt, 1961) coincided with investigations using glycerol and related compounds to cryopreserve red blood cells and other mammalian cells (Lovelock, 1953). Comparative physiologists, who usually measure metabolite levels in millimolar units, were especially intrigued by the accumulation of cryoprotectants to multimolar levels – a larval wasp produces 5 M glycerol equal to 25% of its body mass (Salt, 1961). Cryoprotectants are defined as compounds that enhance cold- and freezing-tolerance; as such they protect against both chilling and freezing injury.

The term cryoprotectant is most frequently used to refer to glycerol and other polyhydric alcohols (sorbitol, mannitol, ribitol, erythritol, threitol and ethylene glycol) and sugars (trehalose and glucose), although amino acids, such as proline and alanine, and other molecules are sometimes included in this category (see Table 2.1 in Lee, 1991; Karow, 1991). Recently, urea was reported to have cryoprotective efficacy equal to that of glucose in the freeze-tolerant wood frog (Costanzo and Lee, 2005), although a similar role has not been demonstrated in insects. Typically, cryoprotectants share the characteristics of being small, stable molecules that are highly soluble in water, and non-toxic even at high concentrations. Carbohydrate cryoprotectants are derived from glycogen stores in the fat body and the biochemistry of their biosynthesis is reviewed by Storey and Storey (1991). Less commonly, antifreeze proteins and ice-nucleating proteins are called high-molecular-mass cryoprotectants.

A primary function of cryoprotectants relates to their colligative effects at high concentration in body fluids (Salt, 1961; Zachariassen, 1985; Lee, 1991). For a freezing-tolerant species, colligative depression of the hemolymph MP results in less ice forming at any given sub-zero temperature, thereby lessening stress due to freeze concentration and extending the lower limit of freezing-tolerance. The maximal amount of survivable ice seems to be ~65% of body water. At high concentrations, glycerol may serve as a solvent that prevents specific ions and other solutes from reaching damaging levels due to freeze concentration (Lovelock, 1953). Glycerol and several other solutes are termed penetrating cryoprotectants because they readily cross the cell membrane, serving to decrease osmotic stress across the cell membrane and limit cellular water loss. A recent report (Kikawada *et al.*, 2007) described a novel trehalose transporter that functions to move trehalose from its site of synthesis in the fat body into the hemolymph. In this case, the transporter functions as a component in a suite of coordinated responses for anhydrobiotic survival in midge larvae. It seems likely that similar transporters will be found promoting uptake of trehalose and other non-penetrating cryoprotectants in freeze-tolerant species.

Cryoprotectants are sometimes referred to as antifreezes because they enhance the supercooling capacity of freezing-susceptible insects. For reasons poorly known, the SCP is depressed by approximately twice as much as the colligative depression of the FP; see Zachariassen *et al.* (2004) for a recent critical analysis of this phenomenon. Cryoprotectants may also promote supercooling by masking the activity of hemolymph ice nucleators and ice-nucleating bacteria through their interaction with antifreeze proteins (Duman, 2002). Coupling of cryoprotectant accumulation with extreme reductions in body water can lead to the avoidance of freezing by vitrification of body fluids (below).

Carbohydrate cryoprotectants also play important specific (i.e. non-colligative) roles in cold-tolerance. Even when there is no change in supercooling capacity, polyol accumulation is associated with increased cold-tolerance in diapausing adults of the bug, *Pyrrhocoris apterus* (Kostal *et al.*, 2001). Injection of ribitol and sorbitol also increased cold-hardiness in this species. Trehalose and glycerol are commonly produced by insects in response to osmotic challenge due to desiccation and freezing (Holmstrup *et al.*, Chapter 7). These compounds serve to stabilize proteins and protect cell membranes (Crowe *et al.*, 1998; Crowe, 2002; Yancey, 2005).

1.7 Water and ice management

Mechanisms of cold-hardening and freezing-tolerance depend fundamentally on altering water properties such as colligative depression of hemolymph osmolality, and the distribution of water and ice formation in body compartments. Reviews of this subject are available elsewhere (Holmstrup *et al.*, Chapter 7; Zachariassen, 1985; Zacharaissen, 1991). Here I focus on selected recent developments in the area.

1.7.1 Intracellular versus extracellular freezing

At temperatures and cooling rates experienced in nature, it is generally believed that ice must be restricted to extracellular compartments for survival (Lee and Costanzo, 1998). Again, based primarily on studies in mammalian systems, it is believed that at slow rates of cooling, intracellular freezing results in cryoinjury. However, recent evidence in some mammalian cells indicates that under certain conditions intracellular freezing is innocuous and may actually cryoprotect cells by preventing cellular dehydration (Acker and McGann, 2003).

Only a few studies have carefully examined intracellular versus extracellular freezing in freeze-tolerant insects. Sinclair and Wharton (1997) concluded that fat body and Malpighian tubule cells from an alpine weta avoid intracellular freezing by ensuring extracellular ice formation and subsequent cellular dehydration. Tolerance of intracellular freezing also does not appear to have evolved in an alpine cockroach (Worland *et al.*, 2004).

In contrast, Salt (1959; Salt, 1962) reported intracellular freeze-tolerance in fat body cells of *E. solidaginis*. Lee et al. (1993b) confirmed Salt's reports using fluorescent vital dyes, and also demonstrated that these cells are susceptible to inoculative freezing from contact with extracellular ice and tolerate extreme hypo- and hyper-osmotic stress. Non-lethal ice formation induces extensive lipid coalescence in the cell's core that displaces the nucleus, mitochondria and cytoplasmic organelles to the periphery (Morason *et al.*, 1994). The fact that the water content of these cells is only 35% (compared to 52–55% for fat body cells from freezing-intolerant larvae from *S. crassipalpis)* may enhance tolerance by reducing mechanical damage as intracellular water freezes (Davis and Lee, 2001). A modeling study of seasonal changes through autumn revealed marked increases in the intrinsic freezing-tolerance of fat body cells that was independent of increases in glycerol and sorbitol levels (Bennett and Lee, 1997). Thus far, the only invertebrate shown to tolerate extensive intracellular freezing in multiple tissues is the Antarctic nematode, *Panagrolaimus davidi* (Wharton and Ferns, 1995). More research is needed to determine whether intracellular freezing plays an important role in insects.

Freezing-tolerant plants and animals reduce mechanical damage in dense organs and freezing sensitive tissues by promoting ice formation in spaces or potential spaces where ice growth may occur innocuously (Lee and Costanzo, 1998). Although the lymphatic sacs of the wood frog *R. sylvatica* are normally collapsed, during freezing, large ice masses form within the lymphatic sacs and the coelomic cavity as body organs lose up to 60% of their initial water content (Lee and Costanzo, 1998). The presence of ice-nucleating agents in the insect hemolymph promotes extra-organ sequestration of ice in the hemocoel. Since many insects are tolerant of wide swings in body water content (and, thus, volume changes in the hemocoel), ice growth can occur innocuously within the hemocoel. The fact that insects are often highly tolerant of water loss may have promoted the evolution of freeze-tolerance.

1.7.2 Cryoprotective dehydration

For many years strategies of winter survival were categorized as either freezing-susceptible or freezing-tolerant. Recently, a third strategy, termed cryoprotective dehydration, was described for several soil-dwelling invertebrates (Holmstrup and Westh, 1994; Holmstrup and Sømme, 1998; Worland *et al.*, 1998; Wharton *et al.*, 2003; Holmstrup *et al.*, Chapter 7). This strategy requires a highly permeable integument that allows rapid dehydration when exposed to even a mildly desiccating environment. When an unfrozen animal is surrounded by frozen substrate it dehydrates owing to the vapor pressure gradient between unfrozen body water and environmental ice (Holmstrup *et al.*, 2002). Water loss continues until, at equilibrium, the vapor pressure of the body fluids equals that of the surrounding ice. At this time, the risk of freezing has been eliminated because the melting point (MP) of the body fluids equals the ambient temperature. In some species, this equilibration of the body fluid MP with that of the environment is facilitated by accumulation of low-molecular-mass cryoprotectants such as glycerol (Holmstrup, 1995; Worland *et al.*, 1998). Thus, with progressive dehydration, both extensive supercooling and freezing are avoided.

Since cryoprotective dehydration was first reported in earthworm egg cases, it has been described in nematodes, enchytraeid annelids and collembolans (Holmstrup *et al.*, 2002; Wharton *et al.*, 2003). An Arctic collembolan, *Megaphorura arctica*, loses approximately 90% of its body water during cryoprotective dehydration. Recently, the potential for cryoprotective dehydration was demonstrated in the first true insect, the chironomid, *Belgica antarctica* (Elnitsky *et al.*, 2008). It seems likely that this distinctive strategy is widespread in polar and alpine invertebrates whose hibernacula are surrounded by frozen substrate.

1.7.3 Vitrification

Another way to avoid damage by internal ice formation is vitrification. Vitrified water exists in an amorphous, metastable state in which its viscosity approaches that of a glass. Vitrification is attractive for cryopreservation because water, ions and other solutes remain in their original intra- and extracellular compartments, and damage due to freeze concentration and mechanical distortion is avoided. In this way, micro-organisms, plant tissues and some mammalian embryos and cells are routinely cryopreserved (Fuller *et al.*, 2004) as well as some insect embryos (Leopold and Rinehart, Chapter 13).

Though rarely documented, vitrification also occurs in natural systems. Hirsh *et al.* (1985) reported intracellular glass formation in winter-hardened twigs of the poplar *Populus balsamifera*, a species whose distribution extends into subarctic regions. Wintering larvae of the Alaskan cucujid beetle, *Cucujus clavipes*, dehydrate extensively to <40% their body water content in summer, causing glycerol levels to reach 7–10 mol l^{-1} (Bennett *et al.*, 2005). They also exhibit the largest amount of thermal hysteresis (−13 °C) ever reported. When some individuals did not freeze, even when cooled to −80 °C, the authors speculated the larvae had vitrified. Further studies demonstrated that larvae did not freeze, even at −150 °C and confirmed vitrification – the first report in an insect (Duman *et al.*, Chapter 3). Interestingly, vitrification is also required for cellular protection during anhydrobiosis in the African midge, *Polypedilum vanderplanki*, a species that accumulates trehalose to ~20% of its dry body mass (Sakurai *et al.*, 2008).

1.8 Role of proteins

Understanding of the roles of various proteins in insect cold-tolerance has advanced substantially in the past 20 years. The development of powerful molecular tools and the increasing ease of their application has facilitated the identification and structural characterization of novel proteins, and progress is being made on determining their function in promoting winter survival in insects. Heat shock proteins (Hsps), also known as stress proteins, play a critical role in protecting organisms from injury due to high or low temperature, anoxia, desiccation and a range of chemical stresses. Recent evidence indicates these proteins play a role in insect cold-tolerance during diapause (Rinehart *et al.*, 2007; Michaud and Denlinger, Chapter 4 in this volume) and may be required for survival in polar environments (Rinehart *et al.*, 2006). In addition to ice-nucleating proteins discussed earlier, progress has been made concerning enzymatic defense against reactive oxygen species (Storey and Storey, Chapter 6), the role of reversible protein phosphorylation in regulating seasonal processes related to metabolic depression and stress-tolerance (McMullen and Storey, 2008), and the involvement of desaturases in membrane restructuring (Kayukawa *et al.*, 2007; Kostal, Chapter 5).

Antifreeze proteins (AFPs), originally discovered in polar fish, are a unique class of proteins that bind to ice crystals and prevent their growth by a non-colligative mechanism (for a detailed review see Duman *et al.*, Chapter 3). Because they create a hysteresis between the equilibrium MP of the hemolymph and the FP at which a small ice crystal can grow, they are also called thermal hysteresis proteins. The difference between the equilibrium MP and FP may reach 8 °C or more. These proteins are widely distributed among diverse terrestrial arthropods, including collembolans, eight orders of insects, spiders and a centipede (Duman *et al.*, 2004).

In freezing-intolerant species AFPs are often produced seasonally and are believed to promote supercooling and inhibit ice formation (Zachariassen and Husby, 1982); sometimes, they also promote supercooling by inhibiting hemolymph ice nucleators (Duman, 2001). The addition of low-molecular-mass compounds, such as glycerol and citrate, enhances their inhibition of bacterial and hemolymph ice nucleators (Duman, 2002). As we strive to understand the functional roles of various proteins in insect cold-tolerance, this discovery suggests that caution should be used whenever isolated proteins are studied outside their normal milieu.

As new roles for AFPs have been identified, terms such as ice-structuring proteins or ice-binding proteins have been suggested to more inclusively describe their diverse functions (Clarke *et al.*, 2002). Olsen *et al.* (1998) demonstrated that AFPs prevent inoculative freezing across the cuticle. Interestingly, these proteins are found in some freezing-tolerant arthropods, where they may cryoprotect by inhibiting recrystallization damage to frozen tissues or intracellular ice formation (Duman, 2001). Attempts to use AFPs to modify ice-crystal growth in frozen foods and in the cryopreservation of mammalian tissues have had mixed results.

1.8.1 Aquaporins

The critical role of water status in cold and desiccation-tolerance has been the focus of numerous reviews (Zachariassen, 1991; Somero, 1992; Storey and Storey, 1996; Danks, 2000; Chown and Nicolson, 2004). Survival of osmotic stress, whether a result of freezing and thawing, desiccation or osmotic challenge, is intimately linked to the movement of water and various small molecules, such as glycerol, between body water compartments and the environment. Aquaporins (AQPs) are a family of transmembrane proteins that facilitate movement of water and other solutes across the cell membrane. Although these proteins are broadly known from yeast, plants and mammals (Borgnia *et al.*,1999), only a few reports have described them in insects (Campbell *et al.*, 2008).

Recent evidence strongly implicates AQPs in the avoidance of freezing injury (Izumi *et al.*, 2006; Philip *et al.*, 2008). Fat body cells of the rice stem borer, *Chilo suppressalis*, require glycerol to survive freezing *in vitro* (Izumi *et al.*, 2006; Izumi *et al.*, 2007). However, the addition of mercuric chloride, a specific inhibitor of certain AQPs, blocks the uptake of glycerol and markedly increases freezing injury. Additional studies with mercuric chloride in *E. solidaginis* larvae demonstrated significant decreases in cell survival in fat body (from 81.6 to 11.6%) and gut (91.4 to 13.5%) following freezing at −20 °C (Philip *et al.*, 2008). The need for these proteins is also evident as overexpression of AQPs in yeast improved their freezing-tolerance (Tanghe *et al.*, 2002), as did the artificial expression of AQPs in cryopreserved mouse oocytes (Edashige *et al.*, 2003).

Furthermore, immunoblotting with antibodies against mammalian AQPs revealed corresponding AQP2, −3 and −4 homologues in *E. solidaginis*, whose expression varied with temperature acclimation (Philip *et al.*, 2008). They also found that AQPs were up-regulated in response to desiccation, which may account, in part, for the remarkable tolerance of these overwintering larvae to desiccation stress (Ramlov and Lee, 2000). Consequently, AQPs appear to play a critical role in facilitating the movement of water out of cells and the uptake of glycerol during freezing.

1.8.2 Dehydrins

Dehydrins are Group II late embryogenesis abundant (LEA) proteins induced by environmental stress associated with low temperature or dehydration, and during seed maturation in plants (Dure, 1993). Their high hydrophilicity and thermostability suggest they function to stabilize protein and membrane structure with surfactant- and possibly chaperone-like properties (Egerton-Warburton *et al.*, 1997; Borovskii *et al.*, 2002). Dehydrins are also thought to influence the availability of free water via their poly-proline II alpha-helical confirmation that may function to decrease water loss and prevent ice crystal formation (Bokor *et al.*, 2005).

Although well known in plants, little is known concerning dehydrin-like proteins in animals. The first LEA protein in an arthropod was reported in the anhydrobiotic larvae of chironomid, *Polypedilum vanderplancki* (Kikawada *et al.*, 2006). Pruitt and colleagues (2007) reported a hydrophilic, heat-stable protein fraction in *E. solidaginis* larvae that protects the catalytic activity of lactate dehydrogenase (LDH) against freezing damage. This protein fraction shares several characteristics with plant dehydrins, and is likely the first report of a dehydrin in a freezing-tolerant insect.

In preliminary studies, we screened for the presence of dehydrin-like proteins in freeze-tolerant larvae of the Antarctic midge, *B. antarctica* (Yi *et al.*, unpublished data). Supernatant from homogenized larvae was boiled, acetone precipitated and the heat-stable protein fraction tested for cryoprotection of lactate dehydrogenase (LDH) activity during freezing (after Pruitt *et al.*, 2007). The heat-stable fraction had significant cryoprotective effects on LDH activity that were greater than those of the heat-shock proteins. We also putatively identified dehydrin-like protein in the heat-stable fraction by immunoblotting with a plant-derived dehydrin antibody. It seems likely that other dehydrins will be discovered to play roles in cold- and freezing-tolerance of insects.

1.9 Future directions

For many years investigations of insect cold-tolerance relied heavily on correlations between organismal measures of cold-hardiness, and seasonal changes in polyhydric alcohols, antifreeze protein levels and supercooling points to probe the nature of insect cold-tolerance. While much progress was made with this approach, recent advances in molecular and cell biology hold the promise of unveiling specific underlying mechanisms through the rapid identification of key metabolites and candidate genes and gene products. Use of gene silencing (RNAi) to knock out specific proteins, such as lipid desaturases, aquaporins and dehydrins, will allow direct assessment of their role in cold- and stress-tolerance. Identification of specific cold-sensing and signal-transduction pathways (Fujiwara and Denlinger, 2007; Teets *et al.*, 2008) leading to cold-hardening or specific types of injury, such as apoptosis (Yi *et al.*, 2007), will provide details into the nature of cold injury and cryoprotection. Once specific molecules are identified, their function can be investigated by such means as vital dyes that can distinguish differences in cold-hardening at the cellular and tissue levels (Yi and Lee, 2004), functional tests of organ performance, such as the Malpighian tubule secretion (Neufeld and Leader,

1997; Yi and Lee, 2005), and direct measures of changes in membrane fluidity (Lee *et al.*, 2006).

These approaches should allow us to address fundamental questions that remain unanswered. What determines the lower limit of chilling and freezing-tolerance? Although cold shock appears to induce apoptosis, is it also associated with injury due to freezing? To what extent, and how, is sub-lethal chilling and freezing injury repaired? A growing body of evidence indicates the response of insects acclimated at constant temperatures does not reflect the response of ones living in fluctuating thermal environments (Colinet *et al.*, 2006; Kostal *et al.*, 2007). Why? What is the basis for cross-tolerance between different types of environmental stresses, including cold, freezing, desiccation and anoxia (Holmstrup *et al.*, Chapter 7)? What environmental factors promote the evolution of freezing-tolerance versus freezing-intolerance (Voituron *et al.*, 2002; Vernon and Vannier, 2002; Sinclair *et al.*, 2003)?

As our understanding of the underlying physiological principles of low temperature biology has advanced and the number and diversity of species studied has grown, it has provided a solid foundation to begin study of their significance to population dynamics and global climate change. Incorporation of life-stage-specific data on SCPs and cold-induced mortality have enhanced models predicting population dynamics and potential range expansions related to climate warming in major forest pests in Europe and North America (Tran *et al.*, 2007; Régnière and Bentz, 2007; Jepsen *et al.*, 2008). Geographical clines in enzyme and inversion polymorphisms have been identified that may serve as biomarkers for populations experiencing climate change (Umina *et al.*, 2005; Dahlhoff *et al.*, 2008). Basic knowledge of thermal tolerance led to the surprising prediction that climate warming poses a greater threat to species living in the tropics than in temperate regions (Huey, Chapter 9). Continued integration of research at molecular, organismal, and ecological levels is needed to provide a deeper and more robust understanding of how the low temperature biology of insects influences their diversity and distribution.

Acknowledgements

I thank Dave Denlinger, Ben Philip, Molly Steinwald and Shu-Xia Yi for their comments on this chapter. Research support from the National Science Foundation is gratefully acknowledged.

References

Acker, J. P. and McGann, L. E. (2003). Protective effect of intracellular ice during freezing? *Cryobiology* **46**, 197–202.

Anderson, J. B. and Brower, L. P. (1996). Freeze-protection of overwintering monarch butterflies in Mexico: critical role of the forest as a blanket and an umbrella. *Ecological Entomology* **21**, 107–116.

Angell, A. (1982). Supercooled water. In *Water: A Comprehensive Treatise*, ed. F. Franks, vol. **7**. New York: Plenum Press, pp. 1–81.

Bennett, V. A. and Lee, R. E. (1997). Modeling seasonal changes in intracellular freeze-tolerance of fat body cells of the gall fly *Eurosta solidaginis* (Diptera, Tephritidae). *Journal of Experimental Biology* **200**, 185–192.

Bennett, V. A., Sformo, T., Walters, K., Toien, O., Jeannet, K., Hochstrasser, R., Pan, Q., Serianni, A. S., Barnes, B. M., and Duman, J. G. (2005). Comparative overwintering physiology of Alaska and Indiana populations of the beetle Cucujus clavipes (Fabricius): role of antifreeze proteins, polyols, dehydration and diapause. *Journal of Experimental Biology* **208**, 4467–4477.

Bokor, M., Csizmok, V., Kovacs, D., Banki, P., Friedrich, P., Tompa, P., and Tompa, K. (2005). NMR relaxation studies on the hydrate layer of intrinsically unstructured proteins. *Biophysical Journal* **88**, 2030–2037.

Borgnia, M., Nielsen, S., Engel, A., and Agre, P. (1999). Cellular and molecular biology of the aquaporin water channels. *Annual Review of Biochemistry* **68**, 425–458.

Borovskii, G. B., Stupnikova, I. V., Antipina, A. I., Vladimirova, S. V., and Voinikov, V. K. (2002). Accumulation of dehydrin-like proteins in the mitochondria of cereals in response to cold, freezing, drought and ABA treatment. *BMC Plant Biology* **2**, 5–12.

Campbell, E. M., Ball, A., Hoppler, S., and Bowman, A. (2008). Invertebrate aquaporins: a review. *Journal of Comparative Physiology B* **178**, 935–955.

Castrillo, L. A., Lee, R. E., Lee, M. R., and Rutherford, S. T. (2000). Identification of ice-nucleating active *Pseudomonas fluorescens* strains for biological control of overwintering Colorado potato beetles (Coleoptera: Chrysomelidae). *Journal of Economic Entomology* **93**, 226–233.

Castrillo, L. A., Lee, R. E., Wyman, J. A., Lee, M. R., and Rutherford, S. T. (2001). Field persistence of ice-nucleating bacteria in overwintering Colorado potato beetles. *Biological Control* **21**, 11–18.

Chen, C. P. and Denlinger, D. L. (1992). Reduction of cold injury in flies using an intermittent pulse of high temperature. *Cryobiology* **29**, 138–143.

Chown, S. L. and Nicolson, S. W. (2004). *Insect Physiological Ecology: Mechanisms and Patterns*. New York: Oxford University Press.

Chown, S. L., Sorensen, J. G., and Sinclair, B. J. (2008). Physiological variation and phenotypic plasticity: a response to 'Plasticity in arthropod cryotypes' by Hawes and Bale. *Journal of Experimental Biology* **211**, 3353–3357.

Chown, S. L. and Terblanche, J. S. (2007). Physiological diversity in insects: ecological and evolutionary contexts. *Advances in Insect Physiology* **33**, 50–152.

Clarke, C. J., Buckley, S. L., and Lindner, N. (2002). Ice structuring proteins: a new name for antifreeze proteins. *CryoLetters* **23**, 89–92.

Colinet, H., Nguyen, T. T. A., Cloutier, C., Michaud, D., and Hance, T. (2007). Proteomic profiling of a parasitic wasp exposed to constant and fluctuating cold exposure. *Insect Biochemistry and Molecular Biology* **37**, 1177–1188.

Colinet, H., Renault, D., Hance, T., and Vernon, P. (2006). The impact of fluctuating thermal regimes on the survival of a cold-exposed parasitic wasp, *Aphidius colemani*. *Physiological Entomology* **31**, 234–240.

Costanzo, J. P., Humphreys, T. L., Lee, R. E., Moore, J. B., Lee, M. R., and Wyman, J. A. (1998a). Long-term reduction of cold hardiness following ingestion of ice-nucleating bacteria in the Colorado potato beetle, *Leptinotarsa decemlineata*. *Journal of Insect Physiology* **44**, 1173–1180.

Costanzo, J. P. and Lee, R. E. (2005). Cryoprotection by urea in a terrestrially hibernating frog. *Journal of Experimental Biology* **208**, 4079–4089.

Costanzo, J. P., Litzgus, J. D., Iverson, J. B., and Lee, R. E. (1998b). Soil hydric characteristics and environmental ice nuclei influence supercooling capacity of hatchling turtles *Chrysemys picta*. *Journal of Experimental Biology* **201**, 3105–3112.

Costanzo, J. P., Moore, J. B., Lee, R. E., Kaufman, P. E., and Wyman, J. A. (1997). Influence of soil hydric parameters on the winter cold hardiness of a burrowing beetle, *Leptinotarsa decemlineata* (Say). *Journal of Comparative Physiology B* **167**, 169–176.

Crowe, J. H., Carpenter, J. F., and Crowe, L. M. (1998). The role of vitrification in anhydrobiosis. *Annual Review of Physiology* **60**, 73–103.

Crowe, L. M. (2002). Lessons from nature: the role of sugars in anhydrobiosis. *Comparative Biochemistry and Physiology A – Molecular and Integrative Physiology* **131**, 505–513.

Dahlhoff, E. P., Fearnley, S. L., Bruce, D. A., Gibbs, A. G., Stoneking, R., McMillan, D. M., Deiner, K., Smiley, J. T., and Rank, N. E. (2008). Effects of temperature on physiology and reproductive success of a montane leaf beetle: implications for persistence of native populations enduring climate change. *Physiological and Biochemical Zoology* **81**, 718–732.

Danks, H. V. (1971). Overwintering of some north temperate and arctic Chironomidae. II. Chironomid biology. *Canadian Entomologist* **103**, 1875–1910.

Danks, H. V. (2000). Dehydration in dormant insects. *Journal of Insect Physiology* **46**, 837–852.

Danks, H. V. (2007). How aquatic insects live in cold climates. *Canadian Entomologist* **139**, 443–471.

Davis, D. J. and Lee, R. E. (2001). Intracellular freezing, viability, and composition of fat body cells from freeze-intolerant larvae of *Sarcophaga crassipalpis*. *Archives of Insect Biochemistry and Physiology* **48**, 199–205.

Denlinger, D. L. and Lee, R. E. (1998). Physiology of cold sensitivity. In *Temperature Sensitivity in Insects and Application in Integrated Pest Management*, ed. G. J. Hallman, and D. L. Denlinger. Boulder: Westview Press, pp. 55–95.

Duman, J. G. (2001). Antifreeze and ice nucleator proteins in terrestrial arthropods. *Annual Review of Physiology* **63**, 327–357.

Duman, J. G. (2002). The inhibition of ice nucleators by insect antifreeze proteins is enhanced by glycerol and citrate. *Journal of Comparative Physiology B* **172**, 163–168.

Duman, J. G., Bennett, V., Sformo, T., Hochstrasser, R., and Barnes, B. M. (2004). Antifreeze proteins in Alaskan insects and spiders. *Journal of Insect Physiology* **50**, 259–266.

Dure, L., III (1993). Structural motifs in LEA proteins of higher plants. In *Response of Plants to Cellular Dehydration during Environmental Stress*, ed. T. J. Close and E. A. Bray. Rockville, MD: American Society of Plant Physiologists, pp. 91–103.

Edashige, K., Yamaji, Y., Kleinhans, F.W., and Kasai, M. (2003). Artificial expression of aquaporin-3 improves the survival of mouse oocytes after cryopreservation. *Biology of Reproduction* **68**, 87–94.

Egerton-Warburton, L. M., Balsamo, R. A., and Close, T. J. (1997). Temporal accumulation and ultrastructural localization of dehydrins in *Zea mays*. *Physiologia Plantarum* **101**, 545–555.

Elnitsky, M. A., Hayward, S. A. L., Rinehart, J. P., Denlinger, D. L., and Lee, R. E. (2008). Cryoprotective dehydration and the resistance to inoculative freezing in the Antarctic midge, *Belgica antarctica*. *Journal of Experimental Biology* **211**, 524–530.

Fields, P. G. (1990). The cold-hardiness of *Cryptolestes ferrugineus* and the use of ice nucleation-active bacteria as a cold-synergist. *Proceedings of the Fifth International Working Conference on Stored-Product Protection*, pp. 1183–1191.

Fields, P. G. (1993). Reduction of cold tolerance of stored-product insects by ice-nucleating-active bacteria. *Environmental Entomology* **22**, 470–476.

Frisbie, M. P. and Lee, R. E. (1997). Inoculative freezing and the problem of winter survival for freshwater macroinvertebrates. *Journal of the North American Benthological Society* **16**, 635–650.

Fujiwara, Y. and Denlinger, D. L. (2007). p38 MAP kinase is a likely component of the signal transduction pathway triggering rapid cold-hardening in the flesh fly, *Sarcophaga crassipalpis*. *Journal of Experimental Biology* **210**, 3295–3300.

Fuller, B. J., Lane, N., and Benson, E. E. (eds.) (2004). *Life in the Frozen State*, Boca Raton: CRC Press.

Gehrken, U. and Southon, T. E. (1992). Supercooling in a freeze-tolerant cranefly larva, *Tipula sp*. *Journal of Insect Physiology* **38**, 131–137.

Gehrken, U., Stromme, A., Lundheim, R., and Zachariassen, K. E. (1991). Inoculative freezing in overwintering tenebrionid beetle, *Bolitophagus reticulatus Panz*. *Journal of Insect Physiology* **37**, 683–687.

Hallman, G. J. and Denlinger, D. L. (1998). Introduction: temperature sensitivity and integrated pest management. In *Temperature Sensitivity in Insects and Application in Integrated Pest Management*, ed. G. J. Hallman and D. L. Denlinger. Boulder: Westview Press, pp. 1–5.

Hawes, T. C. and Bale, J. S. (2007). Plasticity in arthropod cryotypes. *Journal of Experimental Biology* **210**, 2585–2592.

Heinrich, B. (1993). *The Hot-Blooded Insects: Strategies and Mechanisms of Thermoregulation*. Cambridge, MA: Harvard University Press.

Hirsh, A. G., Williams, R. J., and Meryman, H. T. (1985). A novel method of natural cryoprotection. *Plant Physiology* **79**, 41–56.

Holmstrup, M. (1995). Polyol accumulation in earthworm cocoons induced by dehydration. *Comparative Biochemistry and Physiology A* **111**, 251–255.

Holmstrup, M., Bayley, M., and Ramlov, H. (2002). Supercool or dehydrate? An experimental analysis of overwintering strategies in small permeable arctic invertebrates. *Proceedings of the National Academy of Sciences, USA* **99**, 5716–5720.

Holmstrup, M., Costanzo, J. P., and Lee, R. E. (1999). Cryoprotective and osmotic responses to cold acclimation and freezing in freeze-tolerant and freeze-intolerant earthworms. *Journal of Comparative Physiology B* **169**, 207–214.

Holmstrup, M. and Sømme, L. (1998). Dehydration and cold hardiness in the Arctic collembolan *Onychiurus arcticus* Tullberg 1876. *Journal of Comparative Physiology B* **168**, 197–203.

Holmstrup, M. and Westh, P. (1994). Dehydration of earthworm cocoons exposed to cold: a novel cold hardiness mechanism. *Journal of Comparative Physiology B* **164**, 312–315.

Irwin, J. T., Bennett, V. A., and Lee, R. E. (2001). Diapause development in frozen larvae of the goldenrod gall fly, *Eurosta solidaginis* (Fitch) (Diptera: Tephritidae). *Journal of Comparative Physiology B* **171**, 181–188.

Irwin, J. T. and Lee, R. E. (2002). Energy and water conservation in frozen vs. supercooled larvae of the goldenrod gall fly, *Eurosta solidaginis* (Fitch) (Diptera: Tephritidae). *Journal of Experimental Zoology* **292**, 345–350.

Izumi, Y., Sonoda, S., and Tsumuki, H. (2007). Effects of diapause and cold-acclimation on the avoidance of freezing injury in fat body tissue of the rice stem borer, *Chilo suppressalis* Walker. *Journal of Insect Physiology* **53**, 685–690.

Izumi, Y., Sonoda, S., Yoshida, H., Danks, H. V., and Tsumuki, H. (2006). Role of membrane transport of water and glycerol in the freeze tolerance of the rice stem borer, *Chilo suppressalis* Walker (Lepidoptera: Pyralidae). *Journal of Insect Physiology* **52**, 215–220.

Jepsen, J. U., Hagen, S. B., Ims, R. A., and Yoccoz, N. G. (2008). Climate change and outbreaks of geometrids *Operophtera brumata* and *Epirrita autumnata* in subarctic birch forest: evidence of a recent outbreak range expansion. *Journal of Animal Ecology* **77**, 257–264.

Kaneko, J., Kita, K., and Tanno, K. (1991a). INA bacteria isolated from diamondback moth, *Plutella xylostella* L. pupae (Lepidoptera: Yponomeutidae). *Japanese Journal of Applied Entomology and Zoology* **35**, 7–11.

Kaneko, J., Toyohira-ku, H., Owada, T., and Tanno, K. (1991b). Erwinia herbicola: ice nucleation active bacteria isolated from diamondback moth, *Plutella xylostella* L. pupae. *Japanese Journal of Applied Entomology and Zoology* **35**, 247–251.

Karow, A. M. (1991). Chemical cryoprotection of metazoan cells. *BioScience* **41**, 155–160.

Kayukawa, T., Chen, B., Hoshizaki, S., and Ishikawa, Y. (2007). Upregulation of a desaturase is associated with the enhancement of cold hardiness in the onion maggot, *Delia antiqua*. *Insect Biochemistry and Molecular Biology* **37**, 1160–1167.

Kelty, J. D. and Lee, R. E. (2000). Diapausing pupae of the flesh fly Sarcophaga crassipalpis (Diptera: Sarcophagidae) are more resistant to inoculative freezing than non-diapausing pupae. *Physiological Entomology* **25**, 120–126.

Kikawada, T., Nakahara, Y., Kanamori, Y., Iwata, K., Watanabe, M., McGee, B., Tunnacliffe, A., and Okuda, T. (2006). Dehydration-induced expression of LEA proteins in an anhydrobiotic chironomid. *Biochemical and Biophysical Research Communications* **348**, 56–61.

Kikawada, T., Saito, A., Kanamori, Y., Nakahara, Y., Iwata, K., Tanaka, D., Watanabe, M., and Okuda, T. (2007). Trehalose transporter 1, a facilitated and high-capacity trehalose transporter allows exogenous trehalose uptake into cells. *Proceedings of the National Academy of Sciences, USA* **104**, 11585–11590.

Kohshima, S. (1984). A novel cold-tolerant insect found in a Himalayan glacier. *Nature* **310**, 225–227.

Kostal, V., Slachta, M., and Simek, P. (2001). Cryoprotective role of polyols independent of the increase in supercooling capacity in diapausing adults of *Pyrrhocoris apterus* (Heteroptera: Insecta). *Comparative Biochemistry and Physiology Part B* **130**, 365–374.

Kostal, V., Vambera, J., and Bastl, J. (2004). On the nature of pre-freeze mortality in insects: water balance, ion homeostasis and energy charge in adults of *Pyrrhocoris apterus*. *Journal of Experimental Biology* **207**, 1509–1521.

Kostal, V., Renault, D., Mehrabianová, A., and Bastl, J. (2007). Insect cold tolerance and repair of chill-injury at fluctuating thermal regimes: role of ion homeostasis. *Comparative Biochemistry and Physiology, Part A* **147**, 231–238.

Kostal, V., Yanagimoto, M., and Bastl, J. (2006). Chilling-injury and disturbance of ion homeostasis in the coxal muscle of the tropical cockroach (*Nauphoeta cinerea*). *Comparative Biochemistry and Physiology B* **143**, 173–179.

Lacoume, S., Bressac, C., and Chevrier, C. (2007). Sperm production and mating potential of males after a cold shock on pupae of the parasitoid wasp *Dinarmus basalis*. *Journal of Insect Physiology* **53**, 1008–1015.

Larcher, W. (2001). *Physiological Plant Ecology: Ecophysiology and Stress Physiology of Functional Groups*, 4th edn. New York: Springer.

Larsen, K. J. and Lee, R. E. (1994). Cold tolerance including rapid cold-hardening and inoculative freezing in migrant monarch butterflies in Ohio. *Journal of Insect Physiology* **40**, 859–864.

Layne, J. R., Lee, R. E., and Huang, J. L. (1990). Inoculation triggers at high subzero temperatures in a freeze-tolerant frog (*Rana sylvatica*) and insect (*Eurosta solidaginis*). *Canadian Journal of Zoology* **68**, 506–510.

Lee, M. R., Lee, R. E., Strong-Gunderson, J. M., and Minges, S. R. (1995a). Isolation of ice-nucleating active bacteria from the freeze tolerant frog, *Rana sylvatica*. *Cryobiology* **32**, 358–365.

Lee, R. R. (1989). Insect cold-hardiness: to freeze or not to freeze. *BioScience* **39**, 308–313.

Lee, R. E. (1991). Principles of insect low temperature tolerance. In *Insects at Low Temperature*, ed. R. E. Lee and D. L. Denlinger. New York and London: Chapman and Hall, pp. 17–46.

Lee, R. E., Castrillo, L. A., Lee, M. L., Wyman, J., and Costanzo, J. P. (2001). Using ice-nucleating bacteria to reduce winter survival of Colorado potato beetles: development of a novel strategy for biological control. In *Insect Timing: Circadian Rhythmicity to Seasonality*, ed. D. L. Denlinger, J. M. Giebultowicz and D. S. Saunders. Amsterdam: Elsevier, pp. 213–227.

Lee, R. E. and Costanzo, J. P. (1998). Biological ice nucleation and ice distribution in cold-hardy ectothermic animals. *Annual Review of Physiology* **60**, 55–72.

Lee, R. E., Costanzo, J. P., Kaufman, P. E., Lee, M. R., and Wyman, J. A. (1994). Ice-nucleating active bacteria reduce the cold-hardiness of the freeze-intolerant Colorado potato beetle (Coleoptera, Chrysomelidae). *Journal of Economic Entomology* **87**, 377–381.

Lee, R. E., Damodaran, K., Yi, S.-X., and Lorigan, G. A. (2006). Rapid cold-hardening increases membrane fluidity and cold tolerance of insect cells. *Cryobiology* **52**, 459–463.

Lee, R. E. and Denlinger, D. L. (eds.) (1991). *Insects at Low Temperature*. New York: Chapman and Hall.

Lee, R. E., Lee, M. L., and Strong-Gunderson, J. M. (1993a). Insect cold-hardiness and ice nucleating active microorganisms including their potential use for biological control. *Journal of Insect Physiology* **39**, 1–12.

Lee, R. E., Lee, M. R., and Strong-Gunderson, J. M. (1995b). Biological control of insect pests using ice-nucleating microorganisms. In *Biological Ice Nucleations and its Applications*, ed. R. E. Lee, G. J. Warren and L. V. Gusta. St. Paul: APS Press, pp. 257–269.

Lee, R. E. and Lewis, E. A. (1985). Effect of temperature and duration of exposure on tissue ice formation in the gall fly, *Eurosta solidaginis* (Diptera, Tephritidae). *CryoLetters* **6**, 24–34.

Lee, R. E., McGrath, J. J., Morason, R. T., and Taddeo, R. M. (1993b). Survival of intracellular freezing, lipid coalescence and osmotic fragility in fat-body cells of the freeze-tolerant gall fly *Eurosta solidaginis*. *Journal of Insect Physiology* **39**, 445–450.

Lee, R. E., Steigerwald, K. A., Wyman, J. A., Costanzo, J. P., and Lee, M. R. (1996). Anatomic site of application of ice-nucleating active bacteria affects supercooling in the Colorado potato beetle (Coleoptera: Chrysomelidae). *Environmental Entomology* **25**, 465–469.

Lee, R. E., Strong-Gunderson, J. M., Lee, M. R., and Davidson, E. C. (1992). Ice-nucleating active bacteria decrease the cold-hardiness of stored grain insects. *Journal of Economic Entomology* **85**, 371–374.

Lee, R. E., Strong-Gunderson, J. M., Lee, M. R., Grove, K. S., and Riga, T. J. (1991). Isolation of ice nucleating active bacteria from insects. *Journal of Experimental Zoology* **257**, 124–127.

Lee, R. E., Warren, G. J., and Gusta, L. V. (eds.) (1995c). *Biological Ice Nucleation and its Applications*. St. Paul: APS Press.

Leopold, R. A., Rojas, R. R., and Atkinson, P. W. (1998). Post pupariation cold storage of three species of flies: increasing chilling tolerance by acclimation and recurrent recovery periods. *Cryobiology* **36**, 213–224.

Levitt, J. (1980). *Responses of Plants to Environmental Stresses*, 2nd edn. New York: Academic Press, Inc.

Lindow, S. E. (1983). The role of bacterial ice nucleation in frost injury to plants. *Annual Review of Phytopathology* **21**, 363–384.

Lovelock, J. E. (1953). The mechanism of the protective action of glycerol against haemolysis by freezing and thawing. *Biochimica et Biophysica Acta* **11**, 28–36.

Mazur, P. (2004). Principles of cryobiology. In *Life in the Frozen State*, ed. B. J. Fuller, N. Lane and E. E. Benson. Boca Raton: CRC Press, pp. 3–66.

McMullen, D. C. and Storey, K. B. (2008). Suppression of Na^+K^+-ATPase activity by reversible phosphorylation over the winter in a freeze-tolerant insect. *Journal of Insect Physiology* **54**, 1023–1027.

Meryman, H. T. (1968). Modified model for the mechanism of freezing injury in erythrocytes. *Nature* **218**, 333–336.

Moore, M. V. and Lee, R. E. (1991). Surviving the big chill: overwintering strategies of aquatic and terrestrial insects. *American Entomologist* **Summer** 111–118.

Morason, T. R., Allenspach, A., and Lee, R. E. (1994). Comparative ultrastructure of fat body cells of freeze-susceptible and freeze-tolerant *Eurosta solidaginis* larvae after chemical fixation and high pressure freezing. *Journal of Insect Physiology* **40**, 155–164.

Mugnano, J. A., Lee, R. E., and Taylor, R. T. (1996). Fat body cells and calcium phosphate spherules induce ice nucleation in the freeze-tolerant larvae of the gall fly *Eurosta solidaginis* (Diptera, Tephritidae). *Journal of Experimental Biology* **199**, 465–471.

Muldrew, K., Acker, J. P., Elliott, J. A., and McGann, L. E. (2004). The water to ice transition: implications for living cells. In *Life in the Frozen State*, ed. B. Fuller, N. Lane and E. Benson. Boca Raton: CRC Press, pp. 67–108.

Nedved, O. (2000). Snow White and the Seven Dwarfs: a multvariate approach to classification of cold tolerance. *CryoLetters* **21**, 339–348.

Neufeld, D. S. and Leader, J. P. (1997). Freezing survival by isolated Malpighian tubules of the New Zealand alpine weta *Hemideina maori*. *Journal of Experimental Biology* **201**, 227–236.

Oberhauser, K. and Peterson, A. (2003). Modeling current and future potential wintering distributions of eastern North American monarch butterflies. *Proceedings of the National Academy of Sciences, USA* **100**, 14063–14068.

Olsen, T. M., Sass, S. J., Li, N., and Duman, J. G. (1998). Factors contributing to seasonal increases in inoculative freezing resistance in overwintering fire-colored beetle larvae *Dendroides canadensis* (Pyrochroidae). *Journal of Experimental Biology* **201**, 1585–1594.

Oswood, M. W., Miller, L. K., and Irons, J. G. (1991). Overwintering of freshwater benthic marcoinvertebrates. In *Insects at Low Temperature*, ed. R. E. Lee and D. L. Denlinger. New York: Chapman and Hall, pp. 360–375.

Philip, B. N., Yi, S.-X., Elnitsky, M. A., and Lee, R. E. (2008). Aquaporins play a role in desiccation and freeze tolerance in larvae of the goldenrod gall fly, *Eurosta solidaginis*. *Journal of Experimental Biology* **211**, 1114–1119.

Pruitt, N. L., Moqueet, N., and Shapiro, C. A. (2007). Evidence for a novel cryoprotective protein from freeze-tolerant larvae of the goldenrod gall fly *Eurosta solidaginis*. *Cryobiology* **54**, 125–128.

Qi, X.-L., Wang, X.-H., Xu, H.-F., and Kang, L. (2007). Influence of soil moisture on egg cold hardiness in the migratory locust *Locusta migratoria* (Orthoptera: Acrididae). *Physiological Entomology* **32**, 219–224.

Ramlov, H. (1998). Letter to editor. *CryoLetters* **19**, 4.

Ramlov, H. and Lee, R. E. (2000). Extreme resistance to desiccation in overwintering larvae of the gall fly *Eurosta solidaginis* (Diptera: Tephritidae). *Journal of Experimental Biology* **203**, 983–789.

Régnière, J. and Bentz, B. (2007). Modeling cold tolerance in the mountain pine beetle, *Dendroctonus ponderosae*. *Journal of Insect Physiology* **53**, 559–572.

Renault, D., Nedved, O., Hervant, F., and Vernon, P. (2004). The importance of fluctuating thermal regimes for repairing chill injuries in the tropical beetle *Alphitobius diaperinus* (Coleoptera: Tenebrionidae) during exposure to low temperature. *Physiological Entomology* **29**, 139–145.

Rinehart, J. P., Hayward, S. A. L., Einitsky, M. A., Sandro, L. H., Lee, R. E., and Denlinger, D. L. (2006). Continuous up-regulation of heat shock proteins in larvae, but not adults, of a polar insect. *Proceedings of the National Academy of Sciences, USA* **103**, 14223–14227.

Rinehart, J. P., Li, A., Yocum, G. D., Robich, R. M., Hayward, S. A. L., and Denlinger, D. L. (2007). Up-regulation of heat shock proteins is essential for cold survival during insect diapause. *Proceedings of the National Academy of Sciences, USA* **104**, 11130–11137.

Ring, R. A. and Tesar, D. (1980). Cold-hardiness of the arctic beetle, *Pytho americanus* Kirby Coleoptera, Pythidae (Salpingidae). *Journal of Insect Physiology* **26**, 763–774.

Ring, R. A. and Tesar, D. (1981). Adaptations to cold in Canadian Arctic insects. *Cryobiology* **18**, 199–211.

Rojas, R. R. and Leopold, R. A. (1996). Chilling injury in the housefly: evidence for the role of oxidative stress between pupariation and emergence. *Cryobiology* **33**, 447–458.

Sakurai, M., Furuki, T., Akao, K., Tanaka, D., Nakahara, Y., Kikawada, T., Watanabe, M., and Okuda, T. (2008). Vitrification is essential for anhydrobiosis in an African chironomid, *Polypedilum vanderplanki*. *Proceedings of the National Academy of Sciences, USA* **105**, 5093–5098.

Salt, R. W. (1959). Survival of frozen fat body cells in an insect. *Nature* **184**, 1426.

Salt, R. W. (1961). Principles of insect cold-hardiness. *Annual Review of Entomology* **6**, 55–74.

Salt, R. W. (1962). Intracelluar freezing in insects. *Nature* **193**, 1207–1208.

Scholander, P. F., Flagg, W., Hock, R. J., and Irving, L. (1953). Studies on the physiology of frozen plants and animals in the Arctic. *Journal of Cellular Comparative Physiology* **42**, 1–56.

Shimada, K. and Riihimaa, A. (1988). Cold acclimation, inoculative freezing and slow cooling: essential factors contributing to the freezing-tolerance in diapausing larvae of *Chymomyza costata* (Diptera: Drosophilidae). *CryoLetters* **9**, 5–10.

Sinclair, B. J. (1999). Insect cold tolerance: how many kinds of frozen? *European Journal of Entomology* **96**, 157–164.

Sinclair, B., Addo-Bediako, A., and Chown, S. L. (2003). Climatic variability and the evolution of insect freeze tolerance. *Biological Reviews* **78**, 181–195.

Sinclair, B. and Wharton, D. A. (1997). Avoidance of intracellular freezing by the freezing-tolerant New Zealand Alpine weta *Hemideina maori* (Orthoptera: Stenopelmatidae). *Journal of Insect Physiology* **43**, 621–625.

Somero, G. N. (1992). Adapting to water stress: convergence on common solutions. In *Water and Life*, ed. G. N. Somero, C. B. Osmond, and C. L. Bolis. London: Springer-Verlag, pp. 3–18.

Sømme, L. (1982). Supercooling and winter survival in terrestrial arthropods. *Comparative Biochemistry and Physiology* **73A**, 519–543.

Steigerwald, K. A., Lee, M. R., Lee, R. E., and Marshall, J. C. (1995). Effect of biological ice nucleators on insect supercooling capacity varies with anatomic site of application. *Journal of Insect Physiology* **41**, 603–608.

Steponkus, P. L. and Lynch, D. V. (1989). Freeze/thaw-induced destabilization of the plasma membrane and the effects of cold acclimation. *Journal of Bioenergetics and Biomembranes* **21**, 21–41.

Storey, K. B., Baust, J. G., and Storey, J. M. (1981). Intermediary metabolism during low temperature acclimation in the overwintering gall fly larva, *Eurosta solidaginis*. *Journal of Comparative Physiology B* **144**, 183–190.

Storey, K. B. and Storey, J. M. (1988). Freeze tolerance in animals. *Physiological Reviews* **68**, 27–84.

Storey, K. B. and Storey, J. M. (1991). Biochemistry of cryoprotectants. In *Insects at Low Temperature*, ed. R. E. Lee, and D. L. Denlinger. New York and London: Chapman and Hall, pp. 64–93.

Storey, K. B. and Storey, J. M. (1996). Natural freezing survival in animals. *Annual Review of Ecology and Systematics* **27**, 365–386.

Strong-Gunderson, J. M., Lee, R. E., Lee, M. R., and Riga, T. J. (1990). Ingestion of ice-nucleating active bacteria increases the supercooling point of the lady beetle *Hippodamia convergens*. *Journal of Insect Physiology* **36**, 153–157.

Tanaka, K. and Watanabe, M. (2003). Transmission of ice-nucleating active bacteria from a prey reduces cold hardiness of a predator (Araneae: Theridiidae). *Naturwissenschaften* **90**, 449–451.

Tanghe, A., Van Dijck, P., Dumortier, F., Teunissen, A., Hohmann, S., and Thevelein, J. M. (2002). Aquaporin expression correlates with freeze tolerance in baker's yeast, and overexpression improves freeze tolerance in industrial strains. *Applied and Environmental Microbiology* **68**, 5981–5989.

Taylor, M. J., Song, Y. C., and Brockbank, K. G. (2007). Vitrification in tissue preservation: new developments. In *Life in the Frozen State*, ed. B. J. Fuller, N. Lane and E. E. Benson. Boca Raton: CRC Press, pp. 603–642.

Teets, N. M., Elnitsky, M. A., Benoit, J. B., Lopez-Martinez, G., Denlinger, D. L., and Lee, R. E. (2008). Rapid cold-hardening in larvae of the Antarctic midge, *Belgica antarctica*: cellular cold-sensing and a role for calcium. *American Journal of Physiology* **294**, R1938-R1946.

Tran, K., Ylioja, T., Billings, R. F., Regniere, J., and Ayres, M. P. (2007). Impact of minimum winter temperatures on the population dynamics of *Dendroctonus frontalis*. *Ecological Applications* **17**, 882–899.

Tsumuki, H., Konno, H., Maeda, T., and Okamoto, Y. (1992). An ice-nucleating active fungus isolated from the gut of the rice stem borer, *Chilo suppressalis* Walker (Lepidoptera: Pyralidae). *Journal of Insect Physiology* **38**, 119–125.

Turnock, W. J. and Bodnaryk, R. P. (1993). The reversal of cold injury and its effect on the response to subsequent cold exposures. *CryoLetters* **14**, 251–256.

Turnock, W. J. and Fields, P. G. (2005). Winter climates and cold hardiness in terrestrial insects. *European Journal of Entomology* **102**, 561–576.

Tursman, D., Duman, J. G., and Knight, C. A. (1994). Freeze tolerance adaptations in the centipede, *Lithobius forficatus*. *Journal of Experimental Zoology* **268**, 347–353.

Umina, P. A., Weeks, A. R., Kearney, M. R., McKechnie, S. W., and Hoffmann, A. A. (2005). A rapid shift in a classic clinal pattern in *Drosophila* reflecting climate change. *Science* **308**, 691–693.

Vali, G. (1995). Principles of ice nucleation. In *Biological Ice Nucleation and its Applications*, ed. R. E. Lee, G. J. Warren and L. V. Gusta. St. Paul: APS Press, pp. 1–28.

Vernon, P. and Vannier, G. (2002). Evolution of freezing susceptibility and freezing tolerance in terrestial arthropods. *Comptes Rendus Biologies* **325**, 1185–1190.

Voituron, Y., Mouquet, N., de Mazancourt, C., and Clobert, J. (2002). To freeze or not to freeze? An evolutionary perspective on the cold-hardiness strategies of overwintering ectotherms. *American Naturalist* **160**, 255–270.

Walters, K. R., Sformo, T., Barnes, B. M., and Duman, J. G. (2009). Freeze tolerance in an Alaska stonefly. *Journal of Experimental Biology* **212**, 305–312.

Wharton, D. A. and Ferns, D. J. (1995). Survival of intracellular freezing by the antarctic nematode *Panagrolaimus davidi*. *Journal of Experimental Biology* **198**, 1381–1387.

Wharton, D. A., Goodall, G., and Marshall, C. J. (2003). Freezing survival and cryoprotective dehydration as cold tolerance mechanisms in the Antarctic nematode Panagrolaimus davidi. *Journal of Experimental Biology* **206**, 215–221.

Wilson, P. W., Heneghan, A. F., and Haymet, A. D. J. (2003). Ice nucleation in nature: supercooling point (SCP) measurements and the role of heterogeneous nucleation. *Cryobiology* **46**, 88–98.

Worland, M. R., Grubor-Lajsic, G., and Montiel, P. O. (1998). Partial desiccation induced by sub-zero temperatures as a component of the survival strategy of the Arctic collembolan *Onychiurus arcticus* (Tullberg). *Journal of Insect Physiology* **44**, 211–219.

Worland, M. R., Wharton, D. A., and Byars, S. G. (2004). Intracellular freezing and survival in the freeze tolerant alpine cockroach *Celatoblatta quinquemaculata*. *Journal of Insect Physiology* **50**, 225–232.

Yancey, P. H. (2005). Organic osmolytes as compatible, metabolic and counteracting cryoprotectants in high osmolarity and other stresses. *Journal of Experimental Biology* **208**, 2819–2830.

Yi, S.-X. and Lee, R. E. (2004). In vivo and in vitro rapid cold hardening protects cells from cold-shock injury in the flesh fly. *Journal of Comparative Physiology B* **174**, 611–615.

Yi, S.-X. and Lee, R. E. (2005). Changes in gut and Malpighian tubule transport during seasonal acclimatization and freezing in the gall fly *Eurosta solidaginis*. *Journal of Experimental Biology* **208**, 1895–1904.

Yi, S.-X., Moore, C. W., and Lee, R. E. (2007). Rapid cold-hardening protects *Drosophila melanogaster* from cold-induced apoptosis. *Apoptosis* **12**, 1183–1193.

Zachariassen, K. E. (1985). Physiology of cold tolerance in insects. *Physiological Reviews* **65**, 799–832.

Zachariassen, K. E. (1991). The water relations of terrestrial arthropods. In *Insects at Low Temperature*, ed. R. E. Lee and D. L. Denlinger. New York and London: Chapman and Hall, pp. 47–63.

Zachariassen, K. E. (1992). Ice nucleating agents in cold-hardy insects. In *Water and Life*, ed. G. N. Somero, C. B. Osmond and C. L. Bolis. Berlin: Springer-Verlag, pp. 261–281.

Zachariassen, K. E. and Hammel, H. T. (1976). Nucleating agents in the haemolymph of insects tolerant to freezing. *Nature* **262**, 285–287.

Zachariassen, K. E. and Husby, J. A. (1982). Antifreeze effect of thermal hysteresis agents protects highly supercooled insects. *Nature* **298**, 865–867.

Zachariassen, K. E., Kristiansen, E., Perdersen, S. A., and Hammel, H. T. (2004). Ice nucleation in solutions and freeze-avoiding insects: homogeneous or heterogeneous? *Cryobiology* **48**, 309–321.

2

Rapid cold-hardening: Ecological significance and underpinning mechanisms

RICHARD E. LEE, JR. AND DAVID L. DENLINGER

2.1 Introduction

Insects are constantly subjected to changes in environmental temperature. Most studies of insect acclimation to low temperature concern seasonal changes that occur over weeks or months in preparation for winter, and, accordingly, most chapters in this volume focus on seasonal cold-hardening. In contrast, during the past 10 years considerable attention has been paid to rapid acclimatory responses to both high (i.e. induction of heat shock or stress proteins (Feder *et al.*, 2002) and low temperature. This chapter summarizes our current understanding of the rapid cold-hardening (RCH) response. When our previous book (Lee and Denlinger, 1991) was being written, the RCH response had only just been described and merited only a few scattered paragraphs. Indeed, at that time it was unclear whether this response was merely a laboratory artifact or a previously unrecognized type of rapid acclimation. Since then, the RCH response has emerged as a highly conserved trait, allowing diverse insect groups to swiftly adjust their physiological state and organismal performance to match even modest changes in environmental temperature. In this chapter, we summarize evidence supporting the ecological relevance and emerging physiological underpinnings of the RCH response.

The RCH response protects against a form of non-freezing injury known as cold-shock or direct-chilling injury. Cold-shock injury is well known among microbes, plants and animals, and represents a major obstacle for the successful cryopreservation of many types of cells and tissues (Grout, 1987). Injury is not associated with internal ice formation. Rather, it is due to direct effects of cold and often

Low Temperature Biology of Insects, ed. David L. Denlinger and Richard E. Lee Jr. Published by Cambridge University Press.

Table 2.1 *Summary of insects and related arthropod taxa exhibiting RCH in various life stages (E = egg, L = larva/nymph, P = pupae; A = adult)*

Order (Family)	Stage	Selected References
Acari (Phytoseiidae)	A	Broufas and Koveos, 2001
Collembola (Entomobryidae, Onychiuridae, Isotomidae, Hypogastruridae)	A	Bahrndorff *et al.*, 2008
Orthoptera (Acrididae)	L	Wang and Kang, 2003
Hemiptera (Aphidae, Cicadellidae, Lygaeidae)	L, A	Larsen *et al.*, 1993; Lee *et al.*, 1987; Powell and Bale, 2004
Thysanoptera (Tripidae)	L	McDonald *et al.*, 1997
Lepidoptera (Danaidae, Noctuidae)	L, A	Kim and Kim, 1997; Larsen and Lee, 1994
Coleoptera (Cerambycidae, Chrysomelidae, Cucujidae, Curculionidae, Tenebrionidae)	E, A	Burks and Hagstrum, 1999; Lee *et al.*, 1987; Sinclair and Chown, 2006; Shintani and Ishikawa, 2007; Terblanche *et al.*, 2007
Diptera (Agromyzidae, Cecidomyiidae, Ceratopogonidae, Chironomidae, Drosophilidae, Helcomyzidae, Muscidae, Sarcophagidae, Tephritidae)	L, P, A	Coulson and Bale, 1990; Klok *et al.*, 2003; Koveos, 2001; Lee *et al.*, 1987; Lee *et al.* 2006; Li *et al.*, 1999; Nunamaker, 1993; Rosales *et al.* 1994; Terblanche *et al.*, 2007

occurs at temperatures that are 5 °C or more above the supercooling point. The membranes of the cell and its organelles are the primary sites of cold-shock injury, causing loss of selective permeability that results in redistribution of cations, loss of cytoplasmic enzymes and other compounds, damage to mitochondrial function and loss of osmotic responsiveness (Grout, 1987). These effects are believed to be a consequence of thermotropic phase changes in membrane lipids during rapid cooling.

2.2 Taxonomic and life stage diversity

The RCH response is present in diverse insect and related arthropod groups worldwide (Table 2.1). It was first reported in flesh flies, an elm leaf beetle and a milkweed bug (Lee *et al.*, 1987; Chen *et al.*, 1987). Since then the list of taxa exhibiting this trait has grown to eight orders and 26 families. A recent report by Bahrndorff *et al.* (2009) added four families of Collembola. Diptera and Coleoptera are especially well represented by 14 families and numerous species, however not all insects can undergo RCH (e.g. tsetse fly, Terblanche *et al.*, 2008). Notably absent from this list is the Hymenoptera. Whether the absence of representatives from

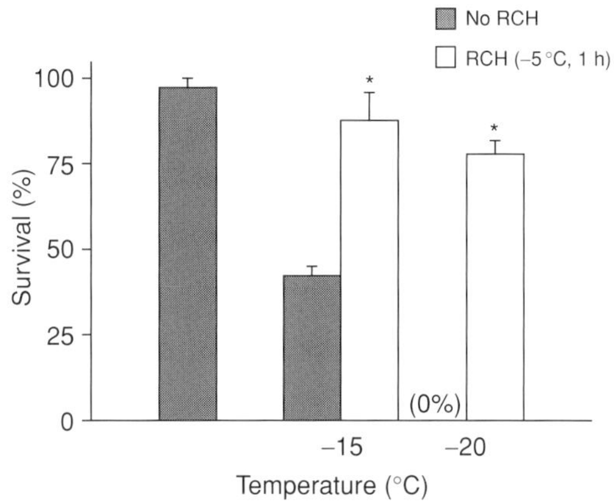

Figure 2.1 Effect of rapid cold-hardening on the freeze-tolerance of cold-acclimated larvae of the Antarctic midge, *Belgica antarctica*. * denotes a significant ($p < 0.05$) difference in survival between groups directly exposed to −15 or −20 °C and those that were first held at −5 °C for 1 h prior to the lower temperature exposure. (From Lee *et al.*, 2006.)

this highly successful, holometabolous order has a physiological basis or, more likely, is simply a result of lack of study is unknown.

Diversity is also apparent in the life stages exhibiting the RCH response (Table 2.1). RCH is found in larvae, pupae and adults of dipterans and for caterpillar and adult stages of the Lepidoptera. RCH is even found in the egg stage. When eggs of the cerambycid beetle *Psacothea hilaris* are directly exposed to −22 °C for 2 h, <30% survive; the survival rate increases to >60% if they are exposed to 0 °C for 4 h before exposure to −22 °C (Shintani and Ishikawa, 2007). Furthermore, within a single species, more than one life stage may exhibit RCH. Czajka and Lee (1990) reported RCH in larvae, pupae and adults of *Drosophila melanogaster*.

2.3 RCH in a freeze-tolerant insect

Until 2005, RCH was only reported in freeze-intolerant insects. However, in 2006 we reported RCH in freezing-tolerant larvae, but not adults, of the Antarctic midge (Lee *et al.*, 2006). The larval supercooling point is ~ −6.6 to −8.6 °C. RCH had no effect on the supercooling point. Fewer than 50% of larvae survived direct transfer to −15 °C and none survived −20 °C (Fig. 2.1). In contrast, 1 h at −5 °C (the lowest temperature reported to induce RCH) increased survival rates to 90% at −15 °C and 75% at −20°C, thus demonstrating an extension of the lower limit of freezing-tolerance. Since these larvae were cold -acclimated prior to these tests and had intrinsic levels of cold-tolerance greater than field-collected summer larvae, these data suggest that the underlying mechanism for RCH is distinct and works in an additive manner to that of winter hardening.

The most common criterion for identifying RCH relies on changes in survival rates at test temperatures above the supercooling point. Although this assessment is useful in identifying RCH in freezing-intolerant species, it is not appropriate for insects that are freezing-tolerant. Future work should carefully examine changes in the lower limit of freezing-tolerance to determine whether other freezing-tolerant species have the capacity for RCH.

2.4 Comparison to other types of cold- and winter-hardening

In several Antarctic microarthropods, swift changes in cold-tolerance are associated with changes in microhabitat temperatures and diurnal thermoperiods (Worland *et al.*, 2000, 2007; Worland and Convey, 2001; Sinclair *et al.*, 2003). However, in these cases cold-hardening involves rapid decreases in the SCP and appears to be distinctly different from the RCH response, which protects against cold-shock injury at temperatures above the SCP. Nonetheless, these Antarctic microarthropods can quickly change their lower lethal temperature (i.e. their SCP) in response to changing environmental conditions.

The RCH response differs from cold-hardening associated with winter in several ways. First, it does not confer extreme cold-tolerance that allows survival of temperatures of −25 °C or below. Rather, it allows survival of relatively mild sub-zero temperatures. Typically insects only overwinter in a single life stage, whereas the RCH response can occur in multiple life stages for a given species. Winter cold-hardening frequently occurs in dormant (diapausing) insects, while RCH occurs in life stages that are actively feeding, growing and reproducing, as well as overwintering stages (Li *et al.*, 1999). Lastly, winter-hardening typically requires weeks or months to achieve maximal cold-tolerance, compared to the RCH response, which can be induced by chilling in hours or even minutes. Remarkably, as little as 30 min at 5 °C increased adult survival from 0% to >50% (Czajka and Lee, 1990). Also, in contrast to winter cold-hardening, the protective effects of RCH are quickly lost upon return to warmer temperatures (also see later section on induction triggers).

The type of cold protection derived from RCH appears to be distinctly different from those associated with longer periods of cold-acclimation or winter-acclimatization in the field. In western flower thrips, RCH increased survival of cold-shock at −11.5 °C, but not chilling-tolerance at −5 °C (McDonald *et al.*, 1997). In contrast, cold-acclimation by rearing at 15 °C increased tolerance of −5 °C but not of −11.5 °C, suggesting that different underlying mechanisms are involved in RCH and cold -acclimation (McDonald *et al.*, 1997). Similar effects are known for other insects including *D. melanogaster* (Rako and Hoffman, 2006), a gall midge (Li *et al.*,

1999), eggs of a cerambycid beetle (Shintani and Ishikawa, 2007) and monarch butterflies (Larson and Lee, 1994). Selection for increased cold-shock-tolerance did not increase indirect chilling-tolerance, thus suggesting that different mechanisms are required for protection against direct versus indirect chilling injury (Chen and Walker, 1994). The diversity of insect responses to different low temperature treatments indicates a high degree of phenotypic plasticity both within and between species and, accordingly, suggests a diverse array of underlying mechanisms (Rako and Hoffman, 2006; Chown and Terblanche, 2007).

2.5 Ecological significance

Shortly after we first reported RCH (Lee *et al.*, 1987) critical questions were raised as to whether this response was, in fact, important to insects in nature (Coulson and Bale, 1990). Since our initial studies used long-established laboratory strains of *Drosophila* and *Sarcophaga* (Lee *et al.*, 1987; Chen *et al.*, 1987; Czajka and Lee, 1990) it was not initially clear whether the responses would match those obtained for field populations. Furthermore, since the extreme experimental treatments (abrupt step transfers of 25 °C or more with accordingly high cooling rates) used to administer cold-shock and to induce RCH were unlikely to occur in nature, would more ecologically relevant rates of cooling and exposure also induce RCH? Lastly, although the RCH response conferred short-term benefits on survival, did these benefits extend to long-term survival and, more importantly, protection of reproductive fitness? And, if so, at what cost?

2.5.1 Induction triggers

Abrupt chilling by step transfer to a lower temperature is the most frequently used method for inducing the RCH response. For example, neither larvae nor adults of *D. melanogaster* survive direct transfer from 23 to −5 °C for 2 h (Fig. 2.2, Czajka and Lee, 1990). However, if allowed to rapidly cold-harden for 2 h at +5 °C prior to 2 h at −5 °C survival increased to >80%. Even very modest cooling (from 21 to 16 °C, Shreve *et al.*, 2004) is sufficient to induce a change in organismal performance. Step transfer to temperatures as low as −2, or even −5 °C, can induce RCH and provide cryoprotection at still lower temperatures (Lee *et al.*, 2006; Terblanche *et al.*, 2007).

The step-transfer method is quick and easily accomplished and serves as a useful tool for rapid screening, however it has significant drawbacks. Following RCH induction, changes in the test temperature of as little as 1 °C near the lower lethal temperature can result in survival rate changes from 100 to 0%. Consequently, with such abrupt transitions, evidence of RCH may be missed unless a range of temperatures is carefully tested. Secondly, the rate of cooling in step transfer is so

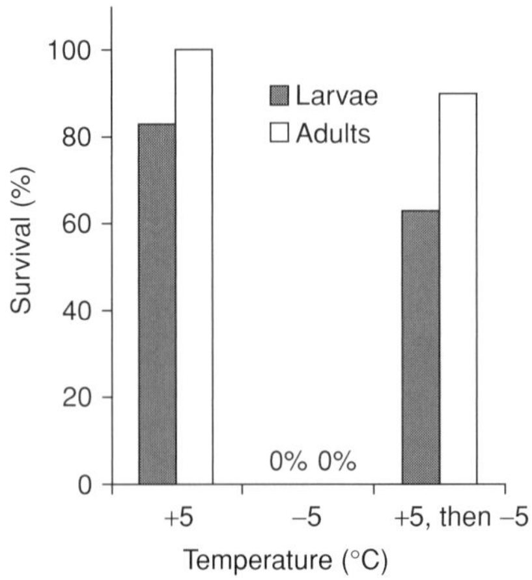

Figure 2.2 Effect of rapid cold-hardening on larvae and five-day-old adults of *Drosophila melanogaster*. Flies were either transferred directly to +5 °C or −5 °C for 2 h or exposed to +5 °C for 2 h prior to sub-zero exposure at −5 °C. (Data from Czajka and Lee, 1990.)

rapid that it would rarely, if ever, occur in nature. Investigations attempting to simulate natural conditions should use natural rates of cooling and thermoperiods to characterize this phenomenon.

Rapid cold-hardening can also be induced by brief exposure to high temperature and anoxia. Exposure of pharate adults of *S. crassipalis* to 36 °C for 2 h significantly increased their tolerance of −10 °C (Chen *et al.*, 1987). Coulson and Bale (1991) reported that exposure of house flies to 40 min of anoxia markedly enhanced their tolerance of −7 °C, but the extent of cryoprotection was not as extensive as that from RCH. Species differences may occur, as indicated by the fact that anoxia blocks RCH in the flesh fly *S. crassipalis* (Yocum and Denlinger, 1994).

While induction of RCH occurs rapidly, it is also quickly lost upon warming. The increased cold-hardiness gained by RCH in house fly pupae is lost within 2 h after return to 27 °C (Coulson and Bale, 1990), and western flower thrips lose their tolerance within 1 h at 20 °C (McDonald *et al.*, 1997). Nymphs of the migratory locust lose RCH-enhanced cold-tolerance within 2 h at 30 °C (Wang and Kang, 2003). Within 15 min after transfer to 24 °C, adults of the olive fruit fly (*Bactrocera oleae*) lose most of the protection gained by RCH (Koveos, 2001).

2.5.2 Short-term organismal effects of RCH

Manifestation of the detrimental effects of chilling/cold-shock may be evident immediately or delayed, i.e. expressed in later developmental stages or in reproductive output. The response is a consequence of the severity and duration

Table 2.2 *Responses to RCH at the organismal level*

Organismal Response	Reference
↑ survival of sub-zero exposure	Lee *et al.*, 1987, and many others
↓ chill-coma temperature	Kelty and Lee, 1999; Powell and Bale, 2006
↓ lower limit of freeze-tolerance	Lee *et al.*, 2006
↓ rate of water loss	Yoder *et al.*, 2006
Preserves reflex behaviors	Kelty *et al.*, 1996
Preserve courtship and mating performance	Shreve *et al.*, 2004
Maintain or enhance fecundity and longevity	Powell and Bale, 2005; Rinehart *et al.*, 2000
Preserves learning/spatial conditioning	Kim *et al.*, 2005
Preserves normal flight behavior	Larsen and Lee, 1994

of exposure to chilling. Accordingly, the consequences may range from minor and inconsequential, to major impairment of the nervous system, to death (Coulson *et al.*, 1992; Kelty *et al.*, 1996).

The organismal effects of RCH are evident soon after the RCH treatment (Table 2.2). With only 1–2 h of RCH, insects can survive sub-zero temperatures that were previously lethal. RCH also causes rapid lowering of the chill-coma temperature, allowing the insect to remain mobile at lower temperatures (Kelty and Lee, 1999; Powell and Bale, 2006). Yoder *et al.* (2006) reported lower rates of water loss following RCH. RCH prevented cold-shock-induced impairment of the proboscis extension reflex and grooming behavior in adult flesh flies (Kelty *et al.*, 1996) and preserved normal flight behavior in monarch butterflies (Larsen and Lee, 1994).

RCH also preserves normal reproductive behaviors in adult flies that are rapidly cooled. When control flies were transferred from 23 to 16 °C, only half of the pairs courted, and no pairs mated during the next hour (Shreve *et al.*, 2004). In contrast, when pairs were allowed 2 h at 16 °C to undergo RCH, 85% of the pairs courted and 11 of 20 pairs mated. Other studies demonstrated that RCH can maintain or enhance fecundity and longevity (Rinehart *et al.*, 2000; Powell and Bale, 2005). Kim *et al.* (2005) found that RCH preserves learning and spatial conditioning that would otherwise be blocked by cold-shock.

2.5.3 Effect of RCH on chill-coma temperature during ecologically relevant cooling

One approach to determining whether RCH occurs under natural conditions is to use cooling regimens that mimic rates and conditions that occur in nature, coupled with ecologically significant measures of organismal performance. The temperature at which insects become immobilized by extreme low

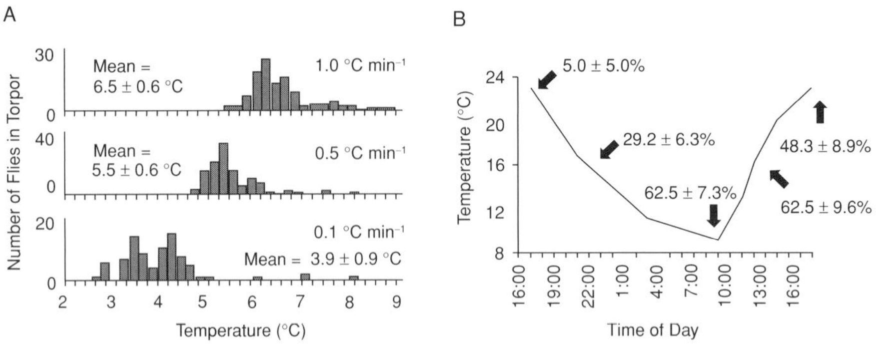

Figure 2.3 (A) Effect of cooling rate on the critical thermal minimum, and (B) on the ability to survive exposure to −7 °C at various times during a natural thermoperiod in *Drosophila melanogaster* adults (from Kelty and Lee, 1999, 2001).

temperature is commonly referred to as the chill-coma or cold-stupor temperature (Kelty and Lee, 1999). Although insects can generally recover from this effect if exposure is not prolonged, it is referred to as the temperature at which "ecological death" occurs, since evasion of predators by running or flying is no longer possible. When adults of *D. melanogaster* are cooled rapidly (1 °C min^{-1}) their chill coma temperature is 6.5 °C (Fig. 2.3A; Kelty and Lee, 1999). However, slow cooling at 0.1 °C min^{-1}, a rate likely to occur in nature, not only increased their tolerance of cold-shock at −7 °C, but significantly lowered their chill coma temperature to 3.9 °C. Similarly, Powell and Bale (2004, 2006) found that slow cooling at 0.1 or 0.05 °C min^{-1}induced RCH and lowered the chill-coma temperature in aphids. Bale (2002) postulated that RCH is significant because it allows insects to quickly reset their thermal activity thresholds, which has important ecological impacts on feeding, dispersal and predator avoidance.

Simulating natural thermoperiods can also induce RCH. When adult fruit flies were exposed to an ecologically based thermoperiod (23 to 9 °C cycle), RCH occurred throughout the cooling phase, increasing the fly's capacity to survive 1 h at −7 °C from 5 to 62% (Fig. 2.3B, Kelty and Lee, 2001). Some, but not all, cold-hardening was lost during the warming phase of the thermoperiod. Flies exposed to repeated thermoperiodic cycling progressively accumulated higher levels of cold-tolerance for at least seven days.

Recovery time from chill coma is another ecologically relevant measure of organismal response to cold (David *et al.*, 1998, 2003). Additionally, selection experiments are possible because it is a non-lethal assessment. Interestingly, this parameter is unaffected by RCH. Despite the fact that RCH enhances cold-shock-tolerance, it does not decrease the recovery time from chill coma in *D. melanogaster* (Rako and Hoffman, 2006).

2.5.4 *RCH under field conditions*

Despite considerable study in the laboratory, few studies have examined RCH under field conditions. Koveos (2001) exposed adult olive fruit flies to field conditions where they experienced daily temperature cycles. Flies tested during the coldest part of the thermoperiod had significantly higher levels of cold-tolerance than ones tested 12 h later that had experienced the warmest part of the cycle.

In 2007, Kelty provided a more extensive field-based validation of RCH using a newly established colony of *D. melanogaster*. Natural diurnal cycles between 22 and 10 °C produced an increase from 10 to 68% survival of cold-shock during the cooling phase and a concomitant decrease in chill-coma values from 8.7 to 6.6 °C. Interestingly, the chill-coma temperatures for RCH in the field closely matched those of long-established laboratory colonies (Czajka and Lee, 1990; Kelty and Lee, 1999). These studies confirm that field-acclimatized insects undergo RCH in response to changes in ambient temperature.

2.5.5 *Effect of RCH on fecundity and longevity*

While there is a growing consensus that RCH confers short-term beneficial effects in response to ecologically relevant cooling, few studies have examined the effects of cold-shock and RCH over extended periods with respect to reproduction and longevity. The possible costs or trade-offs of the RCH response have also received little attention. However, the available reports suggest that effects vary from species to species.

Although RCH increases cold-tolerance in *Musca domestica,* treated females had a shorter life span, a reduced oviposition rate and lower rates of adult emergence among their progeny, compared to control flies that were not cold-shocked (Coulson and Bale, 1992). Rinehart and colleagues (2000) reported that cold-shock reduces fecundity in both sexes of *S. crassipalpis*, but RCH partially prevents this impairment. RCH extends survival time at sub-zero temperatures of diapausing and nondiapausing females of the mite *Euseius finlandicus*, but reduces reproductive output of the nondiapausing females (Broufas and Koveos, 2001). Powell and Bale (2004, 2005) reported that warm-acclimated and cold-acclimated "winter" grain aphids (*Sitobion avenae*) exhibited no ecological costs related to development, fecundity or longevity.

In *D. melanogaster*, costs associated with RCH also appear to vary, but overall appear to be minor. Overgaard *et al.* (2007) found a slight reduction in survival for flies that received an RCH treatment, but were not subjected to cold-shock. Reproductive output was also lower for females during the 8 h immediately after the RCH treatment. However, Kelty and Lee (1999) found that RCH treatment did not affect egg production.

Are gains in low-temperature-tolerance or performance offset by losses at high temperatures? Again, few studies have considered this question. In a field study with *D. melanogaster*, Overgaard and Sorensen (2008) found a significant negative correlation between cold-shock-tolerance and field temperature, suggesting that adults were rapidly acclimating to environmental temperature, but the increased cold-tolerance apparently came at a cost of decreased heat-tolerance. However, in another study using the same species, RCH at 16 °C did not impair courtship or mating at the original rearing temperature of 23 °C (Shreve *et al.*, 2004).

2.6 Underlying mechanisms

It was evident from our first studies on RCH that glycerol levels were boosted during a 2-h exposure of *S. crassipalpis* to 0 °C (Chen *et al.*, 1987; Lee *et al.*, 1987), but the level of elevation was modest (from 28 to 81 mM) and unlikely to significantly alter colligative properties. This result thus suggests that additional biochemical and/or biophysical adjustments are also likely contributors to RCH. In this portion of the chapter we discuss some of the players that have been more recently linked to RCH, as well as a few tantalizing observations that point to signal-transduction pathways that may be involved in initiating this rapid physiological response.

2.6.1 Polyols and sugars

Several independent studies document elevation of glycerol during RCH in different species of flesh flies (Chen *et al.*, 1987; Lee *et al.*, 1987; Yoder *et al.*, 2006), but a metabolomics study using *S. crassipalpis* (Michaud and Denlinger, 2007) indicates that glycerol is not the only polyol that is elevated; sorbitol levels also increase during RCH. Interestingly, glycerol increased during both RCH and during the cold-hardening associated with winter diapause, but during diapause, sorbitol levels decrease rather than increase as they do during RCH, thus we should not assume that cold-hardiness is achieved the same way in both RCH and diapause.

Glucose is elevated in response to RCH in both *S. crassipalpis* (Michaud and Denlinger, 2007) and *D. melanogaster* (Overgaard *et al.*, 2007). Trehalose, the dominant blood sugar in insects, can also function as a cryoprotectant, and thus it would also be a potential candidate for involvement in RCH. But, the trehalose responses differ in these two species. In *D. melanogaster*, trehalose levels do increase in response to RCH (Overgaard *et al.*, 2007), but in *S. crassipalpis* both trehalose and mannose decrease (Michaud and Denlinger, 2007). Interestingly, trehalose alterations have also been noted in other invertebrates; in response to a brief cold-shock an entomopathogenic nematode increases trehalose levels up to sixfold (Jagdale *et al.*, 2005).

Pyruvate, the end product of the glycolytic pathway, also increases quite dramatically in response to RCH in *S. crassipalpis*. These observations are consistent with the idea that RCH may favor the glycolytic pathway, resulting in the transformation of trehalose and mannose to glucose equivalents for use in glycolysis, culminating in the elevation of pyruvate. Lastly, in contrast to winter cold-hardening when massive amounts, sometimes exceeding >1 M, of polyhydric alcohols and sugars are produced, RCH may involve very modest changes in the range of 10–20 mM that function in non-colligative ways to stabilize membranes or function in other roles (Yoder *et al.*, 2006; Overgaard *et al.*, 2007).

2.6.2 Amino acids

Alanine and glutamine levels increase nearly twofold during RCH in *S. crassipalpis*, while β-alanine and ornithine levels drop (Michaud and Denlinger, 2007). Alanine is synthesized directly from pyruvate, thus this increase is likely linked to the elevation in pyruvate noted above, and possibly also to the decrease observed in β-alanine. Alanine up-regulation has been noted in a number of insect cold-hardy states, and this amino acid is similar to glycerol in terms of its colligative properties. Like alanine, glutamine has the potential to increase the overall osmolality of the hemolymph, but it may also work non-colligatively as well, by enhancing the heat-shock response (Phanvijhitsiri *et al.*, 2005) or by suppressing apoptosis (Fuchs and Bode, 2006).

2.6.3 Heat-shock proteins

It has been clear for quite some time that heat-shock proteins (Hsps) are expressed during the recovery phase following low temperature exposure (Denlinger *et al.*, 1991), but what has been less evident is whether the Hsps are actually involved in mounting the RCH response, i.e. during the low-temperature period that generates the protective response. The Hsp that is most highly up-regulated in response to high temperature and recovery from low temperature, Hsp70, does not appear to be expressed during RCH. The mRNA and protein appear only during the recovery phase (Burton *et al.*, 1988; Joplin and Denlinger, 1990; Kelty and Lee, 2001), and knocking down Hsp70 expression with RNAi does not appear to affect the RCH response (Michaud, unpublished observations).

But there are numerous Hsps belonging to different families, and it is becoming evident that the different Hsps can respond differently to distinct stresses. A recent proteomics study of the flesh fly brain demonstrated that one of the small Hsps (smHsp), a stress protein of approximately 23 kDa, increases in abundance during a 2-h exposure to 0 °C (Li and Denlinger, 2008). None of the other Hsps were elevated by RCH, thus underscoring a potentially unique contribution of Hsp23. And, RNAi directed against Hsp23 decreases the protection generated by RCH

(Michaud, unpublished observations). Although these results need to be further verified, we hypothesize a role for Hsps both during RCH, as well as during the recovery period.

2.6.4 *ATP generation*

Several experiments suggest that ATP synthase sub-units increase in response to low temperature (Michaud and Denlinger, Chapter 4) including during RCH (Li and Denlinger, 2008). Low temperature reduces the catalytic activity of ATP synthase, but the insect can compensate for this reduction by increasing the concentration of the enzyme.

2.6.5 *Membrane changes*

In microbes, plants and ectothermic animals acclimation to low temperature is frequently associated with an increase in membrane fluidity, termed homeoviscous adaptation (Hazel, 1995; Koštál, Chapter 5). This increase, coupled with a decrease in the phase-transition temperature, reduces the likelihood of protein and ion leakage across cell membranes damaged by cold-shock.

Most studies of homeoviscous adaptation describe membrane changes that occur after weeks or months of cold-acclimation. However, recently Overgaard *et al.* (2005) found small, but significant, increases in the desaturation level of phospholipid fatty acids in *Drosophila melanogaster* in response to RCH. A similar, but more dramatic response occurs in pharate adults of the flesh fly *S. crassipalpis*, in which oleic acid increases from 30 to 47% in the membrane fatty-acid pool during RCH (Michaud and Denlinger, 2006). Increased levels of oleic acid may be especially important, because oleic acid maintains membrane fluidity and a favorable microenvironment for membrane ATPases, not only at low temperatures, but also over a relatively wide range of temperatures (Michaud and Denlinger, 2006).

Although changes in lipid composition of cell membranes are known to be important events during long-term adaptations to low temperature (Koštál, Chapter 5), it was not clear if short-term responses, such as RCH would also involve changes in the lipid profiles of membranes. It now appears that the answer is yes! In *D. melanogaster*, RCH elicits a significant, but modest increase in linoleic acid (18:2n−6) (Overgaard *et al.*, 2005), and the fatty acid (FA) response is an order of magnitude more robust in the flesh fly, *S. crassipalpis,* a fly that is considerably more cold-hardy than *D. melanogaster* (Michaud and Denlinger, 2006). But, instead of increasing levels of linoleic acid, flesh flies respond to RCH by increasing levels of oleic acid (18:1n-9): within 8 h at 4 °C, oleic acid dramatically increases from 30 to 47% of the total phospholipids FA pool, and it does so at the expense of all other FAs, with the exception of palmitic acid (16:0), which remains at the same proportion. The double bond in oleic acid is located in the center of the acyl chain, a position that maximizes lateral displacement of the chain and thus contributes to overall membrane disorder, i.e. fluidity. Oleic acid and linoleic acid share many biological properties, including the ability to provide a wide temperature window for growth, as demonstrated in bacteria (McElhaney, 1974). By utilizing either oleic acid or linoleic acid in their membranes the insect can quickly adjust to the types of temperature changes that may be regularly encountered during RCH.

In flesh flies, RCH also elicits a restructuring of membrane phospholipids head groups, favoring phosphatidylethanolamines (PE) over phosphatidylcholines (PC) (Michaud and Denlinger, 2006), an observation consistent with similar shifts reported for fish and other poikilotherms (Hazel, 1995). This increase in PE likely functions to reduce torsional stress on the outer layer of the membrane as temperature drops (Thompson, 1983).

While these increases in desaturation level imply increases in membrane fluidity, Lee and colleagues (Lee *et al.*, 2006) directly measured increases in membrane fluidity using ^{31}P solid state NMR spectroscopy. Following RCH at 0 °C for 2 h fat body cells from *S. crassipalpis* markedly increased their membrane fluidity compared to control flies to 25 °C. Furthermore, the survival rate of fat body cells from RCH flies increased by 47% compared to cold-shocked flies that were directly exposed to −8 °C.

The capacity to undergo RCH can also be manipulated using diet to alter membrane composition. When *D. melanogaster* are reared on a diet augmented with cholesterol, membrane levels of cholesterol increase by ~50% (Shreve *et al.*, 2007). This augmentation not only enhances intrinsic cold-tolerance, but also increases the fly's capacity to RCH compared to controls. Increased levels of membrane cholesterol also enhance chilling-tolerance in mammalian sperm (Drobnis *et al.*, 1993), however in other organisms cholesterol may increase or decrease in a tissue-dependent manner with cold-acclimation, thus the mode of action is unlikely to have a single explanation (Crockett, 1998). Collectively, these data indicate that rapid modifications in membrane structure play a vital role in protecting against cold-shock injury.

Lipid remodeling within the cell membrane is accompanied by changes in protein composition as well. Microarray analysis revealed up-regulation of 36 transcripts in adults of *D. melanogaster* following exposure to 0 °C for 2 h plus a 30 min recovery period at 25 °C (Qin *et al.*, 2005). Twelve of these transcripts are classed as membrane-related protein genes. It is still unknown whether all of these protein changes occur only during the recovery phase or if some of these changes in transcript levels occur during the RCH phase.

Table 2.3 *Cell viability of isolated tissues of control (23 °C), cold-shock (−8 °C for 2 h) and rapid cold-hardening (0 °C for 2 h followed by −8 °C, 2 h). (Yi and Lee, 2004)*

Tissue Type	Cell Viability (%, mean ± SEM)		
	Control	Rapidly Cold-hardened	Cold-shocked
Fat body	95.6 ± 1.8	53.2 ± 5.6	26.1 ± 3.6
Gut tissue	96.2 ± 1.2	63.3 ± 5.0	45.9 ± 5.3
Salivary gland	95.1 ± 2.1	46.0 ± 5.2	24.3 ± 2.7
Malpighian tubules	96.5 ± 1.7	50.6 ± 5.5	35.1 ± 4.0

2.6.6 RCH in isolated cells and tissues

Some insect cells can respond directly to environmental stressors. For example, when chilled at 5 °C, fat-body cells incubated *in vitro* secrete glycerol (Yi *et al.*, 1987). Also, Watanabe *et al.* (2002) reported that the induction and termination of cryptobiosis, which confers extreme resistance to cold and desiccation stress, does not involve the brain or thoracic ganglia in the chironomid *Polypedilum vanderplanki*, since isolated abdomens of these larvae synthesize substantial amounts of the cryoprotectant trehalose. However, a paucity of information exists concerning how insects perceive cold and at what level of biological organization this occurs. Is the nervous system required for induction of the RCH response? Or does cold sensing occur at the cellular level?

These questions can be addressed by determining whether RCH occurs in isolated cells and tissues. Yi and Lee (2004) used a fluorescent vital dye assay to examine viability following cold-shock of isolated cells with and without RCH (Table 2.3). For each of the four tissues tested, the RCH group exhibited significantly higher survival following exposure at −8 °C than the cold-shocked group. These *in vitro* results closely match those of intact flies tested in a comparable manner. We obtained similar results with tissues from larvae of *B. antarctica* (Teets *et al.*, 2008).

These results indicate that cold sensing can occur at the cellular level without mediation by the nervous or endocrine systems, a fact that helps to explain the remarkable swiftness of the RCH response (Yi and Lee, 2004). *In vitro* induction of RCH should facilitate future research into the mechanisms underlying this response. Although it is clear that RCH can occur in isolated tissues, the elevation of glycerol associated with RCH in flesh flies is more pronounced when the brain is present (Yoder *et al.*, 2006), thus suggesting that neural or hormonal input from the brain enhances the response.

2.6.7 Cold sensing and signal-transduction pathways

From the results discussed above, it is evident that RCH involves a complex response that includes elevation of several polyols, sugars, amino acids, heat-shock proteins, alterations in the lipid composition of cell membranes, and quite likely shifts in the pools of additional metabolites as well. The question we would now like to address is how are these responses triggered? What is the sensing mechanism that detects "cold" and what signal-transduction mechanism may be involved in orchestrating the response?

In plants, one of several cold-hardening responses is induced directly by low temperature at the cell level and is termed cold sensing (Monroy and Dhindsa, 1995; Smallwood and Bowles, 2002); this response is independent of cold-acclimation, which is mediated by abscisic acid. In this case, chilling reduces membrane fluidity that, in turn, activates calcium channels in the plasma membrane, causing an influx of calcium into the cell and expression of low-temperature responsive genes. Experimental treatment of the cell with a membrane fluidizer or an actin microfilament stabilizer prevents calcium influx and cold-hardening, while a membrane rigidifier promotes cold-acclimation (Murata and Los, 1997; Orvar *et al.*, 2000).

Thus far, two components of the cold-sensing mechanism have been identified in insects: calcium flux and activation of p38 mitogen-activated protein kinase (MAPK). Evidence for the involvement of calcium signaling is based on experiments with the Antarctic midge, *Belgica antarctica* (Teets *et al.*, 2008). Tissues from the midge that are cultured *in vivo* readily undergo RCH when cultured in a medium replete with calcium, but when cultured in a calcium-free medium or a medium containing an intracellular chelator BAPTA-AM, the RCH response is greatly diminished. A calmodulin inhibitor also reduces cell survival, thus further supporting a role for calcium in RCH. Calcium flux is known to trigger cold-acclimation in plants, and these results with the midge suggest that a similar calcium-mediated signaling system may operate for cold detection in insects as well.

Evidence for involvement of p38 MAPK in cold sensing is based on the RCH response of the flesh fly, *S. crassipalpis* (Fujiwara and Denlinger, 2007). Within 10 min after exposure to 0 °C, p38 MAPK is activated, as evidenced by its phosphorylation (Fig. 2.4B). This is the earliest RCH response thus far documented and is likely to be a critical link in initiating this response. Interestingly, the activation of p38 MAPK occurs within a fairly narrow temperature range, 0 to 5 °C (Fig. 2.4A), precisely the temperature range that is most effective in eliciting RCH in this fly. The fact that p38 MAPK is activated, not only in intact flies, but also in isolated abdomens is consistent with our observation that RCH can function independently of the brain. Other MAPKs including extracellular signal-regulated kinase (ERK) and JUN N-terminal kinase (JNK) are not phosphorylated during RCH,

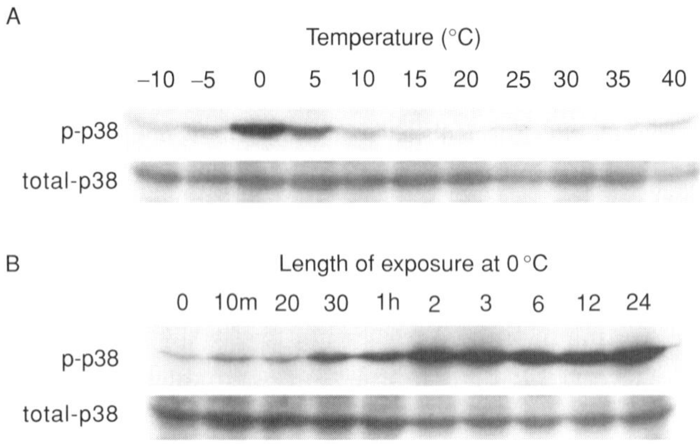

Figure 2.4 Activation of p38 MAPK during rapid cold-hardening in the flesh fly, *Sarcophaga crassipalpis*, indicating that (A) the activation is most pronounced in the temperature range of 0 to 5 °C and (B) the activation can be detected within 10 min of exposure to 0 °C and reaches a maximum in approximately 2 h at 0 °C. (Adapted from Fujiwara and Denlinger, 2007.)

thus indicating that this response is specific to p38 MAPK and does not involve activation of other MAPK family members. The MAPK family is well known to be involved in transmitting stress signals from the environment to the cell nucleus, thus the involvement of p38 MAPK in RCH is consistent with identified functions of this important group of transcription factors. Though it is not yet possible to assemble a definitive pathway linking calcium flux, p38 MAPK activation, and the downstream shifts in metabolites that are associated with RCH, the working model shown in Fig. 2.5, suggests possible interactions that could account for the observations currently available.

2.6.8 RCH blocks cold-induced apoptosis

Injury due to cold-shock has recently been linked to a very specific type of cell death called apoptosis (Yi *et al.*, 2007). Also known as programmed cell death, apoptosis is a highly regulated process by which excess cells during development, or damaged cells, are removed (Fig. 2.5). Using the TUNEL (terminal deoxynucleotidyl transferase-mediated dUTP nick end labeling) assay on cryosections from cold-shocked *D. melanogaster*, Yi *et al.* (2007) found that $>75\%$ of cells tested positive for apoptosis. Further evidence for cell death by cold-induced apoptosis comes from a distinct pattern of DNA fragmentation into short ~180 bp segments and up-regulation of the downstream executioner caspase-3, which promotes the proteolytic cascade culminating in cell death. In contrast, cell survival in RCH flies improves by 38%, caspase-3 levels were markedly down-regulated and levels of the anti-apoptotic protein Bcl-2 increase. It seems likely that thermotropic injury

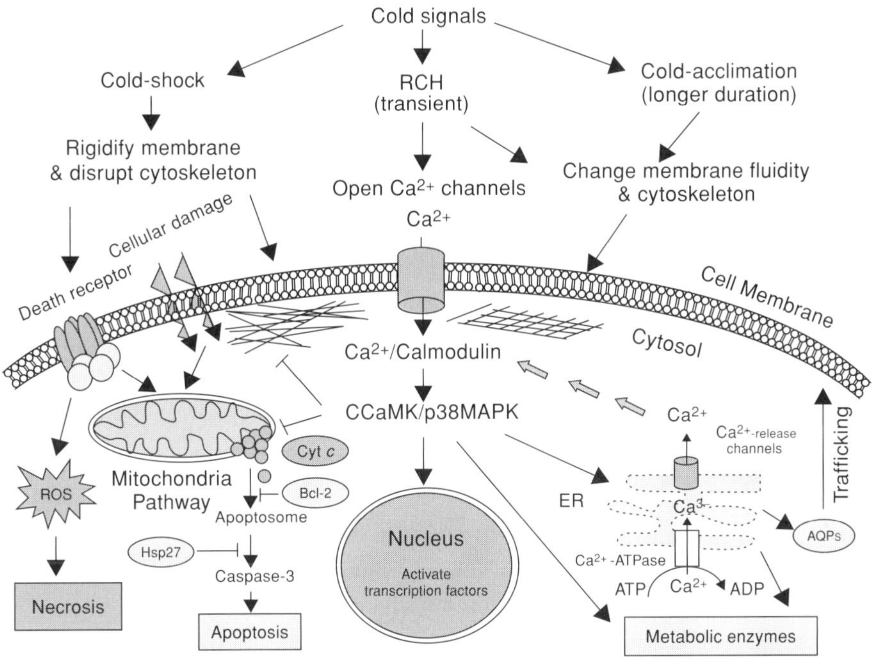

Figure 2.5 Summary of cellular responses to cold-shock and cold-hardening. (Shu-Xia Yi, manuscript submitted.)

caused by cold-shock to the mitochondrial membrane induces the intrinsic or mitochondrial pathway, triggering apoptosis, however this hypothesis requires testing.

2.7 Future directions

Although the importance of the RCH response remains little known to physiological ecologists, we now have evidence that RCH plays a significant role in allowing insects to adjust their physiological function to match rapid changes in environmental temperature. Of special importance is the fact that even modest environmental cooling (e.g. 23 to 16 °C; Shreve *et al.*, 2004), which would not generally be considered cold, can induce RCH and markedly enhance organismal performance. The speed of this response and its widespread taxonomic distribution suggest it is critical for tracking and responding to short-term changes in environmental temperatures caused by unpredictable weather conditions, diurnal thermoperiods or by temperature differences encountered by insects as they visit various microhabitats. However, only a few species have been investigated in any detail, and it is clear that we have only begun to grasp the ecological and evolutionary significance of this acclimatory response. Future studies concerning temperature acclimation should not only consider seasonal and long-term responses, but also acclimatory changes that occur within minutes to hours.

Understanding the RCH response has significant practical implications for cryopreservation. Non-freezing injury due to cold-shock remains a major road-block for cryosurgery (Murakami *et al.*, 1997) and for the cryopreservation of mammalian oocytes, sperm and other cells (Massip *et al.*, 2004). Cold-shock injury in mammalian cells is also linked to apoptosis (Baust *et al.*, 2002). Consequently, study of the mechanisms by which RCH prevents cold-induced apoptosis may provide clues for the development of improved methods for the cryopreservation of *Drosophila* and other insects as well as mammalian tissues (Leopold and Rinehart, Chapter 13).

Understanding the RCH mechanism may also generate strategies for disrupting insect development and stress-tolerance that can be used for pest management and biological control. This information may be especially useful in establishing quarantine procedures for pests of imported agricultural produce. Now that methyl bromide has been taken off the market, high- and low-temperature treatments of fruits and vegetables are being promoted as an environmentally safe technique for eliminating pests in imported produce (Mangan and Hallman, 1998). For temperature treatments to be effective against pests, the RCH response must be understood and circumvented. The increased use of beneficial insects for biological control purposes also demands tools for safely chilling insects for shipment and storage. Again, understanding the RCH response is essential for the effective development of this growing industry (Leopold, 1998).

Acknowledgements

We thank Shu-Xia Yi for Fig. 2.5. Support from the National Science Foundation is gratefully acknowledged.

References

Bahrndorff, S., Loeschcke, V., Pertoldi, C., Beier, C., and Holmstrup, M. (2009). The rapid cold hardening response of Collembola is influenced by thermal variability of the habitat. *Functional Ecology* **23**, 340–347

Bale, J. S. (2002). Insects and low temperatures: from molecular biology to distributions and abundance. *Philosophical Transactions of the Royal Society of London Series B-Biological Sciences* **357**, 849–861.

Baust J. M., Van Buskirk, R., and Baust, J. G. (2002). Gene activation of the apoptotic caspase cascade following cryogenic storage. *Cell Preservation Technology* **1**, 63–80.

Broufas, G. D. and Koveos, D. S. (2001). Rapid cold hardening in the predatory mite *Euseius (Amblyseius) finlandicus* (Acari: Phytoseiidae). *Journal of Insect Physiology* **47**, 699–708.

Burks, C. S. and Hagstrum, D. W. (1999). Rapid cold hardening capacity in five species of coleopteran pests of stored grain. *Journal of Stored Products Research* **35**, 65–75.

Burton, V., Mitchell, H. K., Young, P., and Petersen, N. S. (1988). Heat shock protection against cold stress of *Drosophila melanogaster*. *Molecular and Cellular Biology* **8**, 3550–3552.

Chen, C.-P. and Walker, V. K. (1994). Cold-shock and chilling tolerance in *Drosophila*. *Journal of Insect Physiology* **40**, 661–669.

Chen, C. P., Denlinger, D. L., and Lee, R. E. (1987). Cold-shock injury and rapid cold hardening in the flesh fly *Sarcophaga crassipalpis*. *Physiological Zoology* **60**, 297–304.

Chown, S. L. and Terblanche, J. S. (2007). Physiological diversity in insects: ecological and evolutionary contexts. *Advances in Insect Physiology* **33**, 50–152.

Coulson, S. J. and Bale, J. S. (1990). Characterization and limitations of the rapid cold-hardening response in the house fly *Musca domestica* (Diptera: Muscidae). *Journal of Insect Physiology* **36**, 207–211.

Coulson, S. J. and Bale, J. S. (1991). Anoxia induces rapid cold hardening in the house fly *Musca domestica* (Diptera: Muscidae). *Journal of Insect Physiology* **37**, 497–501.

Coulson, S. C. and Bale, J. S. (1992). Effect of rapid cold hardening on reproduction and survival of offspring in the house fly *Musca domestica*. *Journal of Insect Physiology* **38**, 421–424.

Coulson, S. J., Fisher, J., and Bale, J. S. (1992). A 31P NMR investigation of the energy charge of the house fly *Musca domestica* (Diptera: Muscidae) during rapid cold hardening and cold shock. *CryoLetters* **13**, 183–192.

Crockett, E. L. (1998). Cholesterol function in plasma membranes from ectotherms: membrane-specific roles in adaptation to temperatures. *American Zoologist* **38**, 291–304.

Czajka, M. C. and Lee, R. E. (1990). A rapid cold-hardening response protecting against cold shock injury in *Drosophila melanogaster*. *Journal of Experimental Biology* **148**, 245–254.

David, J. R., Gibert, P., Moreteau, B., Gilchrist, G. W., and Huey, R. B. (2003). The fly that came in from the cold: geographic variation of recovery time from low-temperature exposure in *Drosophila subobscura*. *Functional Ecology* **17**, 425–430.

David, R. J., Gibert, P., Pla, E., Petavy, G., Karan, D., and Moreteau, B. (1998). Cold stress tolerance in *Drosophila*: analysis of chill coma recovery in *D. melanogaster*. *Journal of Thermal Biology* **23**, 291–299.

Denlinger, D. L., Joplin, K. H., Chen, C. P., and Lee, R. E. (1991). Cold shock and heat shock. In *Insects at Low Temperature*, ed. R. E. Lee, and D. L. Denlinger. New York: Chapman and Hall, pp. 131–148.

Drobnis, E. Z., Crowe, L. M., Berger, T., Anchordoguy, T. J., Overstreet, J. W., and Crowe, J. H. (1993). Cold shock damage is due to lipid phase transitions in cell membranes: a demonstration using sperm as a model. *Journal of Experimental Zoology* **265**, 432–437.

Elnitsky, M. A., Hayward, S. A. L., Rinehart, J. P., Denlinger, D. L., and Lee, R. E. (2008). Cryoprotective dehydration and the resistance to inoculative freezing in the Antarctic midge, *Belgica antarctica*. *Journal of Experimental Biology* **211**, 524–530.

Feder, M. E., Bedford, T. C., Albright, D. R., and Michalak, P. (2002). Evolvability of Hsp70 expression under artificial selection for inducible thermotolerance in independent populations of *Drosophila melanogaster*. *Physiological and Biochemical Zoology* **75**, 325–334.

Fuchs, B. C. and Bode, B. P. (2006). Stressing out over survival: glutamine as an apoptotic modulator. *Journal of Surgical Research* **131**, 26–40.

Fujiwara, Y. and Denlinger, D. L. (2007). p38 MAP kinase is a likely component of the signal transduction pathway triggering rapid cold hardening in the flesh fly, *Sarcophaga crassipalpis*. *Journal of Experimental Biology* **210**: 3295–3300.

Grout, B. W. (1987). Direct chilling injury. In *The Effects of Low Temperatures on Biological Systems*, ed. B. W. Grout. and G. J. Morris. London: Edward Arnold, pp. 120–146.

Hazel, J. R. (1995). Thermal adaptation in biological membranes: is homeoviscous adaptation the explanation? *Annual Review of Physiology* **57**, 19–42.

Jagdale, G. B., Parwinder, S. G., and Salmnen, S. O. (2005). Both heat-shock and cold-shock influence trehalose metabolism in an entomopathogenic nematode. *Journal of Parasitology* **91**, 988–994.

Joplin, K. H. and Denlinger, D. L. (1990). Developmental and tissue specific control of the heat shock induced 70kDa related proteins in the flesh fly, *Sarcophaga crassipalpis*. *Journal of Insect Physiology* **36**, 239–249.

Kelty, J. (2007). Rapid cold-hardening of *Drosophila melanogaster* in a field setting. *Physiological Entomology* **32**, 343–350.

Kelty, J. D., Killian, K. A., and Lee, R. E. (1996). Cold shock and rapid cold-hardening of pharate adult flesh flies (*Sarcophaga crassipalpis*): effects on behaviour and neuromuscular function following eclosion. *Physiological Entomology* **21**, 283–288.

Kelty, J. D. and Lee, R. E. (1999). Induction of rapid cold-hardening by cooling at ecologically relevant rates in *Drosophila melanogaster*. *Journal of Insect Physiology* **45**, 719–726.

Kelty, J. D. and Lee, R. E. (2001). Rapid cold-hardening of *Drosophila melanogaster* (Diptera: Drosophilidae) during ecologically based thermoperiodic cycles. *Journal of Experimental Biology* **204**, 1659–1666.

Kim, Y. and Kim, N. (1997). Cold hardiness in *Spodoptera exigua* (Lepidoptera: Noctuidae). *Environmental Entomology* **26**, 1117–1123.

Kim, Y.-S., Denlinger, D. L., and Smith, B. (2005). Spatial conditioning in the flesh fly, *Sarcophaga crassipalpis*: disruption of learning by cold shock and protection by rapid cold hardening. *Journal of Asia-Pacific Entomology* **8**, 345–351.

Klok, C. J., Chown, S. L., and Gaston, K. J. (2003). The geographical range structure of the holly leaf-miner. III. Cold-hardiness physiology. *Functional Ecology* **17**, 858–868.

Koveos, D. S. (2001). Rapid cold hardening in the olive fruit fly *Bactrocera oleae* under laboratory and field conditions. *Entomologia Experimentalis et Applicata* **101**, 257–263.

Larsen, K. J. and Lee, R. E. (1994). Cold tolerance including rapid cold-hardening and inoculative freezing in migrant monarch butterflies in Ohio. *Journal of Insect Physiology* **40**, 859–864.

Larsen, K. J., Lee, R. E., and Nault, L. R. (1993). Influence of developmental conditions on cold-hardiness of adult *Dalbulus* leafhoppers – implications for overwintering. *Entomologia Experimentalis et Applicata* **67**, 99–108.

Lee, R. E., Chen, C. P., and Denlinger, D. L. (1987). A rapid cold-hardening process in insects. *Science* **238**, 1415–1417.

Lee, R. E. and Denlinger, D. L. (1991). *Insects at Low Temperature*. New York: Chapman and Hall.

Lee, R. E., Damodaran, K., Yi, S.-X., and Lorigan, G. A. (2006). Rapid cold-hardening increases membrane fluidity and cold tolerance of insect cells. *Cryobiology* **52**, 459–463.

Leopold, R. A. (1998). Cold storage of insects for integrated pest management. In *Temperature Sensitivity in Insects and Application in Integrated Pest Management*, ed. G. J. Hallman and D. L. Denlinger. Boulder: Westview Press, pp. 235–267.

Li, A. and Denlinger, D. L. (2008). Rapid cold hardening elicits changes in brain protein profiles of the flesh fly, *Sarcophaga crassipalpis*. *Insect Molecular Biology* **17**, 565–572.

Li, Y., Gong, H., and Park, H. Y. (1999). Characterization of rapid cold-hardiness response in the overwintering mature larvae of pine needle gall midge, *Thecodiplosis japonensis*. *CryoLetters* **20**, 383–392.

Mangan, R. L. and Hallman, G. J. (1998). Temperature treatments for quarantine security: new approaches for fresh commodities. In *Temperature Sensitivity in Insects and Application in Integrated Pest Management*, ed. G. J. Hallman and D. L. Denlinger. Boulder: Westview Press, pp. 201–234.

Massip, A., Leibo, S. P., and Blesbios, E. (2004). Cryobiology of gametes and the breeding of domestic animals. In *Life in the Frozen State*, ed. B. J. Fuller, N. Lane and E. E. Benson. Boca Raton: CRC Press, pp. 371–392.

McDonald, J. R., Bale, J. S., and Walters, K. A. (1997). Rapid cold hardening in the western flower thrips *Frankliniella occidentalis*. *Journal of Insect Physiology* **43**, 759–766.

McElhaney, R. N. (1974). The effect of alterations in the physical state of the membrane lipids on the ability of *Acholeplasma laidlawii B* to grow at various temperatures. *Journal of Molecular Biology* **84**, 145–157.

Michaud, M. R. and Denlinger, D. L. (2006). Oleic acid is elevated in cell membranes during rapid cold-hardening and pupal diapause in the flesh fly, *Sarcophaga crassipalpis*. *Journal of Insect Physiology* **52**, 1073–1082.

Michaud, M. R. and Denlinger, D. L. (2007). Shifts in carbohydrate, polyol, and amino acid pools during rapid cold hardening and diapause-associated cold hardening in flesh flies (*Sarcophaga crassipalpis*): a metabolomic comparison. *Journal of Comparative Physiology B* **177**, 753–763.

Monroy, A. F. and Dhindsa, R. S. (1995). Low-temperature signal transduction: induction of cold acclimation-specific genes of alfalfa by calcium at 25 °C. *Plant Cell* **7**, 321–331.

Murakami, M., Kondo, T., Sato, S., Li, Y., and Chan, P. H. (1997). Occurrence of apoptosis following cold injury-induced brain edema in mice. *Neuroscience* **81**, 231–237.

Murata, N. and Los, D. A. (1997). Membrane fluidity and temperature perception. *Plant Physiology* **115**, 875–879.

Nunamaker, R. A. (1993). Rapid cold-hardening in *Culicoides variipennis sonorensis* (Diptera: Ceratopogonidae). *Journal of Medical Entomology* **30**, 913–917.

Orvar, B. L., Sangwan, V., Omann, F., and Dhindsa, R. S. (2000). Early steps in cold sensing by plant cells: the role of actin cytoskeleton and membrane fluidity. *Plant Journal* **23**, 785–794.

Overgaard, J., Malmendal, A., Sørenson, J. G., Bundy, J. G., Loeschcke, V., Nielsen, N. C., and Holmstrup, M. (2007). Metabolomic profiling of rapid cold hardening and cold shock in *Drosophila melanogaster*. *Journal of Insect Physiology* **53**, 1218–1232.

Overgaard, J., Sørensen, J. G., Petersen, S. O., Loeschcke, V., and Holmstrup, M. (2005). Changes in membrane lipid composition following rapid cold hardening in *Drosophila melanogaster*. *Journal of Insect Physiology* **51**, 1173–1182.

Overgaard, J. and Sørensen, J. G. (2008). Rapid thermal adaptation during field temperature variations in *Drosophila melanogaster*. *Cryobiology* **56**, 159–162.

Phanvijhitsiri, K., Musch, M. W., Ropeleski, M. J., and Chang, E. B. (2005). Molecular mechanisms of L-glutamine modulation of heat stimulated Hsp25 production. *FASEB Journal* **19**, A1496–A1497.

Powell, S. J. and Bale, J. S. (2004). Cold shock injury and ecological costs of rapid cold hardening in the grain aphid *Sitobion avenae* (Hemiptera: Aphididae). *Journal of Insect Physiology* **50**, 277–284.

Powell, S. J. and Bale, J. S. (2005). Low temperature acclimated populations of the grain aphid *Sitobion avenae* retain ability to rapidly cold harden with enhanced fitness. *Journal of Experimental Biology* **208**, 2615–2620.

Powell, S. J. and Bale, J. S. (2006). Effect of long-term and rapid cold-hardening on the cold torpor temperature of an aphid. *Physiological Entomology* **31**, 348–352.

Qin, W., Neal, S. J., Robertson, R. M., Westwood, J. T., and Walker, V. K. (2005). Cold hardening and transcriptional change in *Drosophila melanogaster*. *Insect Molecular Biology* **14**, 607–613.

Rako, L. and Hoffman, A. A. (2006). Complexity of the cold acclimation response in *Drosophila melanogaster*. *Journal of Insect Physiology* **52**, 94–104.

Rinehart, J. P., Yocum, G. D., and Denlinger, D. L. (2000). Thermotolerance and rapid cold hardening ameliorate the negative effects of brief exposures to high or low temperatures on fecundity in the flesh fly, *Sarcophaga crassipalpis*. *Physiological Entomology* **25**, 330–336.

Rosales, A. L., Krafsur, E. S., and Kim, Y. (1994). Cryobiology of the face fly and house fly (Diptera: Muscidae). *Journal of Medical Entomology* **31**, 671–680.

Shintani, Y. and Ishikawa, Y. (2007). Relationship between rapid cold-hardening and cold acclimation in the eggs of the yellow-spotted longicorn beetle, *Psacothea hilaris*. *Journal of Insect Physiology* **53**, 1055–1062.

Shreve, S. M., Kelty, J. D., and Lee, R. E. (2004). Preservation of reproductive behaviors during modest cooling: rapid cold-hardening fine-tunes organismal response. *Journal of Experimental Biology* **207**, 1797–1802.

Shreve, S. M., Yi, S.-X., and Lee, R. E. (2007) Increased dietary cholesterol enhances cold tolerance in *Drosophila melanogaster*. *CryoLetters* **28**, 33–37.

Sinclair, B. J. and Chown, S. L. (2006). Rapid cold-hardening in a Karoo beetle, *Afrinus* sp. *Physiological Entomology* **31**, 98–101.

Sinclair, B. J., Klok, C. J., Scott, M. B., Terblanche, J. S., and Chown, S. L. (2003). Diurnal variation in supercooling points of three species of Collembola from Cape Hallett, Antarctica. *Journal of Insect Physiology* **49**, 1049–1061.

Smallwood, M. and Bowles, D. J. (2002). Plants in a cold climate. *Philosophical Transactions of the Royal Society of London Series B-Biological Sciences* **357**, 831–846.

Teets, N. M., Elnitsky, M. A., Benoit, J. B., Lopez-Martinez, G., Denlinger, D. L., and Lee, R. E. (2008). Rapid cold-hardening in larvae of the Antarctic midge, *Belgica antarctica*: Cellular cold-sensing and a role for calcium. *American Journal of Physiology* **294**, R1938-R1946.

Terblanche, J. S., Clusella-Trullas, S., Deere, J. A., and Chown, S. L. (2008). Thermal tolerance in a south-east African population of the tsetse fly *Glossina pallidipes* (Diptera, Glossinidae): implications for forecasting climate change impacts. *Journal of Insect Physiology* **54**, 114–127.

Terblanche, J. S., Marais, E., and Chown, S. L. (2007). Stage-related variation in rapid cold hardening as a test of the environmental predictability hypothesis. *Journal of Insect Physiology* **53**, 455–462.

Thompson Jr., G. A. (1983). Mechanisms of homeoviscous adaptation in membranes. In *Cellular Acclimatisation to Environmental Change*, ed. A. R. Cossins and P. Sheterline. Cambridge: Cambridge University Press, pp. 33–54.

Wang, X. and Kang, L. (2003). Rapid cold hardening in young hoppers of the migratory locust *Locusta migratoria* L. (Orthoptera: Acridiidae). *CryoLetters* **24**, 331–340.

Watanabe, M., Kikawada, T., Minagawa, N., Yukuhiro, F., and Okuda, T. (2002). Mechanism allowing an insect to survive complete dehydration and extreme temperatures. *Journal of Experimental Biology* **205**, 2799–2802.

Worland, M. R. and Convey, P. (2001). Rapid cold hardening in Antarctic microarthropods. *Functional Ecology* **15**, 515–524.

Worland, M. R., Convey, P., and Lukešovà, A. (2000). Rapid cold hardening: a gut feeling. *CryoLetters* **21**, 315–324.

Worland, M. R., Hawes, T. C., and Bale, J. S. (2007). Temporal resolution of cold acclimation and de-acclimation in the Antarctic collembolan, *Cryptopygus antarcticus*. *Physiological Entomology* **32**, 233–239.

Yi, S.-X. and Lee, R. E. (2004). In vivo and in vitro rapid cold hardening protects cells from cold-shock injury in the flesh fly. *Journal of Comparative Physiology B* **174**, 611–615.

Yi, S.-X., Yin, C. M., and Nordin, J. H. (1987). The in vitro biosynthesis and secretion of glycerol by larval fat bodies of chilled *Ostrinia nubilalis*. *Journal of Insect Physiology* **33**, 523–528.

Yi, S.-X., Moore, C. W., and Lee, R. E. (2007). Rapid cold-hardening protects *Drosophila melanogaster* from cold-induced apoptosis. *Apoptosis* **12**, 1183–1193.

Yocum, G. D. and Denlinger, D. L. 1994. Anoxia blocks thermotolerance and the induction of rapid cold hardening in the flesh fly, *Sarcophaga crassipalpis*. *Physiological Entomology* **19**, 152–158.

Yoder, J., Benoit, J. B., Denlinger, D. L., and Rivers, D. B. (2006). Stress-induced accumulation of glycerol in the flesh fly, *Sarcophaga bullata*: Evidence indicating anti-desiccant and cryoprotectant functions of this polyol and a role for the brain in coordinating the response. *Journal of Insect Physiology* **52**, 202–214.

3

Antifreeze and ice-nucleator proteins

JOHN G. DUMAN, KENT R. WALTERS, TODD SFORMO, MARTIN A. CARRASCO, PHILIP K. NICKELL, XIA LIN AND BRIAN M. BARNES

3.1 Introduction

Supercooling ability is a critical component among the suite of adaptations contributing to subzero temperature-tolerance of insects, whether they follow freeze-tolerance or freeze-avoidance strategies. Supercooling points (SCP, nucleation temperature, or crystallization temperature) of insects and other terrestrial arthropods range tremendously, from −2 °C to −100 °C or lower. Supercooling is affected by a number of factors, including the volume and water content of the organism, and the ability of the body surface to prevent inoculative freezing by external ice. However, the topics of this review, ice nucleators and antifreeze proteins, are often of critical importance. Antifreezes can be both small-molecular-mass solutes, such as polyhydroxyl alcohols that depress the freezing point of water on a strictly colligative basis, and high-molecular-mass molecules such as antifreeze proteins that suppress freezing by a non-colligative mechanism. Freeze-tolerant species often exhibit high SCPs (above −10 °C) and have selected *for* extracellular ice nucleators, while freeze-avoiding insects generally have selected *against* ice nucleators and *for* antifreezes, allowing them to supercool below ambient temperatures to which they are exposed over the winter. This review will attempt to provide a broad update on ice nucleators, antifreeze proteins and related adaptations in insects and other arthropods, primarily from the standpoint of how they function in organisms to promote winter survival.

3.2 Protein ice nucleators

Ice nucleators (INs) limit supercooling by organizing water into an ice-like structure, the embryo crystal, that promotes freezing at a temperature higher

Low Temperature Biology of Insects, ed. David L. Denlinger and Richard E. Lee Jr. Published by Cambridge University Press.

than that where ice would otherwise form (Knight, 1967). INs catalyse this heterogeneous nucleation, which can occur at ambient temperatures as high as −2 °C if the IN is sufficiently active to catalyse an embryo crystal to the critical size required to seed the supercooled liquid. In contrast, in the absence of INs, homogeneous nucleation of water has traditionally been thought to take place at approximately −40 °C, as embryo crystals spontaneously reach the critical size for growth. However, recently, Zachariassen *et al.* (2004) argued against heterogeneous nucleation in insects at temperatures lower than −18 °C, based on their studies, which showed that water volume is the critical factor determining nucleation, as shown earlier by Bigg (1953). In contrast, Liu and Du (2004) state that homogeneous nucleation never occurs, even in highly polished water.

Many freeze-tolerant arthropods have selected for "adaptive" (Lundheim, 1996) endogenous protein ice nucleators (PINs) that may have evolved specifically for ice-nucleation purposes. In other cases, insects have selected for nucleating regions of proteins that have other functions, producing dual function PINs. In contrast, freeze-avoiding insects must select against "incidental" INs, such as proteins or motifs of proteins that nucleate ice because of some structural element that cannot be removed without eliminating the normal function of the protein. If this is not possible, then freeze-avoiding species must seasonally remove or inhibit these "incidental" INs.

3.2.1 *Structural characterization*

The structures of insect protein ice nucleators are not well understood. Consequently, it is useful to briefly discuss bacterial PINs, as they are the best-characterized PINs from any organism. These PINs are on the external surface of the outer membranes of ice-nucleating bacteria, such as *Pseudomonas syringae*, that are common epiphytes on the surfaces of plants and in soil (Lindow, 1983, 1995). The 150kDa protein consists mainly (81%) of a central repeat domain containing three regions. One region consists only of highly conserved octapeptide repeats, while the other two contain high-fidelity 48-mer repeats, each of which is composed of 16-mer medium-fidelity repeats, and each of these contain low-fidelity 8-mer repeats (Green and Warren, 1985; Wolber and Warren, 1989). High levels of IN activity of the bacterial PINs require aggregation of several of the proteins, and the proteins must be anchored in the membrane to be active (Govindarajan and Lindow, 1988; Southworth *et al.*, 1988; Mueller *et al.*, 1990).

The only insect PIN that has been reasonably well characterized is from the freeze-tolerant larvae of the crane fly, *Tipula trivittata* (Neven *et al.*, 1989; Duman *et al.*, 1985,1991). This is an 800 kDa hemolymph lipophorin that functions, in addition to ice nucleation, to shuttle lipid from site to site. The lipoprotein ice nucleator (LPIN) contains 45% protein, 51% lipid and 4% carbohydrate. It has two lipid binding apoproteins, 265 kDa apoprotein-1 and 81 kDa apoprotein-2. The lipid components are not unusual, except that one of the phospholipids, which typically form a monolayer on the surface of the spherical insect lipophorins, is phosphotidylinositol (PI), not previously seen in other lipophorins. Various studies, including delipidation and reconstitution of the components into proteoliposomes, showed that only the two apoproteins and PI are essential for IN activity. Also, the *cis*-hydroxyl residues on carbons 4 and 5 of the inositol are required, and may be involved in ordering water into the embryo crystal. Reconstitution of the apoproteins with PI into proteoliposomes resulted in recovery of ice-nucleating ability. Also, immunological evidence indicated similarities between the apoproteins and the bacterial PIN. Antibodies to Apo-1 cross-reacted on Western blots with the bacterial protein, and antibodies to the bacterial protein cross-reacted with the apoproteins, especially Apo-1. In addition, antibodies raised to a synthetic bacterial octapeptide repeat cross-reacted with the apoproteins of the LPIN, indicating the presence of the octapeptide in the apoproteins (Duman *et al.*, 1991). However, the sequence of the apoproteins has never been determined. Also, like the bacterial protein, the LPIN appears to self-aggregate, apparently to permit cooperation between embryo crystal-forming active sites, thereby achieving larger embryo crystals at higher subzero temperatures than could be achieved by LPIN monomers (Yeung *et al.*, 1991; Duman, 2001). The two apoproteins, but not PI, are required for aggregation (Duman *et al.*, 1992).

One other PIN that has been purified from a freeze-tolerant insect is a 70 kDa PIN from overwintering queens of the hornet *Vespula maculata* (Duman and Patterson, 1978), but all that is known about this PIN is its amino-acid composition, that is 20 mol% glutamate/glutamine, suggesting that the water-organizing region may be associated with these residues.

3.2.2 Functions

3.2.2.1 Freeze-tolerance

Zachariassen and Hammel (1976) demonstrated that freeze-tolerant beetles from the mountains of southern California have potent adaptive ice INs in their hemolymph. These PINs function to limit supercooling so that lethal intracellular ice does not form as a result of the rapid ice growth that can accompany freezing that occurs after extensive supercooling (Mazur, 1984). Following extracellular ice formation at higher temperatures the increased osmotic pressure resulting from the freezing-out and concentration of solutes, along with the decreased vapor pressure due to ice, causes an outflux of water from cells that lowers the freezing point of the remaining intracellular water. This produces the paradox of extracellular ice nucleators functioning to prevent intracellular ice. While not all freeze-tolerant insects have extracellular ice nucleators, many do (Zacchariassen, 1982).

Recently, Zachariassen *et al.*, (2008) provided evidence that a function of hemolymph INs in freeze-tolerant species is to initiate extracellular ice at high temperatures to minimize water loss from insects with high rates of cuticular water loss. Overwintering strategies (freeze-tolerant versus freeze-avoiding) and transcuticular water permeability of various species of cerambycid and chrysomelid beetles were compared. The cerambycids were freeze avoiding, with low rates of water loss across the cuticle, while the chrysomelids were freeze-tolerant with high rates of water loss across the cuticle. The authors contend that extracellular freezing in the freeze-tolerant chrysomelids is an adaptation to bring the body fluids into vapor pressure equilibrium with external ice, thereby preventing excessive evaporative water loss across the relatively permeable cuticle. In contrast, low cuticular water permeabilities in the cerambycids preclude the need for extracellular ice nucleation and freeze-tolerance. It will be interesting to see whether this trend holds among other insect groups.

Ice formation in freeze-tolerant species is not always initiated by hemolymph nucleators. Hemolymph INs were initially reported in the freeze-tolerant larvae of the goldenrod gall fly *Eurosta solidaginis* (Sømme, 1978); however Bale attributed the hemolymph activity to detrital contamination (Bale *et al.*, 1989). Later, IN activity coincident with the SCPs of the larvae was identified both in Malpighian tubule fluid, due to calcium phosphate spherules, and fat-body cells (Mugnano *et al.*, 1996). Also, the nucleator activity responsible for the high SCPs (−6 to −10 °C) of arctic wooly bear caterpillars, *Gynaephora groenlandica*, is not in the hemolymph (Kukal *et al.*, 1988).

To exhibit freeze-tolerance, some species require inoculative freezing across the cuticle after minimal supercooling. This is the case with caterpillars of the moth *Cisseps fulvicolis* (Fields and McNeil, 1986) and the centipede *Lithobius forficatus* (Tursman *et al.*, 1994). The latter, if not in contact with ice, supercools to only −7 to −9 °C, but then die when frozen. However, if in contact with ice, as would be the usual situation in their winter microhabitat under logs, they freeze just below the hemolymph freezing point and survive.

It is also important to remember that some freeze-tolerant insects exhibit extensive supercooling and yet survive the rapid freezing that ensues. An extreme example of this is the beetle *Pytho americanus* from the Canadian Rockies (Ring and Tesar, 1980) and the interior of Alaska (Miller, 1982) with SCPs of −50 to −60 °C, but these larvae survive freezing. Also, individual adult fungus gnats, *Exechia nugatoria* (Mycetophilidae) from interior Alaska exhibit two freezing events. Individuals tolerate the first freezing event in the abdomen at −30 °C, but do not survive the second freeze in the head/thorax at −50 °C (Sformo *et al.*, 2009).

Bale and colleagues have identified insects that were initially freeze-tolerant, with high SCPs, but, after thawing, the SCPs of some members of the population decreased and the insects died after being refrozen at the lower SCP (Bale *et al.*, 2000; Brown *et al.*, 2004). This trend continued with multiple rounds of freezing and thawing leading to progressively lower SCPs.

3.2.2.2 Freeze-avoidance

To successfully supercool, freeze-avoiding species must either select against ice-nucleating surfaces on an evolutionary timeframe or remove/mask INs seasonally. An interesting example of the former is the willow cone gall, *Rhabdophaga strobiloides*, which winters on willow branches, where the larvae are exposed to unbuffered air temperatures. In Alaska, the larvae supercool to −56 °C in winter, approximately 37 °C below the hemolymph melting point of −19 °C (Miller and Werner, 1987). One major factor responsible for this extreme supercooling is the high polyol concentrations indicated by the low hemolymph melting point. However, the SCP, even in the summer, is −26.3 °C, approximately 25 °C below the hemolymph melting point. This low summer SCP reflects the absence of IN bacteria, etc. in the phloem on which the larvae feed, but it also indicates an absence of endogenous INs.

On a seasonal timeframe, most freeze-avoiding insects cease feeding and clear their guts to remove IN bacteria, fungi, etc. prior to the onset of subzero temperatures (Salt, 1953; Sømme, 1982; Zachariassen, 1985; Cannon and Block, 1988; Lee *et al.*, 1996). Some species also remove endogenous incidental INs seasonally. Larvae of the stag beetle, *Ceruchus piceus*, remove hemolymph LPINs in winter (Neven *et al.*, 1986). This permits supercooling to −26 °C, even though antifreezes are not produced. The removal of the lipoproteins is probably facilitated by diapause, thereby making the lipid shuttle function of these lipophorins expendable, as metabolism is greatly reduced. Although larvae of the beetle *Dendroides canadensis* also remove some hemolymph PINs and LPINs in winter, not all are removed and consequently AFPs are still required to inhibit the remaining INs (Olsen and Duman, 1997a; Duman, 2002).

3.3 Antifreeze proteins

3.3.1 History

Antifreeze proteins (AFPs) have traditionally been identified by their unique ability to lower the temperature at which an ice crystal will grow, as the temperature is depressed below the equilibrium melting/freezing point to a lower, hysteretic freezing point. The difference between the melting point and the hysteretic freezing point, thermal hysteresis (TH), identifies the presence of AFPs and provides a measure of their activity. Unlike the normal situation where a small crystal just about to melt at the equilibrium freezing/melting point will slowly grow immediately upon depression of the temperature by just a few hundredths of a degree C, in the presence of AFPs crystal size remains constant until the temperature is lowered, typically by 1–10 °C depending on the system, to the hysteretic freezing point at which the crystal grows very rapidly (DeVries 1986). Although the first reports on AFPs are associated with Antarctic fish, the first mention of this unusual melting/freezing behavior was Ramsay's classic report on the physiology of the cryptophridial rectal complex of larvae of the beetle *Tenebrio molitor*. Ramsay measured the melting points of the formative urine, etc. to understand the amazing water reabsorbtion capacity of this complex. He reported in a footnote, the presence of TH in the hemolymph, rectal lumen and especially in the perirectal space (Ramsay, 1964). A subsequent publication (Grimstone *et al.*, 1968) correlated TH with proteins, but the responsible proteins were not identified.

Several prominent physiologists had recognized the need for antifreeze in normally hypo-osmotic non-cartilaginous fish that inhabit oceans where ice forms at approximately −1.9 °C, as these fish would otherwise be supercooled by approximately 1 °C, and would therefore be likely to freeze on contacting ice. In spite of several attempts, the antifreeze could not be identified, mainly because the investigators only searched for low-molecular-mass solutes. Proteins, because of their size, obviously could not be sufficiently concentrated to the near molar concentration required to colligatively depress the freezing point of the fishes' body fluids below that of seawater. DeVries, during his PhD thesis work at Stanford (DeVries, 1969; DeVries and Wohlschlag, 1969), recognized the antifreeze function of certain glycoproteins of Antarctic Nototheniid fishes and provided some structural information. Subsequent studies identified the sequence and further characterized the well-known antifreeze glycoproteins (Komatsu *et al.*, 1970; DeVries *et al.*, 1970). DeVries first recognized the TH activity and continued characterization of the AFGPs, associating the antifreeze activity with certain hydroxyls of the galactose residues. He suggested for the first time that TH might result from the adsorption of the AFGPs to the surface of ice (DeVries 1971). Additional studies showed that AFPs were also present in fish from the east coast of North America (Duman and DeVries, 1974, 1975), and that winter flounder from this region had antifreeze proteins that lacked a carbohydrate component (Duman and DeVries, 1976).

Shortly thereafter AFPs were identified in insects, initially in overwintering larvae of the beetle *Meracantha contracta* (Duman, 1977a,b), and subsequently in other species of insects (Duman, 1979a; Zachariassen and Husby, 1982), including *T. molitor* (Patterson and Duman, 1978), and also in spiders (Duman, 1979b; Husby and Zachariassen, 1980).

3.3.2 Mechanism(s) of freezing inhibition

Ice crystals typically grow by adding water molecules onto various prism faces such that the growth front progresses with a broad, low radius of curvature and low surface free energy. However, as suggested by DeVries (1971), according to the generally accepted adsorption–inhibition mechanism (Raymond and DeVries, 1977; Raymond *et al.*, 1989; Knight *et al.*, 1991), AFPs adsorb onto the surface of ice at preferred growth sites, thereby preventing ice growth in the normal low radius of curvature and restricting growth to regions between the adsorbed AFPs in a high radius of curvature front. Consequently, according to the Kelvin effect, new growth is prevented until the temperature is lowered to the hysteretic freezing point (Raymond *et al.*, 1989; Knight *et al.*, 1991; Wilson, 1993; Kristiansen and Zacheriassen, 2005).

Most fish AFPs bind to various faces of the six prism planes of ice. Consequently, the rapid growth that occurs at the hysteretic freezing point is typically in the non-preferred c-axis (Raymond and DeVries, 1977; Raymond *et al.*, 1989; Knight *et al.*, 1991; Jia and Davies, 2002). However, insect AFPs bind to both the basal and prism planes, and this is probably responsible for the higher TH activities of most insect AFPs (Graether and Sykes, 2004; Pertaya *et al.*, 2008).

The unique repeat structures of many AFPs, with regular spacing of hydroxyl side chains of amino acids, such as threonine, or sugar residues of the AFGPs, may provide a lattice match whereby the hydroxyls hydrogen bond to the oxygens in the ice (DeVries and Cheng, 1992; Sicheri and Yang, 1995). More recently, evidence for hydrophobic and van der Waals forces being the major mechanism of binding has been forwarded for fish AFPs. (See Jia and Davies, 2002 for review.) Burying hydrophobic residues away from water (in ice) while hydrophilic residues project into water makes sense on thermodynamic grounds, but if ice is more hydrophilic than water this argument would not hold. Also, if the binding mechanism is via hydrophobic interactions, then the reason for the evolution of repeat structures of AFPs, resulting in regular spacing of hydroxyls at distances that match the lattice structure of ice, would seem to be unnecessary. Insect AFPs appear to bind via hydrophilic bonds, but this is not certain (Graether and Sykes, 2004). It is likely that the binding mechanism varies between different AFPs as the structures of

AFPs vary dramatically. The binding mechanisms are the subject of considerable research.

3.3.3 Determining TH and the presence of AFPs

Several techniques are used to measure TH, including the capillary technique, nanoliter osmometer and differential scanning calorimeter. Since these may not produce the same results, a discussion of this topic is useful. The capillary technique used by DeVries (1986) to identify fish AFPs is a modification of the method of Ramsay (1964). The method involves placing a few microliters of sample into a glass capillary tube, sealing one end of the tube in a flame and closing the other with mineral oil. Sample volumes down to ~0.5 μl can be reasonably used in a 10 μl capillary, and smaller bore capillaries and samples can be used after some experience. The sample is then partially frozen using an aerosol spray freeze and placed into a cooled viewing bath with fine temperature control. The temperature is slowly raised, while monitoring the ice with a microscope as it melts, until the crystal completely disappears at the melting point. The temperature is then lowered slightly (~0.02 °C) and the sample frozen again and placed back into the bath until the ice melts to a small crystal or group of crystals. The temperature is then lowered by a few hundredths of a degree. If AFPs are not present, the crystal will grow noticeably; however, if AFPs are present, the crystal will not grow until the temperature is lowered to the hysteretic freezing point, where rapid ice growth occurs. The difference between the melting and freezing points is the TH, which has been reported to be between a few tenths of a degree and as much as 13 °C. Slow growth of the ice after initial minor cooling is sometimes observed, but is usually arrested after minimal growth, until the hysteretic freezing point is reached.

A related technique that also relies on visualization of ice as the temperature is raised and lowered, but that uses much smaller sample volumes and crystals, is the nanoliter osmometer. The standard unit has been the Clifton osmometer, but these are no longer available. A related apparatus is supplied from Otago Osmometer Ltd, Dunedin, New Zealand. A drawback of the Clifton unit for some applications is that it can only be used down to −9 °C. The Otago osmometer can be manufactured to operate at much lower temperatures, although some loss of precision results from broadening the range.

For many AFPs the magnitude of TH measured is inversely related to the size of the crystal used (Zachariassen and Husby, 1982; Kristiansen *et al.*, 1999; Zachariassen *et al.*, 2002). Consequently, because the nanoliter osmometer uses a much smaller crystal than the capillary technique (usually ~0.25 mm diameter), the nanoliter osmometer often measures a larger TH than does the capillary technique. Sometimes this difference is considerable. For example, the TH of a hemolymph sample from a transgenic *Drosophila* producing *Dendroides canadensis* AFPs was 0.80 °C using the capillary technique but 4.04 °C using the Clifton osmometer (Nicodemus *et al.*, 2006). This is important when reading the AFP literature and comparing AFP activities from various organisms.

A differential scanning calorimeter (DSC) has been used successfully to determine TH (Hansen and Baust, 1988; Hansen and Baust, 1989; Lu *et al.*, 2002; Zachariassen *et al.*, 2002; Zhang *et al.*, 2004; Ramlov *et al.*, 2005; Amornwittawat *et al.*, 2008). TH values determined with DSC can also vary inversely with the amount of ice in the sample during cooling to determine the hysteretic freezing point (Ramlov *et al.*, 2005). The inability to directly observe ice crystals and the difficulty in melting the sample so that only a small amount of ice remains are potential drawbacks to this method; however, there is low variability of repeated measurements (Amornwittawat *et al.*, 2008).

Because AFPs have a non-colligative mechanism for lowering the freezing point, it should be obvious that a vapor pressure osmometer cannot be used for detection of the hysteretic freezing point, and doing so can result in false negatives (Duman and DeVries, 1974). However, determination of the melting point of a sample with the capillary or nanoliter osmometer techniques can be time consuming, especially in samples with high polyol levels and correspondingly low melting points. The vapor pressure osmometer provides a quick melting point measurement on these samples that then can be used for quicker determination of the TH using either the capillary or nanoliter osmometer.

While freezing point osmometers can generally measure a depressed hysteretic freezing point resulting from AFPs, this is not always the case (Lin *et al.*, 1972), probably due to the inability of some AFPs to stop the growth of ice crystals at high rates of growth (Ramlov *et al.* 2005). Also, freezing point osmometers do not permit TH determinations as they do not measure both melting and freezing points.

Not only do AFPs produce TH, but they also affect ice crystal morphology. This can be used to identify the presence of AFPs and is especially useful when the TH is so low that it cannot confidently be measured (Griffith and Yaish, 2004). The nanoliter osmometer is useful for this purpose as it employs a higher magnification microscope and only one crystal is present, unlike the capillary technique, where polycrystalline ice makes it difficult to observe single crystals. When the ice is melted back to obtain the single crystal for freezing-point determination, the crystal typically has an ovate spheroid shape. However, as it is held just below the melting point, a slight amount of growth generally occurs, resulting in uncharacteristic crystal morphology, such as hexagonal or bipyramidal crystals, that identifies AFP presence. This is particularly obvious, and useful, at low levels of AFP activity.

AFPs have very effective recrystallization inhibition (RI) activity (Knight *et al.*, 1984), even at very low concentrations, as reported later in this chapter. Consequently RI can be a sensitive technique for identifying the presence of AFPs (Knight and Duman, 1986; Griffith and Yaish, 2004)). RI activity of *D. canadensis* hemolymph did not disappear until the hemolymph was diluted 10^5-fold (Knight and Duman, 1986). However, false positives can occur with RI measurements, and high solute concentrations should be used to minimize this problem (Knight *et al.*, 1995).

3.3.4 *Which arthropods have evolved AFPs?*

Over 50 species of insects and Collembola have been reported to have TH. In a survey of 75 species of insects from the interior taiga and tundra regions of arctic and subarctic Alaska, 18 (24%) demonstrated hemolymph TH (Duman *et al.*, 2004a). (This latter publication contains a list of insects and other arthropods known to have AFPs prior to 2004.)

TH-producing species are particularly common in the Coleoptera (beetles). The tenebrionids (darkling beetles) and elaterids (click beetles) are especially well represented. Numerous other insect orders also have TH-producing representatives: the Plecoptera, Orthoptera, Hemiptera, Neuroptera, Mecoptera, Lepidoptera and Diptera. However, only a few species of the large orders Diptera and Lepidoptera are known to have TH, and so far no species of Hymenoptera (bees, wasps and ants). In addition to the known AFP-producing Lepidoptera, adults of the Alaskan moth *M. ciniflonella* previously studied by Miller (1982) have hemolymph TH (K. R. Walters, unpublished). TH is also known to occur in non-insect arthropods. These include Collembola (Zettel, 1984; Meier and Zettel, 1999; Graham and Davies, 2005; Sinclair *et al.*, 2006), spiders (Duman, 1979b; Husby and Zachariassen, 1980; Duman *et al.*, 2004a), mites (Block and Duman, 1989; Sjursen and Sømme, 2000) and centipedes (Tursman *et al.*, 1994; Tursman and Duman, 1995).

Most of the known TH-producing species are from North America and Europe. However, several terrestrial arthropods from Antarctica and the sub-Antarctic Islands have been reported to have TH. The oribatid mite, *Alaskozetes antarcticus*, from Signy Island, South Orkney Islands has TH (Block and Duman, 1989). Also, two species of Collembola from Cape Hallet had hemolymph TH of 0.30 –0.45 °C in summer (Sinclair *et al.*, 2006). Larvae of the ermine moth *Embryononsis halticella* from Marion Island (46° 54′ S, 37° 45′ E) had hemolymph TH of 0.28 °C (Sinclair and Chown, 2002).

To our knowledge, only one Asian species, the pine needle gall midge *Thecodiplosis japonensis* (Diptera: Cecidomiidae) from Korea has been reported to have TH (Li *et al.*, 2000). Recently, a freeze-tolerant alpine cockroach, *Celatoblatta quinquemaculata*, from New Zealand, was shown to have TH in gut and Malpighian tubule extracts (Wharton *et al.*, 2009). No TH-producing species has been reported from Africa, Australia or South America. This may simply be a matter of insects from these regions not having been screened for TH. While most AFP-producing insects reported to date are freeze-avoiding, an increasing number of freeze-tolerant species are known.

3.3.5 Structural characterization of arthropod AFPs

Two basic types of insect AFPs have been characterized, one from beetles and another from a lepidopteran. A third type of AFP has been identified in a Collembolan.

Larvae of the tenebrionid beetle, *T. molitor,* (Graham *et al.*, 1997; Liou *et al.*, 1999) and the pyrochroid beetle, *D. canadensis*, (Duman *et al.*, 1998; Andorfer and Duman, 2000) have nearly identical AFPs. These are families of approximately 7–16 kDa proteins consisting of 12–13-mer repeating units, in which certain positions are highly conserved (Fig. 3.1). NMR and X-ray (Graether *et al.*, 1999, 2000; Leinala *et al.*, 2002; Graether and Sykes, 2004) showed the structure of the *T. molitor* AFP (TmAFP) to be a cylinder that is flattened on one side, formed by a right-handed β-helix. (The general structure of a β-helix is that of a cylindrical helical coil with a flat β-sheet on at least one side of the cylinder.) Every sixth residue of the similar *D. canadensis* AFPs (DAFPs) is a cysteine that forms disulfide bridges (Fig. 3.2) across the interior of the cylinder so as to provide stability (Li *et al.*, 1998a). Threonine residues adjacent to cysteines are spaced to permit their hydroxyl groups to hydrogen bond to oxygen atoms in the ice lattice on both the basal and prism planes. This T–C–T motif forms the β-helical structure that presents a flat ice-binding surface. This surface may also serve to bind water molecules that insert into the ice lattice and assist to bind the AFP to ice. (For reviews of the structure and ice-binding mechanisms, see Jia and Davies, 2002; Graether and Sykes, 2004.) The AFPs of the beetle *Cucujus clavipes* (CcAFPs) (Duman, unpublished) are similar in sequence to both the TmAFPs and DAFPs, and it is likely that both these and the DAFPs have higher-order structures similar to those of the TmAFPs (Jia and Davies, 2002; Duman *et al.* 2004b). AFPs from the beetle *Rhagium inquisitor* were purified and partially characterized (Kristiansen *et al.*, 2005), but sequence information has not yet been published.

AFPs (sbwAFPs) of the spruce budworm, *Choristoneura fumiferana*, a lepidopteran, although lacking sequence homology to the beetle AFPs, also form a β-helical structure (left-handed) with 15-mer repeats and a T–X–T motif, where X may be any of several amino acids (Tyshendo *et al.*, 1997; Gauthier *et al.*, 1998; Graether *et al.*, 1999, 2000). Although the sequences of the sbwAFPs are different than those of the beetles, the budworm T–X–T motif forms a flat β-sheet ice-binding surface that is remarkably similar to that of the beetle AFPs (Graether and Sykes, 2004).

POSITION

			1	2	3	4	5	6	7	8	9	10	11	12	13	
A	1	pQ	C	T	G	G	S	D	C	R	S	C	T	V	S	
	2	pQ	C	T	G	G	S	D	C	R	S	C	T	V	S	
	4	pQ	C	T	G	G	S	D	C	Q	S	C	T	V	S	
	9	QQ	C	T	G	G	S	D	C	S	S	C	T	V	A	
	3	Q	C	T	G	A	P	D	C	S	A	C	T	T	A	
	TEN	Q	C	T	G	G	A	D	C	T	S	C	T	G	A	
B	1		C	T	D	C	Q	N	C	P	N	A	R	T	A	
	2		C	T	D	C	Q	N	C	P	N	A	R	T	A	
	4		C	T	D	C	Q	N	C	P	N	A	R	T	A	
	9		C	T	D	C	Q	N	C	P	N	A	L	I	A	
	3		C	T	D	C	V	N	C	P	N	A	L	T	A	
	TEN		C	T	D	C	G	N	C	P	N	A	V	L	-	
C	1		-	-	-	-	-	-	-	-	-	-	-	-	-	
	2		-	-	-	-	-	-	-	-	-	-	-	-	-	
	4		-	-	-	-	-	-	-	-	-	-	-	-	-	
	9		C	T	D	S	T	N	C	Y	Q	A	M	T	-	
	3		-	-	-	-	-	-	-	-	-	-	-	-	-	
	TEN		-	-	-	-	-	-	-	-	-	-	-	-	-	
D	1		C	T	R	S	S	N	C	I	N	A	L	T	-	
	2		C	T	R	S	S	N	C	N	N	A	L	T	-	
	4		C	T	G	S	S	N	C	I	N	A	L	T	-	
	9		C	T	R	S	T	M	C	N	G	A	L	T	-	
	3		C	T	R	S	T	N	C	Y	K	A	V	T	-	
	TEN		C	T	N	S	Q	H	C	V	K	A	N	T	-	
E	1		C	T	D	S	Y	D	C	H	N	A	E	T	-	
	2		C	T	D	S	Y	D	C	H	N	A	E	T	-	
	4		C	T	D	S	H	D	C	H	N	A	E	T	-	
	9		C	T	D	S	Y	D	C	F	N	A	D	T	-	
	3		C	T	K	S	Y	D	C	Y	K	A	V	T	-	
	TEN		C	T	G	S	T	D	C	N	T	A	Q	T	-	
F	1		C	T	R	S	T	N	C	Y	K	A	K	T	-	
	2		C	T	R	S	T	N	C	Y	K	A	K	T	-	
	4		C	T	R	S	T	N	C	Y	K	A	K	T	-	
	9		C	T	R	S	T	N	C	Y	T	A	K	T	-	
	3		C	T	D	S	T	N	C	Y	E	A	T	A	-	
	TEN		C	T	N	S	K	D	C	F	E	A	N	T	-	
G	1		C	T	G	S	T	N	C	Y	E	A	T	-	A	
	2		C	T	G	S	T	N	C	Y	E	A	T	T	A	
	4		-	-	-	-	-	-	-	-	-	-	-	-	-	
	9		C	T	G	S	T	N	C	Y	E	A	T	-	A	
	3		-	-	-	-	-	-	-	-	-	-	-	-	-	
	TEN		C	T	D	S	T	N	C	Y	K	A	T	A	-	
H	1		C	T	D	S	T	G	C	P	-	-	-	-	-	
	2		C	T	D	S	T	G	C	P	-	-	-	-	-	
	4		C	T	D	S	T	G	C	P	-	-	-	-	-	
	9		C	T	D	S	T	G	C	P	S	S	A	S	V	K Q I K
	3		C	T	N	S	T	G	C	P	P	S	A	Y	I	H Q I N
	TEN		C	T	N	S	S	G	C	P	G	H				

Figure 3.1 Sequences of five mature *Dendroides canadensis* AFPs (DAFPs-1, -2, -4, -9, and -3) (Duman *et al.*, 1998), plus a *Tenebrio molitor* AFP (YL-1) (Graham *et al.*, 1997). Numbers across the top of the figure identify positions within the various repeat units (indicated by letters along the side of the figure. These AFPs were chosen to illustrate the similarities of the beetle-type AFPs, even between species. Note that: (1) several positions are highly conserved between repeats, especially the cysteines at positions 1 and 7, and the threonines at positions 12 (beginning with repeat B) and 2; (2) DAFP-9 is the largest and DAFPs-4 and -3 the smallest (fewest repeats) of the AFPs shown; (3) the hemolymph DAFPs -1, -2, and -4 are more similar to one another than to the gut DAFPs-4 and -9; (4) the N-terminus is typically a blocked pyroglutamate (pQ) in those DAFPs that have been checked, and the C-terminus, or a residue near the terminus is a proline. These terminals probably function to protect the AFPs from exopeptidases.

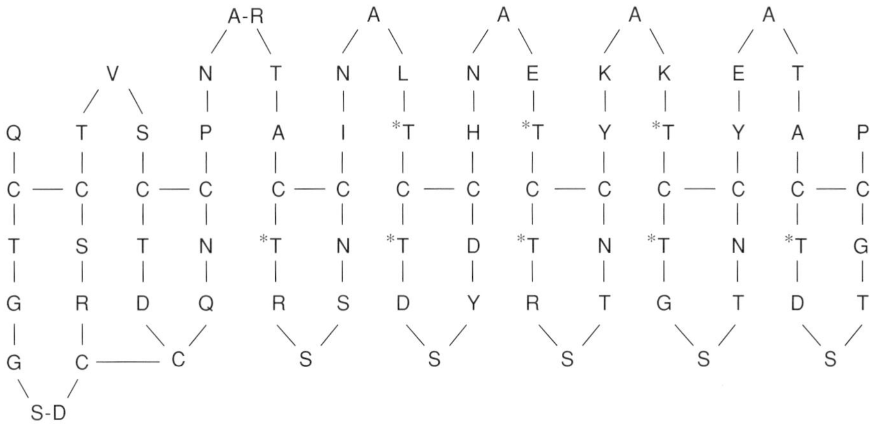

Figure 3.2 Illustrates the disulfide map of DAFP-1 (sequence shown in Fig. 3.1), based on Li *et al.* (1998a). Asterisks identify threonine residues with hydroxyls that provide the probable hydrogen-bonding sites to oxygens in the ice lattice. The DAFP folds into a cylinder such that the T–C–T residues form a flat surface on one side. Note the highly conserved intrachain (vertical) T–C–T residues where the spacing between the threonines is approximately 7.5 Å. The interchain (horizontal) spacing between the threonines is approximately 4.8 Å. These distances approximately match the spacings between oxygens in both the prism and basal planes of ice. Recall that these AFPs bind to both the basal and prism faces. (Distances in this figure are not to scale.) The disulfide bridges are formed across the inside of the cylinder. Also, the serines (at the bottom of this figure), which might be expected to be involved with ice-binding, are actually on the inside of the cylinder where they are removed from the ice. (For details of the structures of these AFPs see Graether and Sykes, 2004).

Also, the sbwAFP binds to both the prism and the basal planes of ice (Graether *et al.*, 2000; Pertaya *et al.*, 2008).

AFPS from the pine needle gall midge, *T. japonensis*, have been purified, but only limited characterization is available (Li *et al.*, 2000). The two AFPs are large, 34.9 and 37.8 kDa, and at very high concentrations (50 mg ml^{-1}) they produce very high TH (11 °C by the capillary method).

A different type of AFP (sfAFP) with TH comparable to the insect AFPs is found in the snow flea *Hypogastrura harveyi* (Graham and Davies, 2005). Two isoforms were purified. Both are glycine rich (∼45.7%) and contain numerous tripeptide repeats consisting of G–X_1–X_2 (X_1 is generally glycine and X_2 is either a charged/hydrophilic residue or hydrophobic alanine or valine). The structure of the smaller 6.5 kDa sfAFP has been modeled (Lin *et al.*, 2007). It consists of a bundle of six polyproline type II helices connected by proline-containing turns that form two sheets of three parallel helices. Like many AFPs, the protein is amphipathic. The hydrophobic side is thought to form the ice-binding surface.

A new, and very different, type of thermal-hysteresis-producing factor has recently been purified from adults of the freeze-tolerant beetle *Upis ceramboides* from Alaska (Walters *et al.*, 2009). Surprisingly, this factor lacks a protein component. It is a glycolipid with the carbohydrate consisting mostly of mannose and xylose with some glucose. The lipid composition is unknown at this time. When concentrated, the glycolipid has 4 °C of TH. This is the first TH factor purified from a freeze-tolerant insect, and the first known to be something other than a protein.

Given the tremendous diversity of insects, they are certain to yield additional surprises, such as new AFPs and other TH factors, that will add greatly to the rapidly expanding body of knowledge in this field.

3.3.6 Enhancement of DAFP activity

TH of individual DAFPs at physiological concentrations (2–4 mg ml^{-1}) is less than the TH measured in winter hemolymph, indicating that some additional factor(s) might be important. A clue to this problem appeared when addition of antibodies raised to the DAFPs resulted in increased TH (Wu *et al.*, 1991). Also, addition of goat anti-rabbit IgG antibodies along with the rabbit anti-DAFP antibody solution further increased TH. The interpretation of these results was that the antibody–DAFP complex was able to bind to ice and, because the complex is larger than the DAFP alone, the complex blocks a larger surface of the crystal. Also, the height of the complex above the ice surface makes it more difficult for ice to overgrow the complex. Consequently, TH is increased. Several other non-endogenous proteins were found to enhance DAFP activity when they bound to the DAFPs (Duman *et al.*, 1993), prompting a search for endogenous proteins that complex with DAFPs and enhance activity. One such protein was purified, but little characterization was accomplished (Wu and Duman, 1991). A second clue emerged when it was noticed that TH was much greater in certain buffers (e.g. malate), leading to the discovery that several, small-molecular-mass solutes greatly increased TH (Li *et al.*, 1998b). The most active was citrate; TH increased from 1.2 °C to 6.8 °C in its presence. However, these small enhancers require a concentration that for most is non-physiological (>0.2M). An exception is glycerol, which is typically present in winter hemolymph at 0.5–1.0M. Endogenous high- and low-molecular-mass enhancers are important to TH of DAFPs (Duman and Serianni, 2002). Also, glycerol enhances the inhibition by DAFPs of the activity of endogenous hemolymph protein and bacterial (*Pseudomonas syringae*) ice nucleators (Duman, 2002).

A yeast-two-hybrid screen was employed, using various DAFPs as "bait", to identify proteins that complex with the four hemolymph DAFPs (1, 2, 4 and 6). Surprisingly, certain hemolymph DAFPs interacted with the bait DAFPs (Wang and Duman, 2005). DAFP-1 interacted with DAFPs-2 and-4, and DAFP-6 with –4. When tested for TH, the combinations identified as interacting exhibited synergistic self-enhancement. Also, glycerol further increased the TH of these combinations, but not of individual DAFPs alone, indicating that the mechanism of glycerol enhancement is to increase the interactions of the self-enhancing DAFPs. Also, a thaumatin-like protein, until recently only known from plants, was identified in *D. canadensis* by the yeast-two-hybrid screen as interacting with DAFP-1 and −2, and was subsequently shown to increase TH, especially in the presence of glycerol. Consequently, combinations of hemolymph DAFPs-1, -2, -4 and -6, along with thaumatin-like protein and glycerol, all at physiological concentrations, resulted in TH of 9–10 °C using the capillary technique (Wang and Duman, 2006).

Clearly, enhancement of DAFPs is critical to their activity, and consequently to the cold-tolerance of *D. canadensis*. In addition, enhancement of DAFPs may be important in various applications of DAFPs. Consequently, low-molecular-mass enhancers that are present in physiological concentrations too low for effective endogenous enhancement still might be useful in certain applications, where increased TH is desired. Recall that citrate was the most active low-molecular-mass enhancer initially identified (Li *et al.*, 1998b). However, a series of polycarboxylates containing 1–6 carboxyl groups, including citrate with three carboxyls, was tested for the ability to enhance DAFP-1 alone (Amornwittawat *et al.*, 2009). Enhancement increased with increasing numbers of carboxyl groups in the compound, with triethylenetetramine-hexaacetate (containing six carboxyl groups) being the best (1.5-fold greater than citrate).

3.3.7 Functions of AFPs in insects and other arthropods

3.3.7.1 Freeze-avoidance

Most TH-producing arthropods are freeze-avoiding, and it is assumed that their AFPs inhibit lethal ice formation. The majority of these species exhibit hemolymph TH only in winter. However, some, such as *D. canadensis,* have low levels of hemolymph activity (0.2–0.4 °C) in summer, but activity is greatly increased in winter. This suggested that AFPs in freeze-avoiding insects function primarily, but perhaps not exclusively, as antifreezes (Duman, 1977a,b). Experimental evidence to this effect was difficult to obtain until sufficient AFPs could be purified, or produced in expression systems, to permit appropriate experiments.

Some 40 different DAFPs are now known to make up the family of DAFPs present in *D. canadensis* (Andorfer and Duman, 2000; Duman, unpublished), and they are expressed differentially in different tissues (Duman *et al.*, 2002). Only four of these are typically present in the hemolymph. Another nine are expressed by the gut epithelium and secreted into the gut fluid. Also, combinations of these DAFPs specific to individuals are expressed by the layer of epithelial cells underlying the cuticle. More recently, numerous DAFP transcripts have been identified in the Malpighian tubule epithelium and are probably responsible for the TH present in the formative urine (P. K. Nickell, E. Blumenthall and J. G. Duman, unpublished). This tissue-specific expression suggests that certain DAFPs and/or combinations of DAFPs are best suited to function in different tissues.

D. canadensis larvae typically have hemolymph TH of 3–5 °C (capillary technique) in winter, although some individuals have 8–9 °C. Comparable TH is present in the gut fluid (Duman, 1984), although the Malpighian tubule fluid has less TH than the other compartments (P. K. Nickell, E. Blumenthall and J. G. Duman, unpublished). The protection afforded by the DAFPs is actually much greater than the 3–5 °C of hemolymph TH. This is at least partially due to the inverse relationship between the size of crystal used in the TH determination and the measured TH (Zachariassen and Husby, 1982). The larvae live under the loose bark of partially decomposed deciduous trees. The larvae can be near the ground and covered by insulating snow in winter, where the microhabitat temperatures are relatively mild, or they can be above snow level and exposed to much colder temperatures. SCPs of the larvae can vary from year to year. In the early 1980s SCPs averaged approximately −28 °C (Horwath and Duman, 1984a): however, more recently, the lowest mean SCPs have been above −20 °C in winter, perhaps as a result of the warmer winters (Duman, unpublished). To promote supercooling in winter, *D. canadensis* accumulate glycerol (0.5–1.0 M), remove ice-nucleating bacteria from the gut, and partially remove ice-nucleating proteins and lipoproteins from hemolymph (Olsen and Duman, 1997a,b). AFPs are also critical components of the suite of winter adaptations, as the DAFPs inhibit inoculative freezing and ice-nucleator activity.

If an insect is to supercool significantly below the freezing point of its hemolymph, it must first prevent inoculative freezing across the cuticle. In the typically damp microhabitat of a decomposing log, *D. canadensis* larvae often are in contact with ice and therefore must solve this problem. The wax coating of the cuticle of insects not only inhibits evaporative water loss, but it also can act as a barrier to inoculative freezing, although this is not always the case. *D. canadensis* larvae are susceptible to inoculative freezing in early autumn, prior to the production of DAFPs. However, later in the season, after accumulation of DAFPs, freezing across the cuticle is inhibited (Olsen *et al.*, 1998). In addition to hemolymph DAFPs, several DAFPs are expressed in the layer of epidermal cells underlying the cuticle (Duman *et al.*, 2002). While some of the epidermal DAFP transcripts lack a signal peptide, indicating that they remain inside the cells, most have signal peptides and are secreted. Immunofluorescence studies using antibodies to DAFPs indicate that DAFPs are present on the membranes of the epidermal cells in winter, where they can inhibit inoculative freezing (Olsen *et al.*, 1998). To test the effectiveness of cuticle and DAFPs in blocking inoculative freezing, patches of cuticle (winter versus summer, with or without the underlying epidermal cells) were positioned in a chamber so as to separate external ice from interior physiological saline (with or without DAFPs), and the temperature at which ice propagated across the cuticle measured (Olsen *et al.*, 1998). Winter cuticle inhibited inoculative freezing better than summer cuticle, indicating seasonal cuticular modifications. The subcuticular epidermal layer added additional protection, but only if DAFPs were in the physiological saline. In fact, winter levels of protection were achieved with summer cuticle and the underlying epidermal layer, if DAFPs were present in the saline.

DAFPs inhibit ice-nucleating bacteria, such as those found in the gut when the *D. canadensis* larvae are ingesting rotted wood (Olsen and Duman, 1997b). Also, DAFPs inhibit the remaining hemolymph ice-nucleating proteins and lipoproteins that are not removed in winter (Olsen and Duman, 1997a). DAFPs are especially active in this regard with glycerol present at physiological concentrations of 250 mM or more (Duman, 2002). As discussed previously, glycerol is a potent enhancer of DAFP activity (Wang and Duman, 2005). TH in Malpighian tubule fluid is slightly less than that in hemolymph or gut fluid, but, as mentioned earlier, several transcripts that encode previously unidentified DAFPs have been identified in the Malpighian tubule epithelium. Since this fluid is not subject to inoculative freezing, these DAFPs presumably function to promote supercooling of the urine.

Cucujus clavipes beetles have AFPs (*Cc*AFPs) similar to those of *D. canadensis* and *T. molitor*. This freeze-avoiding species is found from North Carolina in southeastern United States to the tree line in arctic Alaska, probably as two subspecies, *C. c. clavipes* in eastern North America west to the Great Plains and *C. c. puniceus* in the northwest (Thomas, 2002). Both winter primarily as larvae in standing or fallen, partially decomposed trees (mostly *Populus spp.* in the north), although a few adults winter as well. *C. c. clavipes* from northern Indiana produce AFPs and accumulate glycerol in winter, and the larvae supercool to approximately −25 °C in winter. Alaska *C. c. puniceus* also produce AFPs and glycerol, but in addition they undergo extensive cryoprotective dehydration, from approximately 63% body water in summer to 35% in winter, and some individuals desiccate to 28%. This makes less water available for freezing and concentrates the antifreezes. The desiccation is so extreme that hemolymph samples cannot be taken from winter larvae. Hemolymph from early November larvae (taken prior to desiccation, but after AFP production) and concentrated, as would occur in winter larvae, has TH of nearly 13 °C (measured by the capillary technique), the highest ever reported. Mean winter SCPs from larvae near Fairbanks were −42 °C, but following extensive cold spells, larvae could not be frozen when cooled to as low as −64 °C, the limit of our equipment at that time (Bennett *et al.*, 2005). We now know that when the larvae enter this deep supercooling state they will not freeze when at −150 °C, and they are vitrified at these low temperatures (T. Sformo, K. R. Walters, J. McIntyre, B. Wowk, G. Fahey, B. M. Barnes and J. G. Duman, unpublished). CcAFPs, multimolar glycerol concentrations and the concentrating effects of dehydration promote deep supercooling and vitrification. Recently, using a tandem mass spectrometry approach, a carrot-like AFP was tentatively identified in winter *C. c. clavipes* (M. A. Carrasco and J. G. Duman, unpublished), suggesting the possibility that these larvae have two different types of AFPs.

While AFPs can assist insect survival at extreme low temperatures, they can also be useful at relatively mild subzero temperatures. In northern Indiana *D. canadensis* typically begin to accumulate DAFPs in hemolymph and gut fluid in mid-to-late September and significant TH continues until late April (Duman, 1984). This permits larvae to continue feeding in the autumn well after mild subzero weather has arrived, until lower subzero temperatures occur generally in late November or early December. Likewise, feeding is typically resumed in mid-March, well before the last frosts of the season. Thus, DAFPs significantly extend the feeding period of the larvae. Additional examples of this use of AFPs are winter-active species that remain active and feeding throughout the winter, usually in somewhat thermally protected sites such as the subnivean zone beneath the snow. Husby and Zachariassen (1980) reported examples of winter-active insects and spiders with hemolymph TH from Norway. Also, Zettel (1984) identified TH in the hemolymph of winter-active Collembola from Switzerland that were active on the snow surface at temperatures down to −3 °C. The indication here is that hemolymph AFPs prevent inoculative freezing, but they may also be in the gut, thus permitting ingestion of ice-nucleating micro-organisms by the Collembola.

Another interesting example of the use of AFPs to prevent freezing occurs in the longhorn beetle *R. inquisitor*, where TH occurs not only in the hemolymph and gut, but also intracellularly (Kristiansen *et al.*, 1999). Other than the previously mentioned special case of DAFPs in subcuticular epidermal cells in *D. canadensis*, this is the only known example of intracellular AFPs in insects. This makes good sense if ice-nucleating factors are present inside the cells.

In summary, AFPs provide protection for freeze-avoiding insects by preventing inoculative freezing and by inhibiting ice nucleators in the gut, hemolymph and urine. They lower the temperatures the insects can tolerate, thereby extending their ranges, and AFPs in the gut permit feeding at higher subzero temperatures.

3.3.7.2 Freeze-tolerance

A growing number of freeze-tolerant insects (Duman, 2001; Duman *et al.*, 2004a) and a centipede (Tursman and Duman, 1995) are known to have TH. This should not be surprising, as all of the known plants with TH and/or related activities (recrystallization inhibition, etc.) are freeze-tolerant (Urrutia *et al.*, 1992; Griffith and Yaish, 2004). Since the only TH-producing factor that has been purified and characterized from a freeze-tolerant arthropod is a glycolipid from the freeze-tolerant Alaskan beetle *U. ceramboides*, it is incorrect to call the TH-producing agents from these species AFPs. Also, they may not actually function as antifreezes. Therefore, we will refer to them here as ice-binding-factors (IBFs). The level of TH in most freeze-tolerant species is low, at most only a few tenths of a degree C, although higher activities are occasionally seen. In some cases, measurable TH is not present, but hexagonal crystal growth and/or RI are. Clearly, the usual antifreeze functions of the IBFs are unlikely to be operative, especially in species where there has been selection for endogenous hemolymph ice nucleators that limit supercooling.

If the "normal" antifreeze function of IBFs is not their purpose in freeze-tolerant species, what is their function? Some possibilities are: (1) prevention of intracellular ice, (2) stabilization of membranes and (3) recrystallization inhibition. While the first two possibilities are speculative, recrystallization inhibition (RI) is probable. When an aqueous solution freezes, especially at lower temperature, the resulting ice crystals tend to be small. Although the total amount of ice stays constant, over time there is an increase in crystal size due to migration of water molecules from smaller to larger crystals. This can cause damage to tissues (Tursman and Duman, 1995). Recrystallization occurs because the high radius of curvature of small crystals results in higher surface free energy, and consequently there is a net migration of water from smaller to larger crystals. AFPs stop recrystallization, even at concentrations orders of magnitude lower than required for TH (Knight *et al.*, 1984; Knight and Duman, 1986).

The presence of IBFs on cell membranes and/or inside cells could inhibit the lethal propagation of extracellular ice into the cytoplasm. Also, certain fish AFPs have been demonstrated to protect cell membranes from low-temperature damage at above freezing temperatures (Tomczak and Crowe, 2002). Fish AFPs are structurally quite different from insect IBFs, but membrane protection from cold damage is a possible function that should be investigated, not only in freeze-tolerant arthropods, but also in freeze-avoiding species.

The centipede *L. forficatus* is freeze-tolerant, but the lower lethal temperature is only −6 °C, (Tursman *et al.*, 1994). However, even minimal supercooling (1 °C) prior to freezing is lethal, and consequently for freeze-tolerance to be manifested, the centipedes must be inoculated by external ice at a temperature just below the hemolymph freezing point. Hemolymph TH ranging from 0.33–1.35 °C was measured, but only on rare occasions in late autumn. Generally no TH was seen; however hemolymph RI activity occurred throughout the winter. *In vitro* experiments showed that DAFPs in the bathing medium caused a decrease in LT_{50} of midgut cells taken from both winter and summer centipedes (Tursman and Duman, 1995).

Cells from summer animals incubated with DAFPs and then washed with medium lacking DAFPs, to remove the DAFPs prior to freezing, had increased survivorship after freezing (LT_{50} of −14.5 °C) relative to controls, suggesting that the DAFPs interacted with cell membranes. Immunofluorescence indicated that this was the case, and also demonstrated the presence of endogenous IBFs immunologically similar to DAFPs on the membranes and/or inside the cells in winter. Under conditions designed to promote recrystallization, cells isolated from the midgut were protected from damage by the addition of DAFPs.

Certain features of the described situation of IBFs in *L. forficatus* have been identified in certain freeze-tolerant insects. The stonefly *Nemoura arctica* inhabits small headwater streams in the Brooks Range in arctic Alaska that freeze down to the substratum in winter. The nymphs are freeze-tolerant with an LT_{50} of −15 °C (Walters *et al.*, 2009). While hemolymph TH of 1 °C was measured in some individuals in autumn, most exhibit only RI, a situation reminiscent of the centipede *L. forficatus*. Adults of the beetle *U. ceramboides* from the interior of Alaska are freeze-tolerant, with lethal temperatures near −60 °C (Miller, 1969). This species can exhibit hemolymph TH (range of 0.40–0.70 °C) after cold acclimation (Duman *et al.*, 2004a), but winter individuals generally lack TH and the hemolymph only has RI. As described earlier, a glycolipid IBF capable of producing TH of 4 °C was purified from a homogenate of winter adults. While some TH was present in the soluble portion of the homogenate, a greater amount of TH was associated with the membranes (Walters *et al.*, 2009).

3.3.7.3 Functions not related to cold-tolerance?

Low summer hemolymph TH (generally less than 0.40 °C) in several insects suggests a function for AFPs not related to cold-tolerance. Examples are *D. canadensis* (Duman, 1979) and *T. molitor* (Ramsay, 1964; Patterson and Duman, 1978). Ramsay measured high levels of TH in the perirectal space of the cryptonephridial rectal complex of *T. molitor* larvae acclimated to room temperature. Grimstone *et al.* (1968) conjectured that the proteins responsible for the TH were involved in the water reabsorption ability of the rectal complex. Although no direct experimental evidence has been presented for this possibility, when *T. molitor* larvae were acclimated to low relative humidity at otherwise non-AFP inducing conditions (high temperature, long photoperiod), hemolymph TH increased (Patterson and Duman, 1978), suggesting a function related to dehydration resistance.

While dual functions have not been identified for animal AFPs, this is the normal situation in plants (Griffith and Yaish, 2004). Winter rye AFPs have sequence homology to three classes of pathogenesis-related proteins that function as antimicrobial agents. Also, the bittersweet nightshade, *Solanum dulcamara*, produces a thermal-hysteresis protein that has sequence homology to WRKY proteins, a family of transcription factors involved in regulation of pathogenesis-related proteins in plants. The protein, not produced until late autumn, not only has TH, but specific DNA-binding activity as well. A unique feature of the *S. dulcamara* protein is a C-terminus with 10 consecutive 13-mer repeats, where the TH is thought to reside (Huang and Duman, 2002). Carrot AFP is similar to polygalacturonase-inhibiting proteins (Worral *et al.*, 1998; Meyer *et al.*, 1999; Smallwood *et al.*, 1999). Certain *D. canadensis* DAFPs were tested for both antibacterial and antifungal activity, but no such activity was detected (J. G. Duman, unpublished). Therefore, the potential dual function(s) of insect AFPs, in addition to their antifreeze function, remains unknown.

3.3.8 Control of AFP production and activity

Although some insects have low levels of TH in the summer, TH in these species increases significantly through the autumn and into winter. Also, most exhibit TH only in winter. This raises the question of what factors control the seasonal cycles of AFP production and activity. Short photoperiod and low temperature induce TH in late summer *Meracantha contracta* (Duman, 1977b) and *D. canadensis* (Horwath and Duman, 1982, 1983a) as well as in warm acclimated *T. molitor* (Patterson and Duman, 1978). The circadian system is involved in the photoperiodic timing mechanism inducing TH in *D. canadensis* (Horwath and Duman, 1982, 1983a, 1984b). Also, short thermoperiods are inductive in *D. canadensis* (Horwath and Duman, 1986). Further, as mentioned previously, low relative humidity induces TH in *T. molitor* and *M. contracta* (Patterson and Duman, 1978; Duman, 1977b). Apparently, the central nervous system integrates these environmental signals and stimulates juvenile hormone production that then induces AFP production in *D. canadensis* (Horwath and Duman, 1983b; Xu and Duman, 1991) and *T. molitor* (Xu *et al.*, 1992).

Spruce budworm larvae, *C. fumiferana*, winter as second instars. While a few AFP transcripts are present in the first instar, they increase greatly during the diapausing second instar and then decrease in spring during the third instar, indicating a developmental component of AFP regulation (Doucet *et al.*, 2001). There may also be a developmental effect on AFP accumulation in *T. molitor* (Graham *et al.*, 2000), in addition to the environmental and hormonal regulation (Patterson and Duman, 1978; Xu *et al.*, 1992).

However, recall that TH is dependent, not only on AFP concentration, but also on the specific activity of the AFPs and the presence of enhancers. Also, *D. canadensis* have tissue specific expression of DAFPs. Consequently, the cues for enhancers and/or whether there are different signals controlling DAFP expression in different tissues (fat body, gut epithelium, Malpighian tubule epithelium, etc.) are unknown.

Since photoperiod is the most reliable cue of seasonal change, involvement of photoperiod in AFP induction seems appropriate. However, contrary to the experiments done in the 1980s showing photoperiodic induction of DAFP in *D. canadensis*, as mentioned above, recent experiments failed to demonstrate photoperiodic involvement (P. K. Nickell and J. G. Duman, unpublished). This raises the possibility that *D. canadensis* has responded to the later onset of subzero temperatures over the last 30 years by not using photoperiod as a cue for DAFP production in the autumn. This would save energy by delaying DAFP production until it is needed, but may leave the larvae vulnerable to unusually early cold snaps.

3.4 Future directions

The study of both ice nucleators and thermal-hysteresis factors in insects will be fruitful areas of research far into the future. The tremendous diversity of insects, and their adaptations to cold, insure that novel variations on these themes are awaiting identification. This is especially true of AFPs and other thermal-hysteresis factors.

As regards INs, there is great need for sequence and other structural information. This will permit comparisons to well-studied bacterial and fungal PINs that will be most useful for understanding the structural/ice-nucleating mechanism. Also, experiments on the *in vivo* function of PINs are needed to answer whether PINs are essential for the freeze-tolerance of particular species. Knock-outs/knock-downs of PINs using RNAi, etc. could be particularly useful in this regard.

Additional structural information on AFPs and other thermal-hysteresis factors (such as the glycolipid ice-binding factor) is essential. Only three types of AFPs have been studied to date (beetles, spruce budworm, and Collembola). It is certain that other types are yet to be discovered in insects (probably even within the beetles) and other terrestrial arthropods. This will provide insight into the structure/thermal-hysteresis activity of these factors, something of interest to biologists, biochemists and biophysicists alike. Also, this structural information is important in understanding the *in vivo* functions of AFPs, etc. We need answers to questions such as why so many different isoforms of AFPs are found in insects, such as *D. canadensis*, and why these isoforms tend to be tissue specific, when the differences in sequence are often minimal (often only a few amino acids different). How do these seemingly minor differences apparently favor function of certain DAFPs in certain sites? Likewise, understanding of the enhancement of beetle AFPs by other AFPs, other non-AFP proteins, and by low-molecular-mass solutes, and how this relates to function, is required. Additional *in vivo* functional studies are especially important, but are not easy to properly carry out. How much does the presence of

AFPs extend the supercooling ability of an insect? What are the functions of AFPs and/or other thermal-hysteresis factors, such as the glycolipid ice-binding factor, in freeze-tolerant species? Do the AFPs have a function in summer? Probably the answer is yes, but what is it?

Another area likely to yield important results in the future is the use of AFPs and related factors in applications to cryopreservation, food preservation and agriculture. Insect AFPs, because of their higher specific activities relative to those of other organisms, are likely to prove superior. For example, numerous attempts have been made to use fish AFP genes to transfect other organisms for the purpose of increasing their cold-tolerance. Essentially all of these have failed. In contrast, although the decreases in freezing temperatures (lower SCPs and inhibition of inoculative freezing) have only been a few degrees, they have been significant, both in transgenic *Arabidopsis thaliana* (Huang *et al.*, 2002) and *Drosophila melanogaster* (Nicodemus *et al.*, 2006) expressing a single type of DAFP and in those expressing two self-enhancing types of DAFPs (X. Lin, M. Wisniewski and J. G. Duman, unpublished; X. Lin, J. E. O'Tousa and J. G. Duman, unpublished). Readers interested in the applications of AFPs should consult an excellent recent review, although it sometimes lacks a critical review of the actual efficacy of the AFPs, by Venketesh and Dayanada (2008).

Acknowledgements

Portions of this study were supported by NSF grant IOB06-18342 to JGD and BMB.

References

Amornwittawat, N., Wang, S., Duman, J. G., and Wen, X. (2008) Polycarboxylates enhance beetle antifreeze protein activity. *Biochimica et Biophysica Acta, Proteins and Proteomics* **1784**, 1942–1948.

Amornwittawat, N., Wang, S., Banatlao, J., Chung, M., Velasco, E., Duman, J. G., and Wen, X. (2009). Effects of polyhydroxy compounds on beetle antifreeze protein activity. *Biochimica et Biophysica Acta, Proteins, and Proteomics* **1794**, 341–346.

Andorfer, C. A. and Duman, J. G. (2000). Isolation and characterization of cDNA clones endcoding antifreeze proteins of the pyrochroid beetle *Dendroides canadensis*. *Journal of Insect Physiology* **46**, 365–372.

Bale, J. S., Hansen, T. N., and Baust, J. G. (1989). Nucleators and sites of nucleation in the freeze tolerant larvae of the gallfly *Eurosta solidaginis* (Fitch). *Journal of Insect Physiology* **35**, 291–298.

Bale, J. S., Worland, M. R., and Block, W. (2000). Thermal tolerance and acclimation response of the subAntarctic beetle *Hydromedion sparsutum*. *Polar Biology* **23**, 77–84.

Bennett, V. A., Sformo, T., Walters, K., Toien, O., Jeannet, K., Hochstrasser, R., Pan, Q., Serianni, A. S., Barnes, B. M., and Duman, J. G. (2005). Comparative overwintering physiology of Alaska and Indiana populations of the beetle *Cucujus clavipes* (Fabricus): Roles of antifreeze proteins, polyols, dehydration, and diapause. *Journal of Experimental Biology* **208**: 4467–4477.

Bigg, E. K. (1953). The supercooling of water. *The Physical Society* **66**, 688–691.

Block, W. and Duman, J. G. (1989). The presence of thermal hysteresis producing antifreeze proteins in the Antarctic mite, *Alaskozetes antarcticus*. *Journal of Experimental Zoology* **250**, 229–231.

Brown, C. L., Bale, J. S., and Walters, K. F. A. (2004). Freezing induces a loss of freeze tolerance in an overwintering insect. *Proceedings of the Royal Society*, Series B **271**, 1507–1511.

Cannon, R. J. C. and Block, W. (1988). Cold tolerance of microarthropods. *Biological Reviews of the Cambridge Philosophical Society* **63**, 23–77.

DeVries, A. L. (1971). Glycoproteins as biological antifreeze agents in Antarctic fishes. *Science* **172**, 1152–1155.

DeVries, A. L. (1986). Antifreeze glycopeptides and peptides: Interactions with ice and water. *Methods in Enzymology* **127**, 293–303.

DeVries, A. L. and Cheng, C.-H. C. (1992). The role of antifreeze glycopeptides and peptides in the survival of cold-water fishes. In *Water and Life*, ed. G. N. Somero, C. B. Osmond and C. L. Bolis. Berlin, Heidelberg: Springer-Verlag, pp. 301–315.

DeVries, A. L., Komatsu, S. K., and Feeney, R. E. (1970). Chemical and physical properties of freezing point depressing glycoproteins from Antarctic fishes. *Journal of Biological Chemistry* **245**, 2901–2913.

DeVries, A. L. and Wohlschlag (1969) Freezing resistance in some Antarctic fishes. *Science* **163**, 1073–1075.

Doucet, D., Tyshenko, M. G., Davies, P. L., and Walker, V. K. (2001). A family of expressed antifreeze protein genes from the moth, *Choristoneura fumiferana*. *European Journal of Biochemistry* **269**, 38–46.

Duman, J, G. (1977a). The role of macromolecular antifreeze in the darkling beetle, *Meracantha contracta*. *Journal of Comparative Physiology B* **115**, 279–286.

Duman, J. G. (1977b). Variations in macromolecular antifreeze levels in larvae of the darkling beetle, *Meracantha contracta*. *Journal of Experimental Zoology*, 85–93.

Duman, J. G. (1979a). Thermal hysteresis factors in overwintering insects. *Journal of Insect Physiology* **25**, 805–810.

Duman, J. G. (1979b). Subzero temperature tolerance in spiders: The role of thermal hysteresis factors. *Journal of Comparative Physiology B* **131**, 347–352.

Duman, J. G. (1984). Thermal hysteresis antifreeze proteins in the midgut fluid of overwintering larvae of the beetle *Dendroides canadensis*. *Journal of Experimental Zoology* **230**, 355–361.

Duman, J. G. (2001). Antifreeze and ice nucleator proteins in terrestrial arthropods. *Annual Review of Physiology* **63**, 327–355.

Duman, J. G. (2002). The inhibition of ice nucleators by insect antifreeze proteins is enhanced by glycerol and citrate. *Journal of Comparative Physiology B* **172**, 163–168.

Duman, J. G., Bennett, V., Sformo, T., Hochstrasser, R., and Barnes, B. M. (2004a). Antifreeze proteins in Alaskan insects and spiders. *Journal of Insect Physiology* **50**, 259–266.

Duman, J. G., Bennett, V. A., Li, N., Wang, L., Huang, L., Sformo, T., and Barnes, B. M. (2004b). Antifreeze proteins in terrestrial arthropods. In *Life in the Cold*, ed. B. M. Barnes and H. V. Carey. University of Alaska Press, pp. 527–542.

Duman, J. G. and DeVries, A. L. (1974). Freezing resistance in winter flounder, *Pseudopleuronectes americanus*. *Nature* **247**, 237–238.

Duman, J. G. and DeVries, A. L. (1975). The role of macromolecular antifreeze in cold water fishes. *Comparative Biochemistry and Physiology* **52A**, 193–199.

Duman, J. G. and DeVries, A. L. (1976) The isolation, characterization and physical properties of protein antifreeze from winter flounder, *Pseudopleuronectes americanus*. *Comparative Biochemistry and Physiology* **54B**, 375–380.

Duman, J. G., Neven, L. G., Beals, J. M., Olson, K. R., and Castellino, F. J. (1985). Freeze tolerance adaptations, including haemolymph protein and lipoprotein ice nucleators, in larvae of the cranefly *Tipula trivittata*. *Journal of Insect Physiology* **31**, 1–9.

Duman, J. G., Parmalee, D., Goetz, F. W., Li, N., Wu, D. W., and Benjamin, T. (1998). Molecular characterization and sequencing of antifreeze proteins from larvae of the beetle *Dendroides canadensis*. *Journal of Comparative Physiology B* **168**, 225–232.

Duman, J. G. and Patterson, J. L. (1978). The role of ice nucleators in the frost tolerance of queens of the bald-faced hornet *Vespula maculata*. *Comparative Physiology and Biochemistry* **59A**, 69–72.

Duman, J. G. and Serianni, A. S. (2002). The role of endogenous antifreeze protein enhancers in the hemolymph thermal hysteresis activity of the beetle *Dendroides canadensis*. *Journal of Insect Physiology* **48**, 103–111.

Duman, J. G., Wu, D. W., Yeung, K. L., and Wolf, E. E. (1992). Hemolymph proteins involved in the cold tolerance of terrestrial arthropods; Antifreeze and ice nucleator proteins. In *Water and Life*, ed. G. N. Somero and C. B. Osmond. Berlin: Springer-Verlag, pp. 282–300.

Duman, J. G., Wu, D. W., Olsen T. M., Urrutia, M., and Tursman, D. (1993). Thermal hysteresis proteins. *Advances in Low Temperature Biology* **2**, 131–182.

Duman, J. G., Xu, L. X., Neven, L. G., Tursman, D., and Wu, D. W. (1991). Hemolymph proteins involved in insect subzero temperature tolerance: Ice nucleators and antifreeze proteins. In *Insects at Low Temperatures*, ed. R. E. Lee and D. L. Denlinger. New York and London: Chapman and Hall, pp. 94–127.

Duman, J. G., Verleye, D., and Li, N. (2002). Site specific forms of antifreeze proteins in the beetle *Dendroides canadensis*. *Journal of Comparative Physiology B* **172**, 547–552.

Fields, P. G. and McNeil, J. N. (1986). Possible dual cold-hardiness strategies in *Cisseps fulvicolis* (Lepidoptera: Arctiidae). *Canadian Journal of Entomology* **118**, 1309–1311.

Gauthier, S. Y., Kay, C. M., Sykes, B. D., Walker, V. K., and Davies, P. L. (1998). Disulfide bond mapping and structural characterization of spruce budworm antifreeze protein. *European Journal of Biochemistry* **258**, 445–453.

Govindarajan, A. G. and Lindow, S. E. (1988). Phospholipid requirement for expression of ice nuclei in *Pseudomonas syringae* and *in vitro*. *Journal of Biological Chemistry* **263**, 9333–9338.

Graether, S. P., Kuiper, M. J., Gagne, S. M., Walker, V. K., Jia, Z., Sykes, B. D., and Davies, P. L. (2000). Beta-helix structure of a hyperactive antifreeze protein from an insect. *Nature* **406**, 325–328.

Graether, S. P. and Sykes, B. D. (2004). Cold survival of freeze intolerant insects: The structure and function of beta-helical antifreeze proteins. *European Journal of Biochemistry* **271**, 3285–3296.

Graether, S. P., Ye, Q. L., Davies, P. L., and Sykes, D. B. (1999). Crystallization and preliminary x-ray crystallographic analysis of spruce budworm antifreeze protein. *Journal of Structural Biology* **126**, 72–75.

Graham, L. A. and Davies, P. L. (2005). Glycine-rich antifreeze proteins from snow fleas. *Science* **310**, 461.

Graham, L. A., Liou, Y.-C., Walker, V. K., and Davies, P. L. (1997). Hyperactive antifreeze proteins from beetles. *Nature* **188**, 727–728.

Graham, L. A., Walker, V. K., and Davies P. L. (2000). Developmental and environmental regulation of antifreeze proteins in the mealworm beetle *Tenebrio molitor*. *European Journal of Biochemistry* **267**, 6452–6458.

Green, R. L. and Warren, G. J. (1985). Physical and functional repetition in a bacterial ice nucleation gene. *Nature* **317**, 645–648.

Griffith, M. and Yaish, M. W. (2004). Antifreeze proteins in overwintering plants: A tale of two activities. *Trends in Plant Sciences* **9**, 399–405.

Grimstone, A. V., Mullinger, A. M., and Ramsay, J. A. (1968). Further studies on the rectal complex of the mealworm, *Tenebrio molitor*. *Philosophical Transactions of the Royal Society of London, Series B* **248**, 344–282.

Hansen, T. N. and Baust, J. G. (1988). Differential scanning calorimetric analysis of antifreeze protein activity in the common mealworm, *Tenebrio molitor*. *Biochimica and Biophysica Acta-Protein Structure and Molecular Enzymology* **957**, 217–221.

Hansen, T. N. and Baust, J. G. (1989). Differential scanning calorimetric analysis of *Tenebrio molitor* antifreeze protein activity. *Cryobiology* **26**, 383–388.

Horwath, K. L. and Duman, J. G. (1982). Involvement of the circadian system in photoperiodic regulation of insect antifreeze proteins. *Journal of Experimental Zoology* **219**, 267–270.

Horwath, K. L. and Duman, J. G. (1983a). Photoperiodic and thermal regulation of antifreeze protein levels in the beetle *Dendroides canadensis*. *Journal of Insect Physiology* **29**, 907–917.

Horwath, K. L. and Duman, J. G. (1983b). Induction of antifreeze production by juvenile hormone in larvae of the beetle *Dendroides canadensis*. *Journal of Comparative Physiology B* **151**, 233–240.

Horwath, K. L. and Duman, J. G. (1984a). Yearly variations in the overwintering mechanism of the cold hardy beetle, *Dendroides canadensis*. *Physiological Zoology* **57**, 40–45.

Horwath, K. L. and Duman, J. G. (1984b). Further studies on the involvement of the circadian system in photoperiodic control of antifreeze protein production in the beetle *Dendroides canadensis*. *Journal of Insect Physiology* **30**, 947–955.

Horwath, K. L. and Duman, J. G. (1986). Thermoperiodic involvement in antifreeze protein production in the cold hardy beetle *Dendroides canadensis*: Implications for photoperiodic timer measurement. *Journal of Insect Physiology* **32**, 799–806.

Huang, T. and Duman, J. G. (2002). Cloning and characterization of a thermal hysteresis/antifreeze protein with DNA-binding activity from winter bittersweet nightshade, *Solanum dulcamara*. *Plant Molecular Biology* **48**, 339–350.

Huang, T., Nicodemus, J., Zarka, D. G., Thomashow, M. F., and Duman, J. G. (2002). Expression of an insect (*Dendroides canadensis*) antifreeze protein in *Arabidopsis thaliana* results in a decrease in plant freezing temperature. *Plant Molecular Biology* **50**, 333–344.

Husby, J. A. and Zachariassen, K. E. (1980). Antifreeze agents in the body fluid of winter active insects and spiders. *Experientia* **36**, 963–964.

Jia, X. and Davies P. L. (2002). Antifreeze proteins: An unusual receptor-ligand interaction. *Trends in Biochemical Sciences* **27**, 101–106.

Knight, C. A. (1967). *The Freezing of Supercooled Liquids*. New York: VanNostrand.

Knight, C. A., Cheng, C. C., and DeVries, A. L. (1991). Adsorption of alpha-helical antifreeze peptides on specific ice crystal surface planes. *Biophysical Journal* **59**, 409–418.

Knight, C. A., DeVries, A. L., and Oolman, L. D. (1984). Fish antifreeze protein and the freezing and recrystallization of ice. *Nature* **308**, 295–296.

Knight, C. A. and Duman, J. G. (1986). Inhibition of recrystallization of ice by insect thermal hysteresis proteins: A possible cryoprotective role. *Cryobiology* **23**, 256–262.

Knight, C. A., Wen, D., and Laursen, R. A. (1995). Non-equilibrium antifreeze proteins and the recrystallization of ice. *Cryobiology* **32**, 23–34.

Kristainsen, E., Pedersen, S. L., Ramlov, H., and Zachariassen, K. E. (1999). Antifreeze activity in the cerambycid beetle *Rhagium inquisitor*. *Journal of Comparative Physiology B* **160**, 55–60.

Kristiansen, E., Ramlov, H., Hagen, L., Pedersen, S. L., Andersen, R. A., and Zachariassen, K. E. (2005). Isolation and characterization of hemolymph antifreeze proteins from larvae of the longhorn beetle *Rhagium inquisitor* (L). *Comparative Biochemistry and Physiology B* **142**, 90–97.

Kristiansen, E. and Zachariassen, K. E. (2005). The mechanism by which fish antifreeze proteins cause thermal hysteresis. *Cryobiology* **51**, 262–280.

Kukal, O., Serianni, A. S., and Duman, J. G. (1988). Glycerol production in a freeze tolerant arctic insect, *Gynaephora groenlandica*: An *in vivo* ^{13}C NMR study. *Journal of Comparative Physiology B* **158**, 175–183.

Lee, R. E., Costanzo, J. P., and Mugnano, J. A. (1996). Regulation of supercooling and ice nucleation in insects. *European Journal of Entomology* **93**, 405–418.

Leinala, E. K., Davies, P. L., Doucet, D., Tyshenko, M. G., Walker, V. K., and Jia, Z. (2002). A beta-helical antifreeze protein isoform with increased activity: Structural and functional insights. *Journal of Biological Chemistry* **277**, 33349–33352.

Li, N., Andorfer, C. A., and Duman, J. G. (1998b). Enhancement of insect antifreeze protein activity by low molecular mass solutes. *Journal of Experimental Biology* **201**, 2243–2251.

Li, N., Chibber, B. A. K., Castellino, F. J., and Duman, J. G. (1998a). Mapping of disulfide bridges in antifreeze proteins from overwintering larvae of the beetle *Dendroides canadensis*. *Biochemistry* **37**, 6343–6350.

Li, Y., Gong, H., and Park, H. Y. (2000). Purification and partial characterization of thermal hysteresis proteins from overwintering larvae the pine needle gall midge, *Thecodiplosis japonensis* (Diptera: Cecidomiidae) *CryoLetters* **21**, 117–124.

Lin, F.- H., Graham, L. A., Campbell, R. L., and Davies, P. L. (2007). Structural modeling of snow flea antifreeze protein. *Biophysical Journal* **92**, 1717–1723.

Lin, Y., Duman, J. G., and DeVries, A. L. (1972). Studies on the structure and activity of low molecular weight glycoproteins from an Antarctic fish. *Biochemical Biophysical Research Communications* **46**,87–92.

Lindow, S. E. (1983). The role of bacterial ice nucleation in frost injury to plants. *Annual Review of Phytopathology* **21**, 363–384.

Lindow, S. E. (1995). Control of epiphytic ice-nucleation-active bacteria for management of plant frost injury. In *Biological Ice Nucleation and Its Applications*, ed. R. E. Lee, L. G. J. Warren, and L. V. Gusta. Saint Paul: APS Press, pp. 239–256.

Liou, Y.-C., Thibault, P., Walker, V. K., Davies, P. L., and Graham, L. A. (1999). A complex family of highly heterogeneous and internally repetitive hyperactive antifreeze proteins from the beetle *Tenebrio molitor*. *Biochemistry* **38**, 11415–11424.

Liu, X. Y. and Du, N. (2004). Zero-sized effect of nano-particles and inverse homogeneous nucleation. *The Journal of Biological Chemistry* **279**, 6124–6131.

Lundheim, R. (1996). *Adaptive and Incidental Ice Nucleators*. Doctorate Thesis, Norwegian University of Science and Technology, Trondheim.

Lu, M., Wang, B., Li, Z., Fei, Y., Wei, L., and Gao, S. (2002). Differential scanning calorimetric and circular dicroistic studies on plant antifreeze proteins. *Journal of Thermal Analysis and Calorimetry* **67**, 689–698.

Mazur, P. (1984). Freezing of living cells: mechanisms and implications. *American Journal of Physiology* **247**, C125-C142.

Meier, P. and Zettel, J. (1999). Cold hardiness in *Entomobrya nivalis* (Collembola, Entomobryidae): Annual cycle of polyols and antifreeze proteins, and antifreeze triggering by temperature and photoperiod. *Journal of Comparative Physiology B* **167**, 297–304.

Meyer, K., Keil, M., and Naldrett, M. J. (1999). A leucine-rich repeat protein of carrot that exhibits antifreeze activity. *FEBS Letters* **447**, 171–178.

Miller, L. K. (1969). Freezing tolerance of an adult insect. *Science* **166**, 105–106.

Miller, L. K. (1982). Cold hardiness strategies of some adult and immature insects overwintering in interior Alaska. *Comparative Physiology and Biochemistry* **73A**, 595–604.

Miller, L. K. and Werner, R. (1987). Extreme supercooling as an overwintering strategy in three species of willow gall insects from interior Alaska. *Oikos* **49**, 253–260.

Mueller, G. M., Wolber, P. K., and Warren, G. J. (1990). Clustering of ice nucleation protein correlates with ice nucleation activity. *Cryobiology* **27**, 416–422.

Mugnano, J. A., Lee, R. E., and Taylor, R. T. (1996). Fat body cells and calcium phosphate spherules induce ice nucleation in the freeze tolerant larvae of the gall fly *Eurosta solidaginis* (Fitch). *Journal of Experimental Biology* **199**, 465–471.

Neven, L. G., Duman, J. G., Beals, J. M., and Castellino, F. J. (1986). Overwintering adaptations of the stag beetle *Ceruchus piceus*: Removal of ice nucleators in winter to promote supercooling. *Journal of Comparative Physiology B* **156**, 707–716.

Neven, L. G., Duman, J. G., Low, M. G., Sehl, L. C., and Castellino, F. J. (1989). Purification and characterization of an insect lipoprotein ice nucleator: Evidence for the importance of phosphatidylinositol and apolipoprotein in the ice nucleator activity. *Journal of Comparative Physiology B* **159**, 71–82.

Nicodemus, J., O'Tousa, J. E., and Duman, J. G. (2006). Expression of a beetle, *Dendroides canadensis*, antifreeze protein in *Drosophila melanogaster*. *Journal of Insect Physiology* **52**, 888–896.

Olsen, T. M. and Duman, J. G. (1997a). Maintenance of the supercooled state in overwintering pyrochroid beetle larvae *Dendroides canadensis*: Role of hemolymph ice nucleators and antifreeze proteins. *Journal of Comparative Physiology B* **167**, 105–113.

Olsen, T. M. and Duman, J. G. (1997b). Maintenance of the supercooled state in the gut of overwintering pyrochroid beetle larvae, *Dendroides canadensis*: role of gut ice nucleators and antifreeze proteins. *Journal of Comparative Physiology B*, **167**, 114–122.

Olsen, T. M., Sass, S. J., Li, N., and Duman, J. G. (1998). Factors contributing to seasonal increases in inoculative freezing resistance in overwintering fire-colored beetle larvae *Dendroides canadensis* (Pyrochroidae). *Journal of Experimental Biology* **201**, 1585–1594.

Patterson, J. L. and Duman, J. G. (1978). The role of thermal hysteresis producing proteins in the low temperature tolerance and water balance of the mealworm, Tenebrio molitor. *Journal of Experimental Biology* **74**, 37–45.

Pertaya, N., Marshall, C. B., Celik, Y., Davies, P. L., and Braslovsky, I. (2008). Direct visualization of spruce budworm antifreeze protein interacting with ice: Basal plane affinity confers hyperactivity. *Biophysical Journal* **95**, 333–341.

Ramlov, H., DeVries, A. L., and Wilson, P. W. (2005). Antifreeze glycoproteins from the Antarctic fish *Dissostichus mawsoni* studied by differential scanning calorimetry (DSC) in combination with nanoliter osmometry. *CryoLetters* **26**, 73–84.

Ramsay, R. A. (1964). The rectal complex of the mealworm, *Tenebrio molitor* L. Coleoptera, Tenebrionidae. *Philosophical Transactions of the Royal Society of London, Series B* **248**, 279–214.

Raymond, J. A. and DeVries, A. L. (1977). Adsorption inhibition as a mechanism of freezing resistance in polar fishes. *Proceedings of the National Academy of Sciences, USA* **74**, 2589–2593.

Raymond, J. A., Wilson, P. W., and DeVries, A. L. (1989). Inhibition of growth on nonbasal planes in ice by fish antifreeze. *Proceedings of the National Academy of Sciences, USA* **86**, 881–885.

Ring, R. A. and Tesar, D. (1980). Cold-hardiness of the arctic beetle *Pytho americanus* Kirby Coleoptera, Pythidae (Salpingidae). *Journal of Insect Physiology* **26**, 763–777.

Salt, R. W. (1953). The influence of food on cold-hardiness in insects. *Canadian Entomologist* **85**, 261–269.

Sformo, T., Kohl, F., McIntyre, P., Duman, J. G., and Barnes, B. M. (2009). Simultaneous freeze tolerance and avoidance in individual fungus gnats, *Exechia nugatoria. Journal of Comparative Physiology B*, **179**, 897–902.

Sicheri, F. and Yang, D. S. C. (1995). Ice-binding structure and mechanism of an antifreeze protein from winter flounder. *Nature* **375**, 427–431.

Sinclair, B. J. and Chown, S. L. (2002). Haemolymph osmolality and thermal hysteresis activity in 17 species of arthropods from subAntarctic Marion Island. *Polar Biology* **25**, 928–933.

Sinclair, B. J., Terblanche, J. S., Scott, M. B., Blatch, G. L., Klok, C. J., and Chown, S. L. (2006). Environmental physiology of three species of Collembola at Cape Hallett, North Victoria Land, Antarctica. *Journal of Insect Physiology* **52**, 29–50.

Sjursen, H. and Sømme, L. (2000). Seasonal changes in tolerance to cold and desiccation in *Phauloppia* sp. (Acari, Oribatidae) from Finse, Norway. *Journal of Insect Physiology* **46**, 1387–1396.

Smallwood, M., Worrall, D., Byass, L., Ashford, D., Doucet, C. J., Holt, C., Telford, J., Lilliford, P., and Bowles, D. J. (1999). Isolation and characterization of a novel antifreeze protein from carrot (*Daucus carota*) *Biochemical Journal* **340**, 385–391.

Sømme, L. (1978). Nucleating agents in the haemolymph of the third instar larvae of *Eurosta solidaginis* (Fitch) (Diptera: Tephritidae). *Norwegian Journal of Entomology* **25**, 187–188.

Sømme, L. (1982). Supercooling and winter survival in terrestrial arthropods. *Comparative Physiology and Biochemistry* **73A**, 519–543.

Southworth, M. W., Wolber, P. K., and Warren, G. J. (1988). Nonlinear relationship between concentration and activity of a bacterial ice nucleation protein. *Journal of Biological Chemistry* **263**, 15211–15216.

Thomas, M. C. (2002). Cucujidae (Latreille 1802). In *American Beetles Volume 2*, eds. R. H. Arnett, M. C. Thomas, P. E. Skelley, and J. H. Howard. Boca Raton, London, New York, Washington, D.C.: CRC Press, pp. 329–330.

Tomczak, M. M. and Crowe, J. H. (2002). The interaction of antifreeze proteins with model membranes and cells. In *Fish Antifreeze Proteins*, ed. K. V. Ewart and C. L. Hew, London: World Scientific, pp. 187–212.

Tursman, D. and Duman, J. G. (1995). Cryoprotective effects of thermal hysteresis protein on survivorship of frozen gut cells from the freeze tolerant centipede *Lithobius forficatus. Journal of Experimental Zoology* **272**, 249–257.

Tursman, D., Duman, J. G., and Knight, C. A. (1994). Freeze tolerance adaptations in the centipede *Lithobius forficatus. Journal of Experimental Zoology* **268**, 347–353.

Tyshenko, M. G., Doucet, D., Davies, P. L., and Walker, V. K. (1997). The antifreeze potential of spruce budworm thermal hysteresis protein. *Nature Biotechnology* **15**, 887–890.

Urrutia, M. E., Duman, J. G., and Knight, C. A. (1992). Plant thermal hysteresis proteins. *Biochimica et Biophysica Acta* **1121**, 199–206.

Vanketesh, S. and Dayanada, C. (2008). Properties, potentials and prospects of antifreeze proteins. *Critical Reviews of Biotechnology* **28**, 57–82.

Walters, K. R., Serianni, A. S., Sformo, T., Barnes, B. M., and Duman, J. G. (2009). A novel thermal hysteresis-producing xylomannan antifreeze in a freeze tolerant Alaskan beetle. Proceedings of the National Academy of Sciences, USA (in press).

Walters, K. R., Sformo, T., Barnes, B. M., and Duman, J. G. (2009). Freeze tolerance of an Arctic Alaska stonefly. *Journal of Experimental Biology* **212**, 305–312.

Wang, L. and Duman, J. G. (2005). Antifreeze proteins of the beetle *Dendroides canadensis* enhance one another's activities. *Biochemistry* **44**, 10305–10312.

Wang, L. and Duman, J. G. (2006). A thaumatin-like protein from larvae of the beetle *Dendroides canadensis* enhances the activity of antifreeze proteins. *Biochemistry* **45**, 1278–1284.

Wharton, D. A., Pow, B., Kristensen, M., Ramlov, H. R., and Marshall, C. J. (2009). Ice-active proteins and cryoprotectants from the New Zealand alpine cockroach *Celatoblatta quinquemaculata*. *Journal of Insect Physiology* **55**, 27–31.

Wilson, P. W. (1993). Explaining thermal hysteresis by the Kelvin effect. *CryoLetters* **14**, 31–36.

Wolber, P. K. and Warren, G. J. (1989). Bacterial ice nucleating proteins. *Trends in Biochemical Sciences* **14**, 179–182.

Worral, D., Elias, L., Ashford, D., Smallwood, M., Sidebottom, C., Lilliford, P., Telford, J., Holt, C., and Bowles, D. (1998). A carrot leucine-rich-repeat protein that inhibits ice recrystallization. *Science* **282**, 115–117.

Wu, D. W. and Duman, J. G. (1991). Activation of antifreeze proteins from the beetle *Dendroides canadensis*. *Journal of Comparative Physiology B*, **161**, 279–283.

Wu, D. W., Duman, J. G., and Xu, L. (1991). Enhancement of insect antifreeze protein activity by antibodies. *Biochimica et Biophysica Acta* **1076**, 416–420.

Xu, L. and Duman, J. G. (1991) Involvement of juvenile hormone in the induction of antifreeze protein production by fat body in larvae of the beetle *Dendroides canadensis*. *Journal of Experimental Zoology* **258**, 288–293.

Xu, L., Duman, J. G., Goodman, W. G., and Wu, D. W. (1992) A role for juvenile hormone in the induction of antifreeze protein production by the fat body in the beetle *Tenebrio molitor*. *Comparative Biochemistry and Physiology* **101B**, 105–109.

Yeung, K. L., Wolf, E. E., and Duman, J. G. (1991). A scanning tunneling microscopy study of an insect lipoprotein ice nucleator. *Journal of Vacuum Science and technology B* **9**, 1197–1201.

Zachariassen, K. E. (1982). Nucleating agents in cold-hardy insects. *Comparative Physiology and Biochemistry* **73A**, 557–562.

Zachariassen, K. E. (1985). Physiology of cold tolerance in insects. *Physiological Reviews* **65**, 799–832.

Zachariassen, K. E., DeVries, A. L., Hunt, B., and Kristiansen, E. (2002). Effect of ice fraction and dilution factor on the antifreeze activity in the hemolymph of the cerambycid beetle *Rhagium inquisitor*. *Cryobiology* **44**, 132–141.

Zachariassen, K. E. and Hammel, H. T. (1976). Nucleating agents in the haemolymph of insects tolerant to freezing. *Nature* **262**, 285–287.

Zachariassen K. E. and Husby, J. A. (1982). Antifreeze effects of thermal hysteresis agents protect highly supercooled insects. *Nature* **298**, 865–867.

Zachariassen, K. E., Kristansen, E., Pedersen, S. A., and Hammel, H. T. (2004). Ice nucleation in solutions and freezing in insects – homogeneous or heterogeneous? *Cryobiology* **48**, 309–321.

Zachariassen, K. E., Li, N. G., Laugsand, A. E., Kristiansen, E., and Pedersen, S. A. (2008). Is the strategy for cold hardiness in insects determined by their water balance? A study on two closely related families of beetles: Cerambycidae and Chrysomelidae. *Journal of Comparative Physiology B* **178**, 977–984.

Zettel, J. (1984). Cold hardiness strategies and thermal hysteresis in Collembola. *Revue d'Ecologie de Biologie du Sol* **21**, 189–203.

Zhang, D. Q., Liu, B., Feng, D. R., He, Y. M., and Wang, J. F. (2004). Expression and purification of antifreeze activity of carrot antifreeze protein and its mutants. *Protein Expression and Purification* **35**, 257–263.

4

Genomics, proteomics and metabolomics: Finding the other players in insect cold-tolerance

M. ROBERT MICHAUD AND DAVID L. DENLINGER

4.1 Introduction

Our understanding of the physiological mechanisms of insect cold-hardiness relies on a rich history of discovery that has revealed the presence of cryoprotective agents, antifreeze proteins, ice-nucleating proteins, heat-shock proteins and modifications of the cell membrane. Such discoveries have usually targeted select metabolites, proteins, or fatty acids, and have successfully revealed the identity of many key agents that protect insects from injury at low temperature, yet we suspect that, in many cases, additional protective agents may be involved. In some cases this may mean the potential discovery of entirely new cryoprotective agents or discovery of roles for a multitude of chemical modifications that may accompany a change in a single compound that has already been well documented. This is indeed the power of the "-omics" approach, discovery of large-scale changes at levels of the transcript (genomics or transcriptomics), protein (proteomics), or metabolites (metabolomics). These approaches are especially good for seeing the big picture. What novel genes or gene pathways are turned on or off? What proteins are uniquely synthesized or post-translationally modified in respond to the cold? What is the full range of metabolic change that occurs in response to cold, and what metabolic pathways are affected? These are the sorts of questions that can be uniquely answered by genomics, proteomics and metabolomics. Clearly, all of these approaches mark, not the end of the discovery phase, but the beginning. These approaches thus offer powerful hypothesis-generating tools providing productive leads that can be followed-up with more rigorous and standard procedures

Low Temperature Biology of Insects, ed. David L. Denlinger and Richard E. Lee Jr. Published by Cambridge University Press.

to both verify the results and evaluate functional roles of the gene, gene product, or metabolite.

We are just at the beginning of this phase of discovery for insect cold-hardiness. Though few such studies have thus far been completed, it is already evident from the work on hand that the -omics approach has the potential to dramatically enhance our understanding of cold-tolerance and to especially reveal distinct gene and metabolic pathways that are shifted in response to low temperature. Our goal in this chapter is to review the nascent body of literature that examines large-scale changes in response to insect cold-tolerance, with the hope that this will spur additional interest in these approaches, along with recognition of some of the inherent pitfalls. We are especially intrigued with the obvious conclusion that an insect's response to low temperature seldom involves just the elevation or suppression of a single agent. When we look at the big picture it is evident that low-temperature responses are complex and involve a multitude of gene and metabolic shifts, some of which have already been well studied and others that are novel.

4.2 Genomics

The more that is known about the genetic composition of an organism, the more successful molecular techniques will be when applied to that organism. This is true with targeted analysis, but it is especially true with holistic studies that generate copious data, such as -omics approaches. For example, a proteomic study in the flesh fly, *Sarcophaga crassipalpis*, revealed that a small heat-shock protein is up-regulated during pupal diapause, but the identity of the heat-shock protein was unknown, other than having a high sequence similarity to *Drosphila melanogaster* alpha crystallin, a domain contained in all small heat-shock proteins (Wistow, 1985). Most organisms have multiple small heat-shock proteins, and the functions of each of the genes in this family are thought to be different, so normally the investigator would be forced to identify the protein as "small heat-shock protein-like" and follow-up research would be difficult. However, if an Expressed Sequence Tag (EST) genetic sequencing project were available for this species, which is now the case, the precise identity of the heat-shock protein in question becomes easier to determine, yielding a more definitive identity for the gene. The identities of the genes in the EST project are more definitive because lengths of the fragments used for extra-species comparisons are much longer, often spanning the entire messenger RNA molecule.

A completed EST project also is necessary for the printing of oligonucleotide microarrays, a task that is impossible without sequence information. In this manner, an EST project is a necessary first step for transcriptomic studies, just as it is a valuable tool for the interpretation of proteomic studies. Even for metabolomics, an approach that is not directly connected to sequence information, sequences derived from an EST project may be useful because the enzymes responsible (as well as their isoforms) for controlling the levels of a specific metabolite can be cloned with greater success.

A perusal of publicly available insect EST projects from NCBI reveals 17 insects that are potentially interesting for low-temperature studies (Table 4.1). Of these, only five (*Bombyx mori*, *Leptinotarsa decemlineata*, *Sarcophaga crassipalpis*, *Onychiurus arcticus*, *Folsomia candida*) have thus far been used as models for low-temperature molecular studies. Fortunately, the insect EST projects that are currently available cover species with a wide variety of cold-survival and diapause strategies, allowing for a diversity of biological questions to be pursued.

4.3 Transcriptomics

Over the last decade, a number of emergent technologies have allowed researchers to exploit transcriptomics, a comparison of one pool of mRNA to another, particularly pools derived from different treatment groups, thus making it possible to identify which transcripts are up- or down-regulated in response to a specific treatment. The two main methods currently employed for the transcriptomics approach are suppressive subtractive hybridization and microarray analysis.

Suppressive subtractive hybridization (SSH) involves the tagging of control and treatment RNA pools with different adaptor sequences designed to provide selective polymerase chain reaction (PCR) amplification. The two pools are then combined and allowed to hybridize, and only the single-stranded transcripts that have no corresponding match in the other RNA pool are amplified by PCR. This technique produces PCR products that represent increased and decreased transcripts with respect to the control. The advantage of this technique is that the clones of putatively up- and down-regulated transcripts have already been obtained, making verification of transcriptional change technically a much easier process by eliminating sequence guesswork when designing primers for cloning. The major disadvantage of this technique lies in the fact that there is no quantification of transcript, only its procurement from the greater RNA pool. For this reason, all results obtained by SSH must be verified with some other measure of transcript expression, either through Northern blot analysis or quantitative real-time PCR (qPCR).

To date, only one insect low-temperature study has been published that shows the full results of an SSH experiment, an experiment on the adult diapause of the mosquito, *Culex pipiens* (Robich *et al.*, 2007). In this study, a number of transcripts were up-regulated during diapause, particularly ribosomal sub-units and mitochondrial metabolic genes. In addition, some stress-protein transcripts were

Table 4.1 *Completed EST and genomics projects for insect species with cold-hardiness traits, ranked by number of available ESTs listed in the* National Center for Biotechnology Information (NCBI) *database*

Species	Common name	Cold-hardiness potential	Number of ESTs
Culex quinquefasciatus	Brown house mosquito	Adult reproductive diapause in close relative (*C. pipiens*)	204 742
Bombyx mori	Silk moth	Obligate, cold-hardy, embryonic diapause	184 519
Acyrthosiphon pisum	Pea aphid	Egg diapause, high latitude distribution	167 735
Tribolium castaneum	Red flour beetle	Overwinters all stages, rapid cold-hardening	64 571
Locusta migratoria	Migratory locust	Egg diapause, cold-hardy	45 449
Sarcophaga crassipalpis	Flesh fly	Cold-hardy diapause, rapid cold-hardening	43 155
Nasonia giraulti	Jewel wasp	Cold-hardy, larval diapause	30 060
N. vitripennis	Jewel wasp	Cold-hardy, larval diapauses	30 653
Onychiurus arcticus	Arctic springtail	Arctic habitat, cryoprotective dehydration	16 379
Folsomia candida	White springtail	No diapause, overwinters in all stages	8686
Leptinotarsa decemlineata	Colorado potato beetle	Adult, reproductive, cold-tolerant diapause	8466
Aphis gossypii	Melon aphid	Adult overwinters in mild regions, egg diapause in temperate	8344
Graphocephala atropunctata	Blue-green sharpshooter	Obligate adult diapause, cold tolerant	6488
Manduca sexta	Tobacco hornworm	Pupal diapause, temperate distribution	3317
Culicoides sonorensis	Biting midge	Reproductive diapause, high latitude distribution	2977
Ostrinia nubilalis	European corn-borer	Larval, cold-hardy diapause	1768
Oncopeltus fasciatus	Milkweed bug	Migratory, reproductive diapause	1115
Helicoverpa armigera	Pink bollworm	Cold-hardy, pupal diapause	1055

increased, as well as actin transcripts. Follow-up Northern blots confirmed the up- or down-regulation of 32 of 40 genes, and the failure to confirm results for the remaining eight genes in the follow-up were due to a lack of hybridization or failure to cross the threshold of detection.

SSH was used in two other studies on cold acclimation and recovery from cold-shock in *Drosophila melanogaster*, but the full results were not reported in the literature. However, two genes with altered transcript levels from cold acclimation, *frost* and *senescence marker protein-30*, were subjected to targeted analysis and confirmed to be up- and down-regulated, respectively (Goto, 2001a, 2001b). Quantitative real-time PCR found the same results for the *frost* gene, and also reported that both *frost* and *senescence marker protein-30* transcript levels were up-regulated in response to starvation and desiccation (Sinclair *et al.*, 2007). Another example of partially reported SSH results may be found in a study of diapause in *S. crassipalpis* (Rinehart *et al.* 2007), where only heat-shock proteins were the focus.

The other powerful transcriptomic technique that is widely used is microarray analysis. Microarray analysis involves the hybridization of two mRNA pools (converted to cDNA and labeled with cy3 and cy5 fluorescent dyes) to a "gene chip" containing "spots" of oligonucleotide probes for thousands of genes derived from genomic analysis. Competition of RNAs from the two pools to hybridize with the spots produces a blend of the two fluorescent dye colors for each spot, which can then be read and quantified photometrically. An equal level of transcript from the two pools produces a yellow color, but if one pool of RNA has a higher level of a particular transcript, the corresponding spot on the gene chip will be read as more green or more red (depending on which dye is used). Using this technique, transcript levels of tens of thousands of genes can be measured for any treatment and its corresponding control, producing a holistic view of the entire transcriptome. Microarray analysis, while expensive in terms of start-up cost and limited in terms of model organisms, produces a tremendous amount of data. In fact, the technique produces so much data that a computer must assist in the analysis, and care must be taken to avoid a large number of false positives. Nevertheless, the impact of this powerful technique on hypothesis generation cannot be ignored, and it is therefore anticipated that as more model organisms become available for analysis, microarrays will become more essential as a "first step" for investigating biological responses.

Thus far, few studies in the insect cryobiological literature have used microarrays. Unfortunately, the microarrays that are available commercially do not cover the common insect model systems used for low-temperature research. However, several species such as the parasitoid, *Nasonia virtipennis*, the springtail, *O. arcticus*, and the flesh fly, *S. crassipalpis*, have a considerable number of ESTs present in the

NCBI database, and are excellent candidates for the printing of microarrays for low-temperature experimentation (Table 4.1).

To date, there is only one low-temperature microarray study on an insect (Qin *et al.*, 2005). In this study of *D. melanogaster*, a microarray of 7000 sequences was hybridized with cDNA derived from flies that had been exposed to 0 °C for 2 h and allowed to recover for 30 min, and cDNA from untreated flies kept at 25 °C. The analysis revealed many genes that were up- or down-regulated in response to mild cold-shock. Genes with altered expression profiles included membrane proteins, stress proteins, mitochondrial genes and genes related to protein expression. The distribution of up- and down-regulated genes amongst the gene families was fairly evenly distributed, with the exception of expression-related genes, which were consistently down-regulated. A sub-set of genes with altered transcript levels was chosen for confirmation by qPCR. Of these, the expression profiles of all but one transcript confirmed the results seen in the microarray analysis. Interestingly, *frost* was up-regulated in this study, confirming results originally obtained through SSH (Goto 2001b). It is clear from this *Drosophila* cold-shock study that the results of microarray analyses are highly reproducible and that different transcriptomic methods produce similar results, despite disparate techniques.

Theoretically, cross-species microarrays can be performed with mRNA from a non-model organism hybridized to a chip from a model organism for which a chip is available (Bar-Or *et al.*, 2006). The major constraint to this analysis is establishing that the mRNA successfully hybridizes with the gene chip, but provided there is good sequence identity between the two species, this hurdle can often be overcome by optimization of washes, hybridization temperature, etc. As one might expect, cross-species hybridization becomes less satisfactory as the genetic distance between species increases.

We have performed cross-species microarray analysis using diapausing and rapidly cold-hardened *S. crassipalpis* cross-hybridized with a *D. melanogaster* microarray, but with limited success. A number of putative transcripts were identified as responsive to both treatments, but we were unable to confirm many of the results through Northern blot analysis. The main challenge was the lack of information on sequences in *S. crassipalpis*, which led to cloning difficulties (either the wrong isoform of a gene was cloned or the gene was not cloned at all). The few genes of interest that we did clone either were not detectable through Northern blot hybridization or were not altered by the treatment (unpublished results).

A segment of a cross-species microarray study using the same *D. melanogaster* chip that we used for our experiments in *S. crassipalpis* was published for cold-shock in the gall fly, *Epiblema scudderiana*, and a number of putative gene transcript levels were significantly altered (nearly 10% of the total) (Storey and McMullen, 2004). Of these, a number of membrane transport proteins were identified. However, to date, there is no qPCR or Northern blot confirmation of the results of this experiment. From the limited information available on cross-species microarrays in the insect cryobiological literature, it is clear that cross-species microarrays can generate copious hypotheses, but the results are more tenuous compared to other -omics studies because of non-specific hybridization and experimenters still face the hurdle in cloning genes of interest from non-model species that are not genomic models.

The power of the -omics approach for identifying new players in the field of insect cryobiology can be appreciated from the list of genes and proteins that increase in response to low temperatures (Table 4.2). This list is based on only four species that have been examined thus far. Obviously, this list will expand as more species and types of low-temperature exposure are examined.

4.4 Proteomics

Proteomics seeks to identify the most abundant proteins in a tissue or the whole body by first dispersing these proteins over a two-dimensional matrix, usually polyacrylamide. This dispersion is first achieved across the x-axis by isoelectric point, followed by SDS-PAGE separation on the y-axis. Location and intensity of the "spots" are then compared between gels, using appropriate software, and protein signals are averaged within treatments and compared across treatments to determine which proteins are significantly altered in abundance. The spots of interest are then removed from the matrix, digested with trypsin, and analysed by mass spectrometry to obtain an oligopeptide sequence that can be used to identify the protein.

This form of -omics is particularly effective for examining cold-hardiness in insects because not only does it reveal quantitative information on individual proteins, but it also reveals qualitative information such as phosphorylation state, a modification that is likely quite important for low-temperature responses of insects (Chen and Denlinger, 1990; Muise and Storey, 1997; Fujiwara and Denlinger, 2007). Proteomics can be performed on any organism, even non-model ones, provided the proteins of interest have a high degree of identity with sequences in Genbank.

To date, the literature contains a paucity of proteomic studies focusing on insect responses to cold, but the few studies that have been conducted revealed some expected, as well as unexpected, protein changes (Table 4.2). The brain of the flesh fly, *S. crassipalpis*, showed increased abundance of eight heat-shock proteins (Li *et al.*, 2007), two of which were previously reported from targeted Northern blots (Rinehart and Denlinger, 2000; Hayward *et al.*, 2005) and one-dimensional SDS-PAGE of proteins (Joplin *et al.*, 1990), but the power of proteomics becomes evident

Table 4.2 *Genes/proteins up-regulated by low-temperature exposures as determined by transcriptomics and proteomics (from Colinet* et al., *2007, Li* et al., *2007, Li and Denlinger 2008, Qin* et al., *2005 and Robich* et al., *2007). C-S refers to cold-shock and RCH refers to rapid cold-hardening*

Category	Gene/protein	Treatment	Species
Chaperones	BiP/GRP78	Diapause	*Sarcophaga crassipalpis*
	Hsc 70	Intermittent 4 °C	*Aphidius colemani*
	Hsp 23	Diapause, RCH, C-S	*S. crassipalpis, A. colemani*
	Hsp 70	Diapause	*S. crassipalpis*
	Hsp 90	C-S	*Drosophila melanogaster*
	Hsp70/90 organizing protein	Intermittent 4 °C	*A. colemani*
	Unknown small hsp	Diapause	*Culex pipiens*
Stress-related	Aldehyde oxidase	Diapause	*C. pipiens*
	Frost	Intermittent 4 °C	*A. colemani*
	Selenoprotein	Diapause	*C. pipiens*
	Thiol peroxiredoxin	Diapause	*S. crassipalpis*
	Thioredoxin-dependent peroxide reductase	Diapause	*S. crassipalpis*
	Ubiquitin	Intermittent 4 °C	*A. colemani*
Cytoskeletal	Actin	Diapause	*C. pipiens*
	Profilin	Intermittent 4 °C	*A. colemani*
	Pupal cuticular protein	Intermittent 4 °C	*A. colemani*
	Tropomyosin-1	RCH	*S. crassipalpis*
Metabolic	Aconitase	Intermittent 4 °C	*A. colemani*
	Aldolase	Intermittent 4 °C	*A. colemani*
	Arginine kinase	Intermittent 4 °C	*A. colemani*
	Bellwether (ATP synthase)	Intermittent 4 °C, diapause	*A. colemani, C. pipiens*
	Cytochrome C oxidase	Diapause	*C. pipiens*
	Fumarate hydratase	Intermittent 4 °C	*A. colemani*
	Glyceraldehyde -3-phosphate dehydrogenase	Intermittent 4 °C	*A. colemani*
	Malate dehydrogenase	Intermittent 4 °C, diapause	*A. colemani, C. pipiens*
	Methylmalonate-semialdehyde dehydrogenase	Diapause	*C. pipiens*
	Phosphoglycerate kinase	Intermittent 4 °C	*A. colemani*

Table 4.2 (*cont.*)

Category	Gene/protein	Treatment	Species
Lipids	Enoyl-CoA hydratase	C-S	*D. melanogaster*
	Fatty acid synthase	Diapause	*C. pipiens*
	Membrane-bound fatty acid elongase	C-S	*D. melanogaster*
	Mitochondrial acyl carrier protein	C-S	*D. melanogaster*
Ribosome	23S ribosomal sub-unit from *Wolbachia*	Diapause	*C. pipiens*
	28S large ribosomal sub-unit RNA	Diapause	*C. pipiens*
	Ribosomal protein 27A	Diapause	*C. pipiens*
	Ribosomal protein L18	Diapause	*C. pipiens*
	Ribosomal protein S24	Diapause	*C. pipiens*
	Ribosomal protein S3A	Diapause	*C. pipiens*
	Ribosomal protein S6	Diapause	*C. pipiens*
Cuticle	Dusky	C-S	*D. melanogaster*
	Gasp precursor	Intermittent 4 °C	*A. colemani*
	Pupal cuticular protein	Intermittent 4 °C	*A. colemani*
Membrane	ABC transporter	C-S	*D. melanogaster*
	G-protein	C-S	*D. melanogaster*
	GTP binding protein	C-S	*D. melanogaster*
	SRY interacting protein Y	C-S	*D. melanogaster*
	Unknown membrane protein (6 total)	C-S	*D. melanogaster*
Expression	Methoprene tolerant protein	Diapause	*C. pipiens*
	Multiprotein bridging factor	C-S	*D. melanogaster*
	Receptor for protein kinase C	Intermittent 4°C	*A. colemani*
	RNA binding/splicing factor	C-S	*D. melanogaster*
	Serine/theronine protein kinase	C-S	*D. melanogaster*
	Transcription elongation factor binding protein	Diapause	*C. pipiens*
Other	Glycoside hydrolase	C-S	*D. melanogaster*
	Immunoglobulin heavy chain binding protein	Diapause	*S. crassipalpis*
	Mimo family transposable element	Diapause	*C. pipiens*
	Proteosome sub-unit beta type 6 precursor	C-S	*D. melanogaster*
	Reverse transcriptase	Diapause	*C. pipiens*

by the revelation in Li *et al.* (2007) demonstrating increased abundance of six additional heat-shock proteins not previously known to be influenced by diapause and low temperature. Additionally, an increase in an immune response protein and a decrease in a suite of proteins related to metabolic function and DNA repair were noted. It is not clear if all of the new proteins found are related to cold-hardiness in the flesh fly, but what is clear is that this study suggests a number of new genes that are promising targets for analysis. Similar results were obtained from a rapid cold-hardening study in *S. crassipalpis* that revealed an increase in one heat-shock protein (Hsp23) and an increase in proteins related to energy production and organization of the cytoskeleton (Li and Denlinger, 2008). Unlike the diapause study, all results from the rapid cold-hardening study in flesh flies were novel; no changes in proteins had been previously detected during rapid cold-hardening in this species. Another proteomic study was conducted on the whole body of a parasitic wasp (*Aphidius colemani*) that was exposed to various chilling regimes that were constant or interrupted with periods of warmth (Colinet *et al.*, 2007). Since intermittent periods of warmth greatly enhance survival to chilling injury in insects (Chen and Denlinger, 1992; Colinet *et al.*, 2007), this study was conducted to determine which proteins were synthesized during periods of warmth, with the presumption that these proteins may be responsible for increasing survival during the chilling regime. In addition to an increase in heat-shock proteins that one may have expected, a number of metabolic enzymes involved in glycolysis and the respiratory chain were altered, indicating that chilling causes major disruptions in the cell's central metabolism, necessitating extensive transcriptional/translation control.

These proteomic studies produced results that supported old hypotheses as well as generated new ones, but many more such studies are needed to provide a more comprehensive overview. The proteomic experiments on diapause and rapid cold-hardening of flesh-fly brains allow limited comparisons to be made, e.g. the common detection of a small heat-shock protein (Hsp23), but neither experiment addressed common themes in cold survival, such as long-term acclimation or cold-shock injury. The proteomic study of the parasitic wasp of aphids (Colinet *et al.*, 2007) cannot be compared to most other cold studies because the treatment regime differed from the great majority of insect cryobiological experiments. For proteomics to deliver the greatest contribution to the insect cryobiological community, future studies covering a range of treatments, models and life stages are needed.

4.5 Metabolomics

Metabolomics refers to the study of small molecules found in the tissues and fluids of an organism, with special emphasis on the levels of non-proteinaceous substrates, such as sugars, amino acids, metabolic intermediates and polyols. Because the focus is on substrate, the results of these experiments provide the most direct evidence for biochemical changes elicited by a given treatment. When a study focuses on the protein or transcript level, there is no assurance that changes detected contribute to low-temperature survival because the detected transcript may not be translated and detected protein may not be activated.

Three equally effective, but dissimilar, analytical techniques are used for metabolomics. One uses gas chromatography-mass spectrometry (GC/MS) as the detection method, and the others use either nuclear magnetic resonance (NMR) or liquid chromatography-mass spectrometry (LC/MS). GC/MS offers the advantage of generating more peaks for analysis, but the peaks can only be generated by derivatizing the compounds to improve chromatographic behavior. NMR requires less chemical modification before quantitation, but fewer peaks are generated compared to GC/MS. LC/MS has the advantage of generating a large number of peaks and requiring little chemical modification, but many LC/MS systems do not offer as many searchable spectral libraries as are available for GC/MS, thus requiring multiple comparative standards. Regardless of the technique used, the data is subjected to analysis using univariate (Analysis of Covariance) as well as multivariate statistical analysis (Principal Components Analysis).

Metabolomics data are used to statistically determine the "sameness" of the entire biochemical makeup of the organism in response to different treatments. The metabolomic technique is powerful in its non-discriminatory and holistic approach, but inferences regarding biochemical pathways that result in detected changes must be approached with caution because only a few of the many constituents of any given biochemical pathway are detected, and the cellular compartmentalization of these substrates is not known. Nevertheless, for insect cryobiological studies, metabolomics can eliminate guesswork in identifying compounds that contribute to winter survival, without making the many assumptions necessary when interpreting data from other kinds of -omics studies.

To date, there are three published metabolomic studies related to insect cold-stress (Overgaard *et al.*, 2007; Michaud and Denlinger; 2007, Michaud *et al.*, 2008), all of which soon followed the first insect metabolomics study that examined heat-shock (Overgaard *et al.*, 2005). The first of the cold studies was performed on *D. melanogaster* and focused on adults subjected to cold-shock and rapid cold-hardening (Overgaard *et al.*, 2007). Using NMR-based techniques, 31 compounds were positively identified, and all but two were significantly altered by the experimental treatments. The most significant increases were found among the sugars; three different sugars (glucose, maltose and trehalose) increased in response to both cold-shock and rapid cold-hardening. These results showed that sugars, not polyols, are the metabolites most likely responsible for rapid cold-hardening in this species, a result that was not revealed by a previous, targeted analysis (Kelty and Lee, 2001). In addition to the changes seen for sugars, amino acids were also dramatically altered by both low-temperature treatments. Of the six amino acids that were altered, half increased and half decreased.

Another metabolomic study on rapid-cold hardening, performed on the flesh fly, *S. crassipalpis*, examined changes of metabolites in response to both rapid cold-hardening and diapause (Michaud and Denlinger, 2007). The searchable spectrometric libraries available for GC/MS made it possible to identify nearly twice as many compounds as were identified in the NMR study, with 21 of these compounds displaying some response to cold-hardening. Most notable changes occurred for the polyols, with increases in both glycerol and sorbitol during rapid cold-hardening. These results confirmed previous findings for the elevation of glycerol during rapid cold-hardening (Lee *et al.*, 1987), but the elevation of sorbitol was the first report for this species. Glucose was moderately increased during rapid cold-hardening, but greatly increased during diapause, a result consistent with the activation of phosphorylase A during rapid cold-hardening and diapause (Chen and Denlinger, 1990). As in the *Drosophila* study, amino acids were quite responsive to low temperature, with some increasing and others decreasing. Alanine, an amino acid that also serves as an alternative to lactic acid as an end product of glycolysis during anaerobic conditions, increased in both *S. crassipalpis* and *D. melanogaster*, a finding echoed throughout the literature (e.g. Rivers and Denlinger, 1994; Goto *et al.*, 2001). Finally, changes in intermediates of glycolysis and the TCA cycle demonstrate the possibility that glycolysis continues during both rapid cold-hardening and diapause, while only diapause arrests the TCA cycle. This metabolomics-derived observation helps explain why oxygen consumption is so low during diapause (Slama and Denlinger, 1992).

A third metabolomics study relevant to insects at low temperatures examined freeze-tolerance in the Antarctic midge, *Belgica antartica* (Michaud *et al.*, 2008). As with GC/MS experiments on the flesh fly, roughly twice as many compounds were identified in the samples than in the NMR-based study, but unlike the previous experiment, only 10 metabolites were responsive to low-temperature treatment. The most interesting family of metabolites that responded to freezing was the polyols, represented by glycerol, mannitol and erythritol. It is particularly interesting to note that all three metabolomic low-temperature experiments revealed elevations of multiple cryoprotectants, a condition that may be more common than assumed from previous studies that targeted only a few metabolites (Storey and Storey, 1981; Košťál *et al.*, 2007). Freezing also caused concentrations of four amino acids to change; two were elevated and two were reduced. As in the previous two metabolomics experiments, alanine levels increased. Two components of the TCA cycle were elevated as well, possibly a result of accumulation from low-temperature inhibition of downstream enzymes in this important biochemical pathway.

These studies demonstrate the power of metabolomics as a hypothesis-generating tool as well as an experimental tool. Not only was previous information from the literature confirmed through metabolomics, but several new, low-temperature-induced compounds were identified for future study. All three experiments showed that multiple cryoprotectants are recruited, the amino-acid pool is perturbated, and biochemical pathways of cellular respiration are altered, at least at the substrate level. Not all of these changes are likely to enhance low-temperature survival; in fact, some, such as energy reduction from TCA perturbation, may reflect hurdles to survival that must be overcome.

4.6 Statistical considerations in -omics experiments

All -omics experiments generate vast amounts of data, often from a single sample. This data is divided into "spots" or peaks, each representing a different protein or metabolite, and each must be compared to a corresponding "spot" or peak in another sample. The inevitable result of this type of data set is that comparisons will be the culmination of tens to thousands of ANOVAs. When the number of comparisons increases, the probability for false positives, or Type I errors, concomitantly increases. This uncertainty is further increased by unforeseen interactions in complex mixtures of transcripts, proteins and metabolites. Chemical derivatization of metabolites, cross-hybridization of transcripts and damaged proteins may all contribute to variability in -omics experiments. To correct for these factors, it is imperative that results be verified with a targeted analysis such as Northern or Western blots or that statistical rigor be applied to minimize Type I error. Verification by targeted analysis is the preferred method for transcriptomics; if gene sequences are known, Northern blots and qPCR are fairly easy to perform. Microarray experiments typically do not report results in terms of statistical certainty, but only whether mean expression of a gene is more than a threshold value, typically 50% greater or lesser than what is observed in untreated samples. Statistical reporting is usually reserved for the verification portion of the experiment. For metabolomics and proteomics, statistical rigor is the preferred method to reduce Type I error because independent verification of results is demanding and cannot be done in a short timeframe.

The Bonferroni correction is the best-known method for correcting Type I error in multiple comparisons experiments. The idea is to adjust the *p*-value in such a manner as to effectively render all comparisons performed simultaneously rather than sequentially. Statistically, this is known as a closed testing procedure. To use the Bonferroni correction, the threshold *p*-value is simply reduced to 0.05 divided by the number of hypotheses tested. For a small number of comparisons (2–20), this method works very well, but for the amount of data generated in a typical -omics experiment, the Bonferroni correction may be too stringent, even for the 30 or so metabolites typically identified in NMR metabolomics. The Holm–Bonferroni, or step-wise Bonferroni, is a less-stringent method that is applicable to slightly larger data sets (21–100 comparisons), making it an excellent method for -omics experiments of this scope. This method was applied to the NMR-based metabolomics study reported for *D. melanogaster* (Overgaard *et al.*, 2007). For the largest data sets, a procedure called the False Discovery Rate (Benjamini and Hochberg, 1995) should be applied because it is the least stringent, but still an acceptable, closed testing procedure. The GC/MS-based metabolomics study on *B. antarctica* used this correction (Michaud *et al.*, 2008). One of these statistical procedures must be applied to all -omics experiments that do not include some alternative form of verification.

4.7 Common low-temperature responses revealed by -omics experiments

4.7.1 *Amino acids*

Among the classes of biomolecules detected in metabolomics experiments, none show as much response to low temperature in terms of number of compounds affected, as the amino-acid pool. Freezing (Michaud *et al.*, 2008), cold-shock, rapid cold-hardening (Overgaard *et al.*, 2007; Michaud and Denlinger, 2007) and a cold-hardy diapause induced in the absence of low temperature (Michaud and Denlinger, 2007) all produced a unique profile of amino-acid concentrations, with some interesting overlaps between treatments (Table 4.3). Interestingly, none of the changes in the amino-acid pool were supported by transcriptomic or proteomic studies, suggesting that these changes are the consequence of enzyme kinetics rather than gene induction.

The most notable change is the increase in alanine observed for all three species and all low-temperature treatments tested, raising the possibility that an elevation of this amino acid may be a good marker for low-temperature stress. Alanine elevation in response to low temperatures is well supported in the literature for various insect species and conditions (Storey and Storey, 1981; Goto *et al.*, 1998; Li *et al.*, 2002; Goto *et al.*, 2001), but it is unknown whether alanine elevation is symptomatic of low temperatures or plays a direct role in low-temperature survival. If the former, then elevation of alanine may be the result of low-temperature inhibition or activation of an enzyme immediately upstream or downstream of alanine. If the latter is the case, alanine may contribute to survival by serving as an alternative end-product to lactic acid when the downstream respiratory pathway is slowed or ceases to function (Michaud and Denlinger, 2007). None of the metabolomic studies on insects in the cold have reported an elevation in lactic acid. Alternatively, the insect may gain colligative benefits from elevating alanine, but this is unlikely because amino-acid peaks are not the dominant peaks in

Table 4.3 *Amino-acid changes observed in low-temperature treatments of three dipteran species as determined by metabolomic analysis (from Overgaard* et al., *2007, Michaud and Denlinger 2007, Michaud* et al., *2008)*

Amino acid	Response	Organism
Alanine	↑ cold shock, rapid cold-hardening, diapause, freezing	*Drosophila melanogaster, Sarcophaga crassipalpis, Belgica antarctica*
Arginine	↓ rapid cold-hardening, cold shock	*D. melanogaster*
Asparagine	↓ rapid cold-hardening, cold shock	*D. melanogaster*
Aspartate	↑ freezing	*B. antarctica*
	↓ diapause, cold shock	*S. crassipalpis, D. melanogaster*
Glycine	↑ cold shock	*D. melanogaster*
	↓ freezing, diapause	*B. antarctica, S. crassipalpis*
Glutamine	↑ rapid cold-hardening	*S. crassipalpis*
Leucine	↓ rapid cold-hardening	*D. melanogaster*
	↑ diapause	*S. crassipalpis*
Lysine	↑ cold shock	*D. melanogaster*
Ornithine	↓ rapid cold-hardening	*S. crassipalpis*
Phenylalanine	↓ diapause	*S. crassipalpis*
Proline	↓ diapause	*S. crassipalpis*
Serine	↓ freezing	*B. antarctica*
Tyrosine	↓ diapause	*S. crassipalpis*

metabolomics experiments when compared to sugars and polyols, at least among the results published thus far.

Other amino acids affected by more than one low-temperature condition and more than one species include aspartate, glycine and leucine, but none of these amino acids showed consistent directional change, i.e. they were elevated in one study, but reduced in another. For aspartate, levels were elevated due to cold acclimation in the flat grain beetle (Fields *et al.*, 1998) but reduced in freezing frogs and snails (Storey and Storey, 1986; Churchill and Storey, 1989). Glycine was elevated in response to cold acclimation in *Arabidopsis* (Cook *et al.*, 2004), but reduced in freezing snails (Churchill and Storey, 1989). Low-temperature reduction of leucine was observed for both cold-acclimated flat grain beetles (Fields *et al.*, 1998), cold-shocked *Arabidopsis* (Kaplan *et al.*, 2007) and rapidly cold-hardened *D. melanogaster*, but was elevated in diapausing *S. crassipalpis*. However, we should note that the diapause of *S. crassipalpis* is primarily dependent on photoperiod, and therefore the elevated leucine levels were observed in the absence of a low-temperature stimulus.

Table 4.4 *Cryoprotective compounds elevated by low temperatures in three dipteran species as determined by metabolomics (from Overgaard* et al., *2007; Michaud and Denlinger 2007; Michaud* et al., *2008)*

Species	Treatment	Cryoprotective compounds
Drosophila melanogaster	Rapid cold-hardening, cold-shock	Glucose, maltose, trehalose
Sarcophaga crassipalpis	Rapid cold-hardening	Glycerol, sorbitol
Belgica antarctica	Freezing	Glycerol, mannitol, erythritol

4.7.2 Cryoprotectants

As previously mentioned, all insects that have been subjected to low temperatures and analysed using metabolomic techniques displayed elevation of multiple compounds that have cryoprotective characteristics (Table 4.4). However, no genes directly related to these cryoprotectants were elevated in response to low temperatures in either transcriptomic or proteomic studies. When all of the results of -omics studies are considered, it is apparent that elevation of polyols does not have an appreciable gene-induction component and most likely occurs at the level of protein activation or inhibition. Certainly, the only evidence to date on cryoprotective compound recruitment at low temperatures in insects is consistent with this view (Storey and Storey, 1981; Li *et al.*, 2002).

4.7.3 Heat-shock proteins

Targeted analysis demonstrated the up-regulation of several heat-shock proteins (Hsps) during recovery from cold-shock as well as during diapause, as reviewed earlier (Denlinger *et al.*, 1991; Denlinger *et al.*, 2001). But what -omics approaches have contributed is the discovery of a wealth of additional Hsps that are responsive to both low temperature and diapause, as well as the discovery that at least one Hsp is produced during rapid cold hardening.

While Hsp70 and Hsp23 were known previously to be up-regulated during the cold-hardy pupal diapause of *Sarcophaga crassipalpis*, proteomics (Li *et al.*, 2007) and transcriptomics (Rinehart *et al.*, 2007) added six additional members of the Hsp family to this list. This response is not restricted to *S. crassipalpis*. Hsps are up-regulated during overwintering diapause in species and developmental stages as diverse as the embryonic diapause of the gypsy moth *Lymantria dispar*, larval diapause of the European corn borer *Ostrinia nubilalis*, pupal diapause of the tobacco hornworm *Manduca sexta* (Rinehart *et al.*, 2007) and adult diapause of the Colorado potato beetle *Leptinotarsa decemlineata* (Yocum, 2000). Several additional studies reporting Hsp-related responses to low temperature and/or overwintering are included in Table 4.5.

Table 4.5 *Heat-shock proteins (Hsps) and related molecules elevated due to diapause and low-temperature treatments in insect -omics experiments*

Gene	Treatment	Technique	Species	References
Unknown small Hsp	Diapause	Transcriptomics	*Sarcophaga crassipalpis*	Rinehart *et al.*, 2007
Hsp18	Diapause	Transcriptomics	*S. crassipalpis*	Rinehart *et al.*, 2007
Hsp23	Diapause, rapid cold-hardening	Proteomics	*S. crassipalpis*	Li *et al.*, 2007; Li and Denlinger 2008
	Diapause	Transcriptomics	*S. crassipalpis*	Rinehart *et al.*, 2007
	Mild cold-shock	Transcriptomics	*Drosophila melanogaster*	Sinclair *et al.*, 2007
Hsp23 pseudogene	Diapause	Transcriptomics	*S. crassipalpis*	Rinehart *et al.*, 2007
Hsp25	Diapause	Transcriptomics	*S. crassipalpis*	Rinehart *et al.*, 2007
Hsp60	Diapause	Transcriptomics	*S. crassipalpis*	Rinehart *et al.*, 2007
Hsp70	Diapause	Proteomics,	*S. crassipalpis*	Li *et al.*, 2007
		transcriptomics	*S. crassipalpis*	Rinehart *et al.*, 2007
Hsp70/90 organizing protein	Intermittent 4 °C	Proteomics	*Aphidius colemani*	Colinet *et al.*, 2007
Hsc70	Intermittent 4 °C	Proteomics	*A. colemani*	Colinet *et al.*, 2007
	Diapause	Transcriptomics	*S. crassipalpis*	Rinehart *et al.*, 2007
Hsp90	Mild cold-shock	Transcriptomics	*D. melanogaster*	Qin *et al.*, 2005
	Diapause	Transcriptomics	*S. crassipalpis*	Rinehart *et al.*, 2007

These Hsps contribute to cold-tolerance during diapause. When genes encoding either Hsp70 or Hsp23 were suppressed by RNA interference (RNAi), pupae of *S. crassipalpis* lost their cold tolerance (Rinehart *et al.*, 2007). Additionally, suppression of Hsp70 transcripts and protein by RNAi drastically reduced survival from chilling injury in *Pyrrhocoris apterus* (Koštál and Tollarová-Borovanská, 2008). Presumably the presence of Hsps protects the integrity of essential proteins during the overwintering diapause, thus preventing damage at low temperatures.

A proteomics study that focused on fluctuating temperatures to enhance the shelf-life of a commercially important parasitoid, *A. colemani,* reported an increase of an Hsp cognate protein Hsc70–5, as well as an Hsp70/90 organizing protein, during periods of warming that were interspersed between chilling periods (Colinet *et al.*, 2007). Hsc70 was also up-regulated during recovery from cold-shock in *S. crassipalpis* (Rinehart *et al.*, 2000), as are several other Hsps (Joplin *et al.*, 1990).

Table 4.6 *Insect cytoskeletal and structural genes up-regulated by low temperature exposures as determined by -omics (from Colinet* et al., *2007; Robich* et al., *2007; Li and Denlinger, 2008)*

Gene	Treatment	Technique	Species
Profilin	Fluctuating exposures to 4°C	Proteomics	*Aphidius colemani*
Pupal cuticular protein	Fluctuating exposures to 4°C	Proteomics	*A. colemani*
Actin	Diapause	Transcriptomics	*Culex pipiens*
Tropomyosin-1	Rapid cold-hardening	Proteomics	*Sarcophaga crassipalpis*

Though it has been appreciated for some time that Hsps can be up-regulated during recovery from low temperature (Burton *et al.*, 1988), the common assumption has been that this response is restricted to the recovery phase and is not likely to occur during low-temperature exposure. This assumption was challenged by the increased abundance of one of the small Hsp, Hsp23, during rapid cold-hardening (see Chapter 2). A proteomics study revealed that Hsp23 increases in the brain during a 2 h exposure to 0 °C, conditions that generate cryoprotection during a subsequent exposure to −10 °C (Li and Denlinger, 2008).

4.7.4 Cytoskeleton

When microtubules are exposed to low temperatures, the components of the αβ tubulin system tend to disaggregate, resulting in depolymerization. This property of tubulin is used routinely in the laboratory for purification purposes, but in a living organism, cold exposure may disrupt the cytoskeleton, rendering intracellular transport and organization more difficult. Certainly, if insects are to survive the inimical conditions of winter, the cytoskeletal system needs to be fortified against depolymerization or the damage from cytoskeletal disruption must be minimized.

To date, most information pertaining to cytoskeletal arrangement in insects originated from an -omics experiment. SSH transcriptomics was used in *Culex pipiens* to discover an actin gene that is up-regulated during adult diapause (Robich *et al.*, 2007). Further analysis revealed that not one, but two actins are up-regulated in this species during diapause, and that cold exposure causes the entire actin matrix to reorganize, especially in the midgut (Kim *et al.*, 2006). The -omics experiments suggest the possibility that additional players are involved in structural rearrangement in response to low temperature, but most of those results await experimental verification (Table 4.6).

4.7.5 Glycolysis and cellular respiration

No biochemical process is more central to life than cellular respiration. The enzymes involved must be carefully coordinated in a synchronous manner to ensure efficient energy production to power the cell. For this reason, enzymes involved in central cellular respiration tend to be highly conserved, and alterations in the kinetics of these enzymes have the potential to seriously disrupt homeostasis. Low temperatures disrupt enzymes of cellular respiration by direct temperature inhibition or by the introduction of torsional strain that disrupts the active site of the enzyme. It is for this reason that dramatic alterations in glycolysis and the TCA cycle are expected when insects are exposed to low temperatures, and physiological responses that mitigate these changes or initiate changes to produce cryoprotectants favor survival.

Evidence that disruption occurs in cellular respiration as a result of low temperature comes from studies using both cold-susceptible and cold-hardy species of insects. In the chill-susceptible silkmoth, *Philosamia ricini*, exposure to a mild low temperature (2 °C) resulted in major disruption of the overall chemistry of the insect, causing a number of enzymes to be shut down and glycolytic products such as pyruvate to be greatly elevated (Pant and Gupta, 1979). Since these exposures caused mortality, it is presumed that these alterations in the moth's biochemical make-up are detrimental to survival. Low temperature in the freeze-tolerant arctic caterpillar, *Gynaephora groenlandica*, induces a breakdown of mitochondria, which would presumably alter the level of TCA metabolites in the overall organism, but mitochondrial breakdown coincides with freezing survival and successful overwintering in this species (Levin *et al.*, 2003).

Disruption of central metabolism by cold may produce substances that promote cold survival. Targeted metabolite analysis of central metabolism in *Eurosta solidiginis* revealed that this freeze-tolerant fly larva manufactures two polyols during the winter, by using the hexose monophosphate shunt. The type of polyol produced depends on whether the cold exposure takes place in an aerobic or anaerobic environment; sorbitol is generated in an aerobic environment, while glycerol is generated in an anaerobic environment (Storey and Storey, 1990). In addition, a study of the gall moth, *Epiblema scudderiana*, shows that UDP-glucose is recruited from glycogen reserves by direct low-temperature activation of glycogen phosphorylase (Holden and Storey, 1993). This glucose may be used as the source material for production of glycerol from the glycolytic pathway.

The -omics studies that have been published for low-temperature biology in insects all show some form of metabolic alteration, mostly in lipid or carbohydrate-based metabolic pathways. Changes in substrates, genes, or proteins of central metabolism are observed across all three -omics fields (Fig. 4.1). Metabolomic

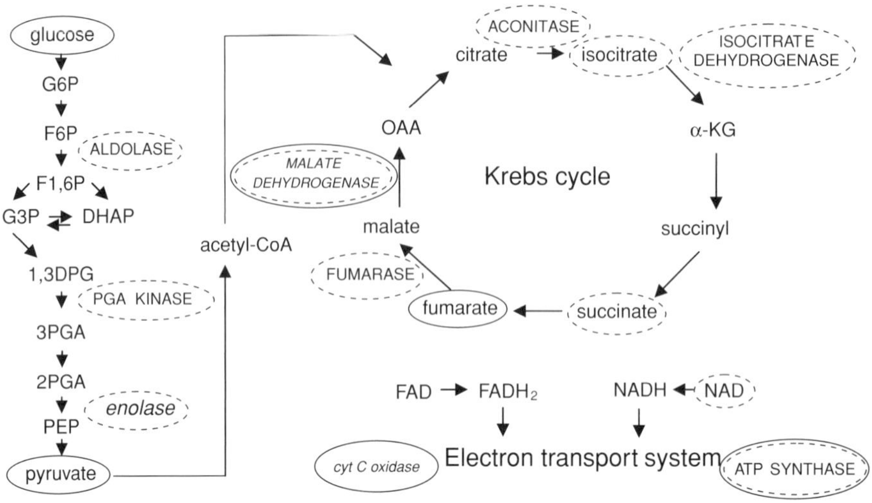

Figure 4.1 Alteration of insect substrates and enzymes due to low-temperature treatments as detected by -omics experiments. This schematic diagram of the tripartite system of cellular respiration highlights the substrates, genes and enzymes detected as significantly changed compared to controls due to low-temperature exposure or during cold-hardy states. Solid circles indicate up-regulation and dotted circles represent down-regulation (or both if both circles are indicated). Data derived from proteomics are capitalized while transcriptomics results are in italics (data derived from both are formatted both ways). (Source data: Robich *et al.* 2007; Overgaard *et al.*, 2007; Michaud and Denlinger, 2007; Michaud *et al.*, 2008; Li *et al.*, 2007; Li and Denlinger, 2008; Colinet *et al.*, 2007).

studies show increases in both source material and end-products of glycolysis, with some metabolites of the TCA cycle altered as well (Overgaard *et al.*, 2007; Michaud and Denlinger, 2007; Michaud *et al.*, 2008). A proteomic experiment in the wasp, *A. colemani*, showed that recovery from chilling involves an increase in many metabolic enzymes. Failure to increase these enzymes by intermittent warm periods is correlated with mortality for this species, suggesting that cold-exposure disrupts the functions of some or all of these enzymes (Colinet *et al.*, 2007). Perhaps this is also the case for isocitrate dehydrogenase, which is less abundant following rapid cold-hardening in the flesh fly (Li and Denlinger, 2008). Transcriptomic studies have shown similar changes in central metabolism, with an increase of malate dehydrogenase transcript observed for the cold-hardy diapause of *C. pipiens* (Robich *et al.*, 2007), an observation mirrored by the proteomic observation of an increase of this enzyme during intermittent chilling for *A. colemani* (Colinet *et al.*, 2007). The *C. pipiens* study also revealed an increase in cytochrome C oxidase, the only change in the electron transport chain reported from a cold-related -omics study.

It is clear that disruption of central metabolism is a common physiological response in insects exposed to low temperatures. So many substrates, products, enzymes, transcripts and sub-cellular compartments are involved in central metabolism that initial study of these pathways lends itself to a holistic approach such as an -omic study. Caution must be taken, however, when drawing conclusions, because -omics techniques are broad in scope and do not have the resolution to draw specific conclusions about pathway directionality. But, the time saved in identifying excellent targets for detailed analysis cannot be overemphasized.

4.8 Future directions for -omics studies in insect cryobiology

Given the tremendous amount of data garnered from relatively few -omics studies, it is clear that the three main branches of experimental -omics (transcriptomics, proteomics and metabolomics) offer unique tools for insect cryobiology by simultaneously highlighting low-temperature processes for future study, as well as by identifying the individual molecular players involved. For the potential of -omics to be fully realized, -omics experiments need to be performed extensively on a variety of insects at the tissue and whole-body level using a wide array of low-temperature treatments. Insects chosen for such experimentation should cover the gamut of cold-survival phenotypes (freeze-tolerant versus freeze-intolerant, chill susceptible versus chill-resistant, etc.). Casting a wide net generates copious and varied data with its inevitable amount of "noise", but such "noise" can be filtered out by using follow-up methods of verification and by identifying similarities between low-temperature physiological events using comparisons across species, habitats, phenotypes and tissues.

The hypothesis-generating character of -omics studies suggests that such approaches should be used as early as possible on new model systems to quickly and accurately identify the major players and to compare responses to what is known in the literature. Conversely, insect model systems that are already well known in the cryobiological literature can also benefit from -omics studies, especially species for which there is an EST project. Thus, processes that may have been overlooked by targeted analysis can be discovered and used to augment our understanding of established cold-survival models.

References

Bar-Or, C., Czosnek, H., and Koltai, H. (2006). Cross-species microarray hybridizations: a developing tool for studying species diversity. *Trends in Genetics* **23**, 200–207.

Benjamini, Y. and Hochberg, Y. (1995). Controlling the false discovery rate: a practical and powerful approach to multiple testing. *Journal of the Royal Statistical Society, Series B (Methodological)* **57**, 289–300.

Burton, V., Mitchell, H. K., Young, P., and Peterson, N. S. (1988). Heat shock protection against cold stress of *Drosophila melanogaster*. *Molecular and Cellular Biology* **8**, 3550–3552.

Chen, C.-P. and Denlinger, D. L. (1990). Activation of phosphorylase: response to cold and heat stress in the flesh fly, *Sarcophaga crassipalpis*. *Journal of Insect Physiology* **36**, 549–554.

Chen, C.-P. and Denlinger, D. L. (1992). Reduction of cold injury in flies using an intermittent pulse of high temperature. *Cryobiology* **29**, 138–143.

Churchill, T. A. and Storey, K. B. (1989). Intermediary energy metabolism during dormancy and anoxia in the land snail, *Otala lactea*. *Physiological Zoology* **62**, 1015–1030.

Colinet, H., Nguyen, T. T. A., Cloutier C., Michaud, D., and Hance, T. (2007). Proteomic profiling of a parasitic wasp exposed to constant and fluctuating cold exposure. *Insect Biochemistry and Molecular Biology* **37**, 1177–1188.

Cook, D., Fowler, S., Fiehn, O., and Thomashow, M. F. (2004). A prominent role for the CBF cold response pathway in configuring the low temperature metabolome of *Arabidopsis*. *Proceedings of the National Academy of Sciences, USA* **101**, 15243–15248.

Denlinger, D. L., Joplin K. H., Chen, C.-P., and Lee, R. E. (1991). Cold shock and heat shock. In *Insects at Low Temperature*, ed. Lee, R. E. Jr., and D. L. Denlinger: New York: Chapman and Hall, pp. 131–148.

Denlinger, D. L., Rinehart, J. P., and Yocum, G. D. (2001). Stress proteins: a role in diapause? In *Insect Timing: Circadian Rhythmicity to Seasonality*, ed. Denlinger, D. L., Giebultowicz J., and Saunders, D. S. Amsterdam: Elsevier Science, pp. 155–171.

Fields, P. G., Fleurt-Lessard, F., Lavenseau, L., Febvay, G., Peypelut, L., and Bonnot, G. (1998). The effect of cold acclimation and deacclimation on cold tolerance, trehalose, and free amino acid levels in *Sitophilus granaries* and *Cryptolestes ferrugineus* (Coleoptera). *Journal of Insect Physiology* **44**, 955–965.

Fujiwara, Y. and Denlinger, D. L. (2007). p38 MAPK is a likely component of the signal transduction pathway triggering rapid cold hardening in the flesh fly *Sarcophaga crassipalpis*. *Journal of Experimental Biology* **210**, 3295–3300.

Goto, M., Fujii, M., Suzuki, K., and Sakai, M. (1998). Factors affecting carbohydrate and free amino acid content in overwintering larvae of *Enosima leucotaeniella*. *Journal of Insect Physiology* **44**, 87–94.

Goto, M., Sekine, Y., Out, H., Hujikura, M., and Suzuki, K. (2001). Relationships between cold hardiness and diapause, and between glycerol and free amino acid contents in overwintering larvae of the oriental corn borer, *Ostrinia furncalis*. *Journal of Insect Physiology* **47**, 157–165.

Goto, S. G. (2001a). Expression of *Drosophila* homologue of senescence marker protein-30 during cold acclimation. *Journal of Insect Physiology* **46**, 1111–1120.

Goto, S. G. (2001b). A novel gene that is up-regulated during recovery from cold shock in *Drosophila melanogaster*. *Gene* **270**, 259–264.

Hayward, S., Pavlides, S. C., Tammariello, S. P., Rinehart, J. P., and Denlinger, D. L. (2005). Temporal expression patterns of diapause-associated genes in flesh fly pupae from the onset of diapause through post-diapause quiescence. *Journal of Insect Physiology* **51**, 631–640.

Holden, C. P. and Storey, K. B. (1993). Purification and characterization of glycogen phosphorylase A and B from the freeze-avoiding gall moth larvae *Epiblema scudderiana*. *Journal of Comparative Physiology B* **163**, 499–507.

Joplin, K. H., Yocum, G. D., and Denlinger, D. L. (1990). Cold shock elicits expression of heat shock proteins in the flesh fly, *Sarcophaga crassipalpis*. *Journal of Insect Physiology* **36**, 825–834.

Kaplan, F., Kopka, J., Sung, D. Y., Zhao, W., Popp, M., Porat, R., and Guy, C. L. (2007). Transcript and metabolite profiling during cold acclimation of *Arabidopsis* reveals an intricate relationship of cold-regulated gene expression with modifications in metabolite content. *Plant Journal* **50**, 967–981.

Kelty, J. D. and Lee, R. E. (2001). Rapid cold-hardening of *Drosophila melanogaster* (Diptera: Drosophilidae) during ecologically-based thermoperiodic cycles. *Journal of Experimental Biology* **204**, 1659–1666.

Kim, M., Robich, R. M., Rinehart, J. P., and Denlinger, D. L. (2006). Up-regulation of two actin genes and redistribution of actin during diapause and cold stress in the northern house mosquito, *Culex pipiens*. *Journal of Insect Physiology* **52**, 1226–1233.

Koštál, V., Zahradníčková, H., Simek, P., and Zelený, J. (2007). Multiple component system of sugars and polyols in the overwintering spruce bark beetle, *Ips typographus*. *Journal of Insect Physiology* **53**, 581–586.

Koštál, V. and Tollaro-Borovanská (2008). The 70kDa heat shock protein assists during the reparation of chilling injury in the insect, *Pyrrhocoris apterus*. *Journal of Insect Physiology* **53**, 581–586.

Lee, R. E., Chen, C.-P., and Denlinger, D. L. (1987). A rapid cold-hardening process in insects. *Science* **238**, 1415–1417.

Levin, D., Danks, H., and Barber, S. (2003). Variations in mitochondrial DNA and gene transcription in freezing-tolerant larvae of *Eurosta solidaginis* (Diptera: Tephritidae) and *Gynaephora groenlandica* (Lepidoptera: Lymantriidae). *Insect Molecular Biology* **12**, 281–289.

Li, A. Q., Popova-Butler, A., Dean, D. H., and Denlinger, D. L. (2007). Proteomics of the flesh fly brain reveals an abundance of up-regulated heat shock proteins during pupal diapause. *Journal of Insect Physiology* **53**, 385–391.

Li, A. Q. and Denlinger, D. L. (2008). Rapid cold hardening elicits changes in the brain protein profiles of the flesh fly, *Sarcophaga crassipalpis*. *Insect Molecular Biology*, **17**, 565–572.

Li, Y-.P., Ding, L., and Goto, M. (2002). Enzyme activities in overwintering larvae of the shonai ecotype of the rice stem borer, *Chilo suppressalis Walker*. *Archives of Insect Biochemistry and Physiology* **50**, 53–61.

Michaud, M. R. and Denlinger, D. L. (2007). Shifts in the carbohydrate, polyol and amino acid pools during rapid cold-hardening and diapause-associated cold-hardening in flesh flies (*Sarcophaga crassipalpis*): a metabolomic comparison. *Journal of Comparative Physiology B* **177**, 753–763.

Michaud, M. R., Benoit, J. B., Lopez-Martinez., G., Elnitsky, M. A., Lee, R. E., and Denlinger, D. L. (2008). Metabolomics reveals unique and shared metabolic changes in response to heat shock, freezing, and desiccation in the Antarctic midge, *Belgica antarctica*. *Journal of Insect Physiology* **54**, 645–655.

Muise, A. M. and Storey, K. B. (1997). Reversible phosphorylation of fructose 1,6-bisphosphatase mediates enzyme role in glycerol metabolism in the freeze-avoiding gall moth *Epiblema scudderiana*. *Insect Biochemistry and Molecular Biology* **27**, 617–623.

Overgaard, J., Jacob, G., Sorensen, J. G., Nielsen, N. C., Loeschcke, V., and Holmstrup, M. (2005). Metabolomic profiling of heat stress: hardening and recovery of homeostasis in *Drosophila*. *American Journal of Physiology – Regulatory, Integrative, and Comparative Physiology* **291**, R205–R212.

Overgaard, J., Malmendal, A., Sorensen, J. G., Bundy, J. G., Loeschcke, V., Niels, N. C., and Holmstrup, M. (2007). Metabolomic profiling of rapid cold hardening and cold shock in *Drosophila melanogaster*. *Journal of Insect Physiology* **53**, 1218–1232.

Pant, R. and Gupta, D. K. (1979). The effect of exposure to low temperature on the metabolism of carbohydrates, lipids and protein in the larvae of *Philosamia ricini*. *Journal of Biosciences* **1**, 441–446.

Qin, W., Neal, S. J., Robertson, R. M., Westwood, J. T., and Walker, V. K. (2005). Cold hardening and transcriptional change in *Drosophila melanogaster*. *Insect Molecular Biology* **14**, 607–613.

Rinehart, J. P. and Denlinger D. L. (2000). Heat-shock protein 90 is down-regulated during pupal diapause in the flesh fly, *Sarcophaga crassipalpis*, but remains responsive to thermal stress. *Insect Molecular Biology* **9**, 641–645.

Rinehart, J. P., Yocum, G. D., and Denlinger, D. L. (2000). Developmental upregulation of inducible hsp70 transcripts, but not the cognate form, during pupal diapause in the flesh fly, *Sarcophaga crassipalpis*. *Insect Biochemistry and Molecular Biology* **30**, 515–521.

Rinehart, J. P., Li., A. Q., Yocum, G. D., Robich, R. M., Hayward, S. A. L., and Denlinger, D. L. (2007). Up-regulation of heat shock proteins is essential for cold survival during insect diapause. *Proceedings of the National Academy of Sciences, USA* **104**, 11130–11137.

Rivers, D. B. and Denlinger, D. L. (1994). Redirection of metabolism in the flesh fly, *Sarcophaga bullata*, following envenomation by the ectoparasitoid *Nasonia vitripennis* and correlation of metabolic effects with diapause status of the host. *Journal of Insect Physiology* **40**, 207–215.

Robich, R. M., Rinehart, J. P., Kitchen, L. J., and Denlinger, D. L. (2007). Diapause-specific gene expression in the northern house mosquito, *Culex pipiens* L., identified by suppressive subtractive hybridization. *Journal of Insect Physiology* **53**, 235–245.

Sinclair, B. J., Gibbs, A. G., and Roberts, S. P. (2007). Gene transcription during exposure to, and recovery from, cold and desiccation stress in *Drososphila melanogaster*. *Insect Molecular Biology* **16**, 435–443.

Slama, K. and Denlinger, D. L. (1992). Infradian cycles of oxygen consumption in dispausing pupae of the flesh fly, *Sarcophaga crassipalpis*, monitored by scanning microrespirographic method. *Archives of Insect Biochemistry and Physiology* **20**, 135–143.

Sonoda, S., Fukumoto, K., Izumi Y., Yoshida, H., and Tsumuki, H. (2006). Cloning of heat shock protein genes (hsp90 and hsc70) and their expression during larval diapause and cold tolerance acquisition in the rice stem borer, *Chilo suppressalis Walker*. *Archives of Insect Biochemistry and Physiology* **63**, 36–47.

Storey, K. B. and Storey, J. M. (1981). Biochemical strategies of overwintering in the gall fly larva, *Eurosta solidiginis*: effect of low temperature acclimation on the activities of enzymes of intermediary metabolism. *Journal of Comparative Physiology B* **144**, 191–199.

Storey, K. B. and Storey, J. M. (1986). Freeze tolerant frogs' cryoprotectants and tissue metabolism during freeze-thaw cycles. *Canadian Journal of Zoology* **64**, 49–56.

Storey, J. M. and Storey, K. B. (1990). Carbon balance and energetics of cryoprotectant synthesis in a freeze-tolerant insect: responses to perturbation by anoxia. *Journal of Comparative Physiology B* **160**, 77–84.

Storey, K. B. and Churchill, T. A. (1995). Metabolic responses to anoxia and freezing by the freeze tolerant marine mussel *Geukensia demissus*. *Journal of Experimental Marine Biology and Ecology* **188**, 99–114.

Storey, K. B. and McMullen, D. C. (2004). Insect cold hardiness: new advances using gene-screening technology. In *Life in the Cold: Evolution, Mechanisms, Adaptation, and Application*, ed. Barnes, B. M., Carey, H. V., Biological Papers of the University of Alaska, number 27. Fairbanks, Alaska, USA. pp. 275–281.

Wistow, G. (1985). Domain structure and evolution in alpha-crystallins and small heat shock proteins. *FEBS Letters* **181**, 1–6.

Yocum, G. D. (2000). Differential expression of two HSP70 transcripts in response to cold shock, thermoperiod, and adult diapause in the Colorado potato beetle. *Journal of Insect Physiology* **47**, 1139–1145.

5

Cell structural modifications in insects at low temperatures

VLADIMÍR KOŠTÁL

5.1 Introduction

Cells of poikilotherm organisms are subjected to a whole range of variation in environmental temperature and have evolved powerful responses to cope with daily and seasonal temperature fluctuations (Lee, 1991). This chapter deals with the acclimatory changes of the basic cell structural components, such as biological membranes, cytoskeleton, organelles (mitochondria) and large (nucleo) protein complexes, which were observed in preparation for, or in a direct response to, a decline in environmental temperature in insects. The main focus will be on biological membranes because this information is by far the most complete, and the situation in insects may be compared to knowledge on fish and other poikilotherms (Cossins and Sinensky, 1984; Cossins, 1994; Hazel, 1989; 1995; Hazel and Williams, 1990).

Considering the adaptive meaning of acclimatory responses, two broad categories can be theoretically distinguished: (i) compensation of physiological function (capacity adaptation) and (ii) preservation of biological structure (resistance adaptation) (Cossins and Bowler, 1987). Some insects inhabiting temperate zones remain fully active in buffered microhabitats throughout the cold season (Aitchison, 1979a,b), and such insects probably need a certain level of physiological compensation. Most species, however, spend winter in dormancy (Koštál *et al.*, 2006), and those insects probably have to rely more on resistance mechanisms. Strong resistance mechanisms can be expected in those insects that undergo freezing or dehydration, sometimes reaching the cryptobiotic state (*sensu* Clegg, 2001). The enormous diversity of insects, together with their ability to survive cold periods

Low Temperature Biology of Insects, ed. David L. Denlinger and Richard E. Lee Jr. Published by Cambridge University Press.

in different states (activity versus dormancy; chilling versus supercooling; freezing versus dehydration) gives little opportunity for drawing general conclusions. When we add the relative paucity of experimental data, it is clear that the ambition of this review cannot be much more than to summarize, and briefly comment, on the available literature.

5.2 Biological membranes

5.2.1 Basic concepts

Biological membranes play essential roles in cellular physiology. They form a barrier for unregulated diffusion, mediate regulated transport of solutes, store and utilize the energy of ion electrochemical gradients, provide organization and a catalytic matrix for membrane proteins, participate in cell–cell recognition and interaction and supply precursors for signal-transduction cascades. The physical properties of lipidic bilayers, i.e. the *phase* state and the *fluidity* (the *order*, allowing a specific rate of molecular motions), are acutely sensitive to temperature. Membrane lipids can assume one of three basic *phase* states: (i) a highly ordered bilayer formed by lipids in a lamellar gel phase (L_{β}); (ii) a fluid bilayer, liquid crystalline phase (L_{α}); and (iii) a non-bilayer, reversed hexagonal phase (H_{II}) (Chapman, 1975). The phase of a particular membrane is dictated by its chemical composition, degree of hydration, pressure and temperature. Generally, at low temperatures, the membrane lipids become closely packed in a gel phase. The rates of molecular motion and lateral diffusion are slow. As the temperature increases, it reaches a specific phase-transition temperature (T_m), at which the gel phase "melts" to form a fluid phase, where the bilayer thickness is reduced, its volume increases and individual lipid molecules may rapidly move and laterally diffuse. Further increases of temperature may cause the transition into a non-bilayer, hexagonal phase at a specific temperature (T_h), where the bilayer losses its integrity. Transition to the H_{II} phase is favored by low hydration rates (Kirk *et al.*, 1984), and, thus, those insects that overwinter in a frozen or dehydrated state may need specific adaptations to avoid transition of their membranes into the hexagonal phase (Pruitt and Lu, 2008). Membrane functions, such as permeability and activities of membrane-bound enzymes, are also directly influenced by bilayer fluidity. Functions tend to decrease gradually with decreasing temperature, even when the "functional" fluid phase is maintained, i.e. without phase transition (Cossins and Macdonald, 1989; Hazel, 1989).

Our knowledge of acclimatory alteration of membrane composition has been emerging gradually, originating from early observations a century ago (Henriques and Hansen, 1901) and culminating in 1974 when the theory of *homeoviscous adaptation* (HVA) was formulated by Sinensky (1974). According to this theory, temperature-induced restructuring of membrane lipidic composition aims to maintain a specific level of membrane fluidity, at which membrane functions are optimal. Later, this theory was extended to a variety of organisms from bacteria to vertebrates, and its validity was tested at both organismal (acclimatory) and evolutionary (adaptational) levels (Cossins and Prosser, 1978; Behan-Martin *et al.*, 1993). Although HVA is the most often-used paradigm to interpret the temperature-induced restructuring of membranes, some observations are difficult to explain solely in terms of HVA. Thus, McElhaney (1984) coined the term *homeophasic adaptation* (HPA), which stresses the significance of regulating the lipid bilayer phase rather than its fluidity. Hazel (1995) further developed and broadened the concept of HPA to include, not only the static description of lipid phase, but also the dynamism of phase changes. In his model of *dynamic phase behavior* (DPB), the relationship between body temperature and transition temperatures of the lipid phases are conserved by acclimatory and adaptational adjustments of membrane composition. The adaptive meaning of DPB is to keep the membrane lipids sufficiently "far" from the deleterious transition to gel phase, and optimally "close" to the transition to hexagonal phase, in order to permit regulated, and prevent unregulated, membrane fusions (regulated membrane fusions occur frequently during vesicle trafficking).

The phase state of membrane lipids critically influences various membrane-associated processes. The fluid/gel transition drastically reduces the activities of membrane-bound enzyme and transport systems (for examples see: Hazel, 1989). Moreover, as the membranes are composed of many diverse lipid species (Dowhan, 1997), the phase transitions may span relatively broad temperature ranges. Thus, the gel and fluid domains may coexist at certain temperatures and areas of *phase separation* are formed between them, resulting in rapid loss of barrier function. Formation of a gel phase is thus believed to directly threaten cell functionality and survival. Similarly, unregulated transitions in the H_{II} phase are considered incompatible with life processes.

5.2.2 *Commonly observed patterns of cold-induced membrane restructuring*

Several alterations of membrane lipid composition (schematically depicted in Fig. 5.1) were repeatedly reported in various poikilotherms in response to low temperatures:

- Desaturation: increasing the relative proportion of unsaturated fatty acids (UFAs) at the expense of saturated fatty acids (SFAs) in the membrane glycerophospholipids (GPLs)
- Shortening the average fatty acyl (FA) chain length

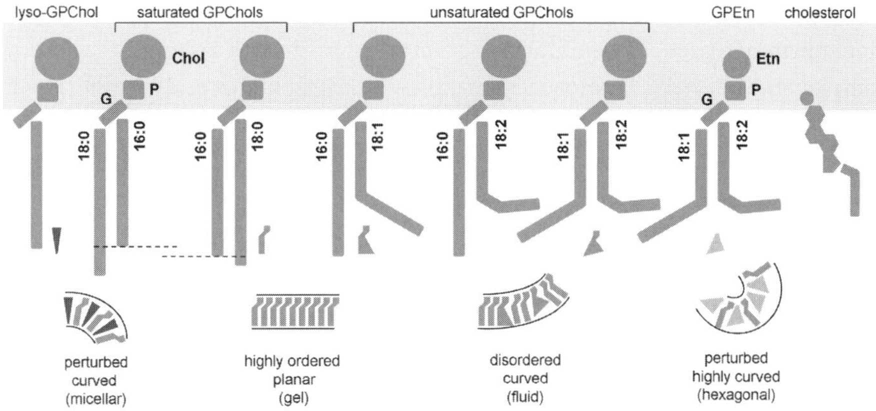

Figure 5.1 Schematic representation of basic structural features of lipidic molecules in biological membranes of animals. Seven different molecules of glycerophospholipids (GPLs) are shown on the left and one molecule of cholesterol is shown on the right. GPLs are composed of a glycerol backbone (G), to which fatty acyls (identified by numbers of carbon atoms:numbers of double bonds) are bound by esteric or etheric (plasmalogenes) bonds to the positions *sn1* and *sn2*. A phosphate group (P) is esterified at position *sn3* and binds various hydrophilic moieties (here either choline (Chol) or ethanolamine (Etn)). Phospholipase A may cleave one of the fatty acyls producing lyso-GPL. This molecule has the shape of an inverted cone, which disrupts tight packing of GPChols into planar membranes, decreases membrane order (increases its fludity) and may lead to formation of isolated lipidic micelles. Saturated GPChols assume a cylindrical shape and form highly ordered planar membranes, which are relatively prone to the transition to a gel phase at low temperatures. Length and position of fatty acyls in GPChols influence the hydrophobic interactions between them (depicted by dashed lines) and, consequently, physical features of the membrane (explained in text). Unsaturated GPChols assume a conical shape, pack less efficiently into a bilayer and decrease its order. Replacing choline by a smaller and less hydrated ethanolamine has a similar effect. If the number of non-bilayer GPLs is high, the membrane may undergo a transition from the fluid to a reversed hexagonal phase. Probability of such a transition increases with increased temperature and decreased hydration. Insertion of cholesterol molecules increases the order of fluid membranes through hydrophobic interaction between the rigid steran structure and the neighboring fatty acyls.

- Reshuffling of fatty acyls resulting in formation of new molecular species without changing the relative proportions of individual FAs
- Head-group restructuring: increasing the relative proportions of glycerophosphoethanolamines (GPEtns) at the expense of glycerophosphocholines (GPChols)
- Changing the GPL/cholesterol ratio.

Desaturation of fatty-acyl chains in membrane GPLs is the most thoroughly documented aspect of membrane restructuring in diverse poikilotherm organisms, such as bacteria, protozoa, yeast, plants, invertebrates and ectotherm vertebrates (Nozawa *et al.*, 1974; Sinensky, 1974; Murata and Yamaya, 1984; Hazel, 1989; Cossins, 1994; Tasaka *et al.*, 1996; Suutari *et al.*, 1997). Direct genetic proof for the adaptive role of desaturation in enhancing survival at low temperatures has been obtained in plants, cyanobacteria and nematodes (Browse *et al.*, 1994; Allakhverdiev *et al.*, 1999; Murray *et al.*, 2007). The introduction of a *cis* double bond into fatty-acyl chain results in formation of a bend of about 30°. Desaturated FA tends to occupy a greater area and increases the conical character of the GPL molecule, which packs less efficiently into the bilayer. Consequently, the temperature of fluid/gel transition (T_m) is markedly reduced (Li *et al.*, 1998; Wang *et al.*, 1999). The adaptive meaning of desaturation thus presents itself: it enhances membrane fluidity, thus counteracting the ordering effects of low temperatures (in accordance to HVA theory), and it also extends the range over which membranes remain fluid to lower temperatures (in accordance to HPA theory).

Thanks to the greater area of hydrophobic interactions, the GPLs containing long-chain FAs have higher melting temperatures than their short-chain counterparts (Huang *et al.*, 1997). Thus, regulation of FA chain length provides another potential strategy for controlling membrane fluidity. Shortening of FA chain length in response to decreasing environmental temperature has been demonstrated in some bacteria, yeast and plants (Harwood *et al.*, 1994). The two positions of the glycerol backbone, *sn-1* and *sn-2*, to which FA chains can be esterified in GPL molecules are not equivalent. If two identical FAs are attached, the one at the *sn-1* position will penetrate deeper into the bilayer. When two FAs have different lengths, positioning of a shorter chain to the *sn-2* will lower the temperature of gel/fluid transition (T_m) and vice versa. The mechanistic explanation of this phenomenon lies again in changes of the area of hydrophobic interactions between the two FA chains. Rapid reshuffling of existing FA chains through deacylation/reacylation of GPL molecules was first observed in a protozoan, *Tetrahymena* (Ramesha and Thompson, 1983). The composition of GPL molecular species may be altered also through slower changes in their *de novo* synthesis and incorporation into the GPL molecules (Hazel, 1989; Brooks *et al.*, 2002).

The area occupied by the ethanolamine moiety is smaller than the area occupied by choline. Moreover, choline tends to be more hydrated and, as a result, GPEtns assume a more conical conformation than GPChols, which are more cylindrical. Conical forms of lipids pack less efficiently into the bilayer, thus decreasing its order, which may explain why it has been observed that membranes of cold-acclimated fish tend to have higher ratios of GPEtn/GPChol (Hazel, 1989). The conical shape of the GPEtn molecule is even more pronounced when it contains unsaturated FAs. Increasing the relative proportion of such molecules, together with decreasing hydration rate, may lead to transition into the H_{II} phase at physiological temperatures (Kirk *et al.*, 1984). Other non-bilayer lipids assuming conical molecular shapes, such as lyso-GPLs, may have a similarly disruptive influence on membrane order.

The impact of cholesterol introduction into the membrane depends on the phase state. A fluid membrane becomes more ordered with increasing cholesterol content and, consequently, its permeability for solutes decreases and the activities of some membrane-bound enzymes may also decrease (Haines, 2001). On the other hand, the conical shape of the cholesterol molecule tends to disrupt tight packing of GPLs into the gel phase. The changes in cholesterol content may thus depend on specific requirements of different membrane fractions at different conditions (Cossins, 1977; Sørenson, 1993; Crockett and Hazel, 1995).

5.2.3 Patterns of membrane restructuring observed in insects

5.2.3.1 Lipid composition of insect membranes

Insects and poikilotherm vertebrates share qualitatively similar compositions of their biological membranes. Despite a broad similarity of the main constituents (lipid classes, FAs and sterols), quantitative proportions of individual molecular species are very diverse (Thompson, 1973; Stanley-Samuelson *et al.*, 1988; Canavoso *et al.*, 2001). Two important features of lipid metabolism differ between insects and other animals: (i) insects are absolutely dependent on a dietary supply of sterols (either of plant or animal origin); and (ii) insects are able to convert oleic FA, 18:1 (*n*-6) to linoleic FA, 18:2 (*n*-3), which makes them nutritionally independent of plant-derived polyunsaturated fatty acids (PUFAs) (Stanley-Samuelson *et al.*, 1988; Blomquist *et al.*, 1991; Canavoso *et al.*, 2001). Saturated and unsaturated fatty acids with 16 or 18 carbons (C16 or C18, respectively) are the most abundant. The content of C20 PUFAs is usually relatively low in total GPLs extracted from terrestrial insects, with one notable exception of Collembola, where it may reach over 40% (Bayley *et al.*, 2001; Holmstrup *et al.*, 2002; Chamberlain and Black, 2005; Bahrndorf *et al.*, 2007). Specific membrane fractions of different organs, however, are highly enriched with PUFAs (Stanley-Samuelson and Dadd, 1983) and PUFAs are also frequently represented in aquatic insect species (Hanson *et al.*, 1985).

Although it is known that FA compositions of various insect tissues and GPL fractions differ, and change flexibly depending on diet quality and developmental stage (Stanley-Samuelson *et al.*, 1988), such knowledge has seldom been taken into account in studies on cold-induced membrane restructuring in insects. Most often, the total polar lipid fraction is extracted from whole specimens or tissues, and FA composition is analysed. Lipid classes are seldom separated and detailed information at a level of lipid molecular species is available for only three insects. Thus, in comparison to the situation in fish, plants, or micro-organisms, our knowledge of insects suffers from a relative paucity of data and a low resolution level. In this chapter, results of 24 papers published on temperature-related changes of membrane lipid composition in 21 different insect species will be reviewed in Sections 5.2.3.2–5.2.3.6. Most of the included species overwinter in a supercooled state and only two are freeze-tolerant, larvae of flies *Chymomyza costata* and *Eurosta solidaginis*. Most of the papers were published within the period 1997–2008 (one in 1981). Some earlier papers, where total lipids were analysed (containing a large fraction of triacylglycerols, TGs) are not included here. For the purpose of this review, original results were sometimes re-calculated to obtain comparable indices of FA unsaturation level (UFA/SFA), chain length (16C/18C), or GPL headgroup structure (GPEtn/GPChol).

5.2.3.2 UFA/SFA ratio

Relevant information on unsaturation rates of membrane fatty acids can be derived for 20 insect species. Only in about half of them, however, could a clear positive correlation between the change of UFA/SFA ratio and low-temperature acclimation/survival be found (Table 5.1). Cursory inspection of UFA/SFA values shows that both the values and magnitude of changes are quite similar to those found in various cold-acclimated fish (Hazel, 1989). An increase in the relative proportion of palmitoleic (16:1) and oleic (18:1) FAs, countered with a decrease of their saturated counterparts, is the most often reported cause of the UFA/SFA increase. Taking the six drosophilid species individually, the data by Ohtsu *et al.* (1998; 1999) suggest a clear acclimatory compensation of membrane fluidity at two rearing temperatures, 23 and 15 °C. However, no correlation between UFA/SFA and cold-tolerance was found in the same data over the latitudinal cline from the sub-tropics to the cold temperate zone. Although the proportions of UFA were constant over the latitudinal cline, a highly significant increase of monounsaturates (MUFAs, 16:1. 18:1), countered by a decrease of polyunsaturates (18:2, 18:3) was observed with increasing cold-tolerance (Ohtsu *et al.*, 1998; 1999). Data for the fly, *Eurosta solidaginis*, and two adult heteropterans, *Dolycoris baccarum* and *Piezodorus lituratus*, exemplify the autumnal acclimatization changes observed in the field over several months (Bennett *et al.*, 1997; Bashan and Cakmak, 2005; Pruitt and Lu, 2008). While larvae of the fly survive winter in the frozen state, the adults of the two bugs supercool. Despite that big difference in overwintering strategy, similar adjustments were seen in their membranes. Changes observed in the collembolan, *Folsomia candida*, were triggered by desiccation. Although they occurred at

Table 5.1 *Examples of insects showing a positive correlation between change of the UFA/SFA ratio and low-temperature acclimation/survival*

Species	Stage[1]	Analysis[2]	Experiment[3]	Change in[4] UFA/SFA	Reference
Drosophila subauraria	A	FA, body	L, 23 vs. 15°C	5.1 → 5.9	Ohtsu *et al.*, 1998
Drosophila triauraria				3.6 → 6.0	Ohtsu *et al.*, 1998
Drosophila rufa				3.3 → 6.3	Ohtsu *et al.*, 1998
Drosophila lutescens				4.4 → 6.0	Ohtsu *et al.*, 1998
Drosophila takahashi				4.8 → 6.0	Ohtsu *et al.*, 1998
Drosophila curviceps	A	FA, body	L, 21 vs. 15°C	5.2 → 6.2	Ohtsu *et al.*, 1999
Eurosta solidaginis	L	FA, body	F, Sep → Dec	2.8 → 4.2	Bennett *et al.*, 1997
Eurosta solidaginis			F, Aug → Nov	2.3 → 3.5	Pruitt and Lu, 2008
Dolycoris baccarum	A	FA, body	F, May → Feb	2.7 → 8.5	Bashan and Cakmak, 2005
Piezodorus lituratus				3.1 → 9.9	Bashan and Cakmak, 2005
Cymbalophora pudica	L	FA, midgut	L, cold-acclimation	1.7 → 2.0	Koštál and šimek, 1998
Folsomia candida	A	FA, body	L, desiccation	1.9 → 2.4	Bayley *et al.*, 2001; Holmstrup *et al.*, 2002

[1] L, larva; P, pupa; A, adult.

[2] FA, fatty acid; GPL, glycerophospholipid; GPEtn, glycerophosphoethanolamine; GPChol, glycerophosphocholine.

[3] L, laboratory; F, field; RCH, Rapid cold-hardening.

[4] UFA, unsaturated fatty acid; SFA, saturated fatty acid.

a constant temperature of 20 °C, they positively correlated with cold-tolerance (Bayley *et al.*, 2001; Holmstrup *et al.*, 2002). In addition to the data listed in Table 5.1, Kayukawa *et al.* (2007) reported that cold-acclimation of non-diapause pupae of the fly, *Delia antiqua*, increased the relative proportions of 16:1 and 18:1 FAs in brain GPEtns and GPChols, respectively.

The list of species where no, or even a negative, correlation between UFA/SFA and cold-acclimation/survival was observed, is almost equally long. Thus, cold-acclimation of aestivating mature prepupae of the moth, *Cymbalophora pudica*, resulted in a moderate enhancement of its cold-tolerance. Although this was accompanied with significant restructuring of the midgut FA composition (Table 5.1), the UFA/SFA ratio remained constant in four other tissues (brain, fat body, silk gland and body wall) (Koštál and Šimek, 1998). Similar lack of correlation between cold acclimation and FA unsaturation rate was observed in: three drosophilids,

Table 5.2 *Examples of insects showing a positive correlation between a change in the 16C/18C ratio and low-temperature acclimation/survival*

Species	Stage	Analysis	Experiment	Change in 16C/18C	Reference
Drosophila immigrans	A	FA, body	L, 21 vs. 15 °C	0.33 → 0.40	Ohtsu *et al.*, 1999
Drosophila albomicans				0.31 → 0.38	
Sarcophaga crassipalpis	P	FA, body	RCH, diapause	0.49 → 0.57	Michaud and Denlinger, 2006
Pyrrhocoris apterus	A	FA, muscle	F, Sep → Jan	0.18 → 0.23	Hodková *et al.*, 1999
Pyrrhocoris apterus	A	GPEtn, muscle	L, 26 °C vs. F, Feb	0.16 → 0.24	Hodková *et al.*, 2002
		GPEtn, fat body		0.21 → 0.34	
		GPchol, muscle		0.11 → 0.15	
Pyrrhocoris apterus	A	GPL, muscle	F, Oct → Dec	0.16 → 0.20	Tomčala *et al.*, 2006
		GPL, fat body		0.16 → 0.20	
Eurygaster integriceps	A	FA, body, male	F, May → Feb	0.33 → 0.46	Bashan *et al.*, 2002
		FA, body, female		0.30 → 0.77	
Folsomia candida	A	FA, body	L, desiccation	0.22 → 0.26	Holmstrup *et al.*, 2002
Megaphorura arctica	A	FA, body	L, cold-acclimation	0.19 → 0.24	Bahrndorf *et al.*, 2007

See Table 5.1 for explanation of abbreviations used.

D. imigrans, *D. albomicans* (Ohtsu *et al.*, 1999) and *D. melanogaster* (Overgaard *et al.*, 2005, 2006, 2008); GPChols extracted from muscles and fat bodies of larvae of the temperate fly, *Chymomyza costata* (Koštál *et al.*, 2003); pupae of the fly, *Sarcophaga crassipalpis* (Michaud and Denlinger, 2006); GPChols and GPEtns extracted from muscle and fat body tissues of the adult bugs, *Pyrrhocoris apterus* (Hodková *et al.*, 1999; 2002; Šlachta *et al.*, 2002; Tomčala *et al.*, 2006). In three cases, significantly negative correlations between UFA/SFA ratio and cold acclimation were reported: GPEtns extracted from muscles and fat bodies of larvae of the temperate fly, *Chymomyza costata* (Koštál *et al.*, 2003); adults of the bug, *Eurygaster integriceps* (Bashan *et al.*, 2002); and the Spitsbergen springtails, *Megaphorura arctica*, acclimated to −10 and −20 °C (Bahrnodrf *et al.*, 2007).

5.2.3.3 Fatty-acyl chain length

Data on cold-related changes in fatty-acyl chain length are available for 14 insect species. Because 16C and 18C fatty acids predominate in the membrane lipids of most insects studied, the changes in average chain length were estimated based on the ratio of 16C/18C. First, at an evolutionary level, a strong positive correlation between the 16C/18C ratio and cold-tolerance was found by comparing eight different *Drosophila* species sampled over a latitudinal cline from the sub-tropics to the temperate zone of Japan (Ohtsu, 1998; 1999). Second, a significant increase in the 16C/18C ratio was observed at an acclimatory level in several cases listed in Table 5.2. The two sub-tropical drosophilid flies, in which no correlation of UFA/SFA with cold-acclimation was observed, i.e. *D. immigrans* and *D. albomicans*, did show a significant shortening of FA chains (Ohtsu *et al.*, 1999). Similarly, in two heteropterans where no (*P. apterus*) or a negative (*E. integriceps*) correlation of cold-acclimation with UFA/SFA was found, significant shortening of FA chain length was apparent (Bashan *et al.*, 2002). And finally, a similar trend was seen in the springtail, *M. arctica*: decreasing the UFA/SFA countered with increasing the 16C/18C ratio during cold-acclimation (Bahrndorf *et al.*, 2007). These observations suggest that these two different adaptive responses, FA desaturation and FA shortening, may be alternatively employed by different species to reach a similar goal. In support of this view, in two other heteropterans, *D. baccarum* and *P. lituratus*, where the 16C/18C ratio significantly *decreased* with cold-acclimation (0.19 → 0.14 and 0.20 → 0.14, respectively), the positive change of UFA/SFA ratio was by far the most pronounced (see Table 5.1). Decreases of the 16C/18C ratio with cold-acclimation were also observed in freeze-tolerant fly larvae of *E. solidaginis* (Bennett *et al.*, 1997; Pruitt and Lu, 2008) and *C. costata* (Koštál *et al.*, 2003). In two cases, *C. pudica* and *D. curviceps*, no significant changes in 16C/18C were reported (Koštál and Šimek, 1998; Ohtsu *et al.*, 1999).

5.2.3.4 GPL headgroups

Of the five insects for which data are available, three showed a significant increase in the GPEtn/GPChol ratio with cold-acclimation (Table 5.3) and two, the freeze-tolerant larvae of *C. costata* and *E. solidaginis*, showed no change (Koštál *et al.*, 2003; Pruitt and Lu, 2008).

5.2.3.5 Molecular species remodeling

Three insect species were studied with respect to cold-related remodeling of their membranes at the level of individual GPL molecular species. Analysis of lipid molecules helped to reveal details that are not accessible by standard FA analysis. For instance, none of the most frequently used general descriptors of

Table 5.3 *Examples of insects showing a positive correlation between a change of the GPEtn/GPChol ratio and low-temperature acclimation/survival*

Species	Stage	Analysis	Experiment	Change in GPEtn/GPChol	Reference
Drosophila melanogaster	A	Body	L, 25 vs. 20, 15 °C	1.1 → 1.4	Overgaard *et al.*, 2008
Sarcophaga crassipalpis	P	Body	RCH, diapause	1.9 → 2.2	Michaud and Denlinger, 2006
			RCH, non-diapause	0.8 → 1.2	
Pyrrhocoris apterus	A	Muscle	F, Sep → Jan	1.1 → 1.3	Hodková *et al.*, 1999
Pyrrhocoris apterus	A	Muscle	L, 26 °C vs. F, Feb	1.1 → 1.4	Hodková *et al.*, 2002
		Fat body		0.8 → 1.1	
Pyrrhocoris apterus	A	Muscle	L, cold accl.,	1.0 → 1.1	Šlachta *et al.*, 2002
		Fat body	non-diapause	0.7 → 1.0	
Pyrrhocoris apterus	A	Muscle	F, Oct → Dec	0.8 → 1.2	Tomčala *et al.*, 2006
		Fat body		0.8 → 1.0	

See Table 5.1 for explanation of the abbreviations used

membrane restructuring, i.e. UFA/SFA, 16C/18C, GPEtn/GPChol, changed significantly during cold-acclimation in larvae of *C. costata*. At the level of molecular species, however, significant remodeling in muscle and fat-body membranes was observed. Most changes were relatively small, and their adaptive value, if any, remains unexplored. Some changes, however, were highly pronounced and tightly correlated with freeze-tolerance. Thus, relative proportions of GPEtn 16:0/18:2 and GPChol 16:0/18:2 increased during cold-acclimation by about 15–20% in muscles and fat-body tissue (Koštál *et al.*, 2003). The same two molecular species showed the highest increases, ranging between 5 and 10%, in the muscles and fat bodies of diapausing adults of *P. apterus* during their autumnal cold-acclimatization in two independent studies (Hodková *et al.*, 2002; Tomčala *et al.*, 2006). In adults of *D. melanogaster* that were raised as larvae and pupae at three different temperatures of 25, 20 and 15 °C, GPEtn 16:0/18:2 was among the three molecular species that showed a significant increase with decreasing temperature (Overgaard *et al.*, 2008). And, finally, the same lipid molecule showed the most prominent increase in relative abundance in overwintering water striders, *Aquarius palludum* (Hodková, unpublished results). Such observations of a common biochemical response to cold in phylogenetically unrelated insect species, which use very different overwintering strategies, strongly suggest that an increase in relative proportion of GPEtn 16:0/18:2 has some adaptive meaning. It has been proposed (Hodková *et al.*, 2002) that the adaptive value may be related to the fact that the overwintering insects are usually exposed to two environmental stressors, low temperatures and dehydration, at the same time (Block, 1996). It is known that the range of temperatures at which membranes composed of GPEtns remain fluid (i.e. $T_h - T_m$) markedly increases with a decrease in acyl chain length. For instance, the $T_h - T_m$ range is approximately 20 °C wider in GPEtns pairing 16C and 18C FAs than in those with two 18C FAs (Lewis *et al.*, 1989). The presence of two double bonds in the unsaturated linoleic FA (18:2) dramatically reduces the T_m of the whole GPL molecule by about 60–70 °C in comparison to GPL 16:0/18:0 (Huang *et al.*, 1997). Thus, the specific pairing of two common fatty acids, 16:0 with 18:2, may serve two purposes: (i) reduction of T_m and thus avoidance of fluid/gel transition at low temperatures, and (ii) maintenance of a sufficiently wide $T_h - T_m$ interval, so that unregulated transitions into the reversed H_{II} phase are less probable when the ambient temperature suddenly increases or when the hydration rate suddenly decreases. Both threats are common in overwintering insects.

5.2.3.6 Cholesterol

A positive correlation between cholesterol content and increasing ambient temperature was observed in flight-muscle mitochondria of the orthopteran, *Schistocerca gregaria*, acclimated at 31 and 45 °C for 30 days (Downer and Kallapur, 1981). Because cholesterol increases the order of fluid membranes, its role in this case could be to stabilize membrane structure at elevated temperature. No temperature-induced change of cholesterol content, however, was detected in the Malpighian tubules of *E. solidaginis* (Yi and Lee, 2005). Shreve *et al.* (2007) raised larvae of *D. melanogaster* on a cholesterol-augmented diet and managed to elevate cholesterol content in the membranes of adult flies ca. 1.6-fold. Cholesterol-augmented flies exhibited significantly higher cold-tolerance and greater capacity to rapidly cold-harden (see Chapter 2).

5.2.4 Metabolic pathways involved in membrane restructuring

The most consistently observed response of poikilotherms to cold is enrichment of their membranes with UFAs. Consequently, the highest attention was paid to the study of integral membrane desaturases, which occur ubiquitously in eukaryotic cells and are believed to play a primary role in the homeostatic regulation of membrane physical properties (Macartney *et al.*, 1994; Knipple *et al.*, 2002). Cytosolic fatty acyl synthase (FAS) of animals produces predominantly the 16C saturated FAs with smaller amounts of 14C and 18C SFAs. All fatty acids, either from *de novo* synthesis by FAS or from dietary input, first enter the pool of fatty-acyl CoA, where they can be modified by microsomal desaturases and elongases before they are selectively incorporated into various membrane lipids. The first double bond is usually introduced into the Δ9 position (between carbons 9 and 10, counted from the carboxyl end) by a Δ9 desaturase. Second and subsequent double bonds may be introduced by Δ5 and Δ6 terminal desaturases, only between the existing double bond and the carboxyl end of the FA chain in most animals. In insects and some other invertebrates, however, double bonds may be inserted also on the methyl side of the FA chain by Δ12 and some other desaturases (Jurenka *et al.*, 1987; Blomquist *et al.*, 1991; Cook and McMaster, 2002; Knipple *et al.*, 2002).

Because of its primary role, Δ9 desaturase has been most often implicated in cold-induced membrane restructuring (Macartney *et al.*, 1994; Cossins *et al.*, 2002). Activation of Δ9 desaturase at low ambient temperatures has often been reported in fish (Schunke and Wodtke, 1983; Wodtke and Cossins, 1991; Tiku *et al.*, 1996). Although Δ9 desaturase has been characterized and cloned in several insects (Wicker-Thomas *et al.*, 1997; Gonzales and Brenner, 1999; Liu *et al.*, 1999; Eigenheer *et al.*, 2002; Riddervold *et al.*, 2002), there is little information on its potential role during cold-induced membrane restructuring. Kayukawa *et al.* (2007) observed, for the first time in insects, that the abundance of mRNA transcripts of Δ9 desaturase increases 4.5-fold in the brain, midgut and Malpighian tubules of cold-acclimated pupae of the fly, *Delia antiqua*. In parallel, the authors reported a moderate increase in relative proportions of 16:1 and 18:1 FAs in some GPL fractions extracted from pupal brains (Kayukawa *et al.*, 2007). Six desaturase-encoding sequences were found in the genome of *D. melanogaster* (Knipple *et al.*, 2002). While no functions have been ascribed to four of them so far, the *desat1* and *desat2* genes code for two Δ9 desaturases with different substrate preferences: 16:0 > 18:0 in Desat1 and 14:0 in Desat2. Both Δ9 desaturases are used for biosynthesis of sex pheromones, and *desat2* expression was detected only in females (Wicker-Thomas *et al.*, 1997; Dallerac *et al.*, 2000). It has been shown by Greenberg *et al.* (2003) that flies carrying a fully functional Z-allele of *desat2* gene (ancestral tropical populations) not only produce a different sexual pheromone than flies with a non-functional M-allele (temperate populations), but are also less cold-tolerant. This result suggested a system where insect cold-tolerance is markedly influenced by a single gene/enzyme. Unfortunately, Coyne and Elwyn (2006) were not able to replicate the results of Greenberg *et al.* using transgenic flies differing in the nature of their *desat2* alleles. Although they confirmed the role of Desat2 in sexual isolation, carriers of the M allele were *not* more cold-tolerant than carriers of the Z allele. Perhaps, it would be more rewarding to test the role of the *desat1* product on cold-tolerance, as this gene is expressed in both sexes and also in immature stages of the fly (Knipple *et al.*, 2002). In a recent study of *Caenorhabditis elegans*,

Murray *et al.* (2007) used a combination of genetic knockout and RNAi to suppress expression of *fat-5* and *fat-7* FA-CoA desaturases. This way, they managed to abolish cold-induced changes in lipid unsaturation. Although the degree of *C. elegans* cold-tolerance was also influenced, the authors calculated that only 16% of the cold-tolerance response could be attributed to changes in lipid saturation. Such a result clearly documents the fact that cold-acclimation is a highly complex process and that cold-tolerance must be explained as a combination of different adaptive mechanisms (Hayward *et al.*, 2007).

In insects, low temperature is not the exclusive trigger of membrane restructuring. Similar changes were stimulated by cold and by exposure to a short-day photoperiod (Košťál *et al.*, 2003; Michaud and Denlinger, 2006; Kayukawa *et al.*, 2007). Photoperiodically driven changes were simulated by surgical removal of the *corpora allata*, the source of juvenile hormone (Hodková *et al.*, 2002). In addition to long-term seasonal changes, it was shown that lipid composition may be altered also within a much shorter time course that is typical of diurnal temperature fluctuations (Overgaard *et al.*, 2005, 2006; Michaud and Denlinger, 2006). Desiccation and cold-acclimation were shown to have similar effects on collembolan membrane composition (Bayley *et al.*, 2001; Holmstrup *et al.*, 2002; Bahrndorf *et al.*, 2007).

5.2.5 Membrane physical properties and functions

Studies on how animal membrane properties change with cold-induced restructuring of their composition are almost exclusively limited to fish (Cossins and Macdonald, 1989; Hazel, 1989; Hazel and Williams, 1990). Only one report is available for insects so far. Using ^{31}P solid-state NMR spectroscopy, Lee *et al.* (2006) found that the fluidity of membrane lipids extracted from fat-body tissue of the adult fly, *Sarcophaga bullata*, substantially increased within a few hours of rapid cold-hardening (RCH) (Lee *et al.*, 1987). Although no parallel analysis of membrane lipid composition was performed in this study, the results suggest that: (i) membrane properties may be modified very rapidly, within a timeframe of hours, and (ii) the change in membrane properties is in a direction predicted by HVA (or HPA) theory and thus could be causally related to the parallel increase in cold-tolerance.

Supercooled insects, i.e. those maintaining a liquid water phase at temperatures below the melting point of their body fluids, are capable of maintaining their transmembrane electrochemical potentials. It was also shown that such ability is acquired during cold-acclimation and correlates well with the level of cold tolerance (Košťál *et al.*, 2004; Zachariassen *et al.*, 2004). In contrast, membranes of insects surviving in a frozen state are not "functional". For instance, the transmembrane gradients of sodium and potassium completely dissipate within one day in larvae of the fly, *Xylophagus cincta*, when frozen at −10°C (Kristiansen and Zachariassen, 2001).

5.3 Cold-related changes in the cytoskeleton and other structural elements of cells

The cytoskeleton creates and maintains the gross morphology and internal organization of eukaryotic cells, allows movement, division and internal trafficking, participates in signal transduction and is thus essential for proper cell function (Amos and Amos, 1991). Destabilization and depolymerization of actin microfilaments and tubulin microtubules, and loss of associated functions, have been recognized recently as important factors of cold-injury in plants and animals (Pucciarelli *et al.*, 1997; Egiersdorff and Kacperska, 2001; Garlick and Robertson, 2007). Little is known about the fate of cytoskeleton structural elements of insects in response to low temperatures. Studies on gene expression showed that *actin* and *tubulin* mRNA abundance changes specifically with insect diapause. A full range of responses was observed, however, including down-regulation, up-regulation, and no change (Lee *et al.*, 1998; Yocum *et al.*, 2005; Robich *et al.*, 2007; Colinet *et al.*, 2007). A recent study by Kim *et al.* (2006) gives a first hint of possible cold-rearrangement of actin microfilaments in insects. In this study, relatively high expression of two *actin* genes during early diapause in the mosquito, *Culex pipiens*, was confirmed. Fluorescent staining revealed that the polymerized *F* actin also changes its distribution (in the midgut cells) upon entry into diapause and in response to a low-temperature stimulus. While in non-diapausing mosquitoes reared at 25 °C the *F* actin was preferentially clustered along the intersections of muscle fibers, in diapause cold-acclimated mosquitoes, it was evenly distributed along the muscle fibers, probably fortifying their structure. Cytoskeletal elements closely cooperate with various families of molecular chaperones (HSPs, heat-shock proteins). HSPs participate in the folding of nascent cytoskeletal polypeptides, in their assembly and disassembly and, perhaps most importantly for this chapter, in their stabilization during and/or refolding after thermal stress (Liang and MacRae, 1997; Mounier and Arrigo, 2002; Kayukawa *et al.*, 2005; Garlick and Robertson, 2007). As HSPs accumulate in various diapausing insects and are essential for their cold survival (Li *et al.*, 2007; Rinehart *et al.*, 2007), the study of interactions between the cytoskeleton and HSPs emerges as a promising new field of study.

Considering the dramatic differences between insects' active lifestyle during summer and dormancy during winter, some changes in the number and structure/function of various cell organelles likely accompany entrance into diapause and cold-acclimation. Nevertheless, very little work has been done on this topic. Gradual degradation of mitochondria was detected in fat-body and brain tissue of the freeze-tolerant arctic moth larvae, *Gynaephora groenlandica*, during a 3-month-long cold-acclimation from +15 to −15 °C (Kukal *et al.*, 1989). The number of mitochondria decreased 10 000-fold! The pre-adult development of *G. groenlandica* in high arctic regions may extend to 14 years and most of the time each year (11 months) is spent in a deeply hypometabolic dormant state (Kukal and Kevan, 1987). High predictability and extreme conditions of the arctic habitat may explain why this species can "afford" such a drastic reduction of its cell energetic machinery. Approximately a two-fold reduction in the amount of mitochondrial DNA was also observed in larvae of the goldenrod gall fly, *Eurosta solidaginis*, overwintering in Canada (45°N; 75°W) (Levin *et al.*, 2003). Larvae of *E. solidaginis* spend winter in a frozen state, with metabolism deeply depressed for several months. In general, evolution of similar responses is less probable in less predictable/severe habitats and in insects that overwinter in a supercooled state. Indeed, no reduction of mitochondrial DNA content was found in larvae of the moth, *Epiblema scudderiana*, which overwinter in a supercooled state in the same microhabitat as the freeze-tolerant larvae of *E. solidaginis* (McMullen and Storey, 2008).

In addition to changes at the level of organelles, insect dormancy and cold-acclimation likely involves capacity and resistance adaptations at the level of cell structural elements, such as large multienzyme and nucleoprotein complexes participating in chromatin structure and function, gene transcription, mRNA splicing, ribosomal translation, protein synthesis, folding, trafficking and proteasome degradation (Storey and Storey, 2007). Although acclimatory changes in these processes are currently almost totally unexplored, recent transcriptomic and proteomic studies identified some elements that are differentially regulated. For example, several genes coding for ribosomal subunits are up-regulated in diapausing mosquitoes, *C. pipiens* (Robich *et al.*, 2007); a proteasome β-subunit is more abundant in the parasitic wasp, *Aphidius colemani* exposed to fluctuating thermal regimes than in wasps exposed to constant temperature (Colinet *et al.*, 2007); the transcripts of several genes coding for proteins involved in the regulation of gene transcription were differentially expressed after RCH treatment of the fly, *D. melanogaster* (Qin *et al.*, 2005).

5.4 Future directions

So far, our knowledge has been extended mostly in a "horizontal" direction. Additional insect species have been studied using similar approaches. Most often, the composition of fatty acids in the polar lipid fraction was correlated with the level of cold-tolerance. The diversity of responses observed in various insects, however, does not allow any easy generalizations, thus suggesting the utility of additional horizontal studies. Important contributions may be obtained when resourceful experimental design (separating the effects of various stressors) is combined with detailed chemical analysis (distinguishing lipid classes, GPL analysis, extracting lipids from tissues, characterizing sub-cellular fractions).

Research in a "vertical" direction is in its infancy in insects. Direct proof of causality and studies on mechanisms/regulations are needed along the axis: environment → cellular structure → function → stress-tolerance. It is not easy to find a suitable model organism that both shows a robust acclimatory response as well as allows usage of a whole array of sophisticated methods of analysis. For instance, the power of genetics may be fully exploited in *D. melanogaster*, but this species displays a relatively weak capacity for cold-acclimation/tolerance, and its small size makes application of some techniques difficult. Such disadvantages may be overcome by employing other species of *Drosophila* and other flies, such as *C. costata* and *S. crassipalpis*, for research.

Studies of the "-omic" type (see Chapter 4) help to detect pathways and structures that might participate in the insect's response to cold. They provide a starting point, which must be verified and extended by detailed studies. One outcome from these studies is relatively clear, however: the number and variety of genes (proteins, structures, processes) involved in thermal acclimation and adaptation of insects is formidable and illustrate the magnitude of the challenge involved in understanding the physiology of a complex adaptive trait such as cold-acclimation.

References

Aitchison, C. W. (1979a). Winter-active subnivean invertebrates in Southern Canada. II. Coleoptera. *Pedobiologia* **19**, 1121–128.

Aitchison, C. W. (1979b). Winter-active subnivean invertebrates in Southern Canada. IV. Diptera and Hymenoptera. *Pedobiologia* **19**, 176–182.

Allakhverdiev, S. I., Nishiyama, Y., Suzuki, I., Tasaka, Y., and Murata, N. (1999). Genetic engineering of the unsaturation of fatty acids in membrane lipids alters the tolerance of *Synechocystis* to salt stress. *Proceedings of National Academy of Sciences USA* **96**, 5862–5867.

Amos, L. A. and Amos, W. G. (1991). *Molecules of Cytoskeleton*. New York: Guilford Press.

Bahrndorf, S., Petersen, S. O., Loeschke, V., Overgaard, J., and Holmstrup, M. (2007). Differences in cold and drought tolerance of high arctic and sub-arctic populations of *Megaphorura arctica* Tullberg 1876 (Onychiuridae: Collembola). *Cryobiology* **55**, 315–323.

Bashan, M., Akbas, H., and Yurdakoc, K. (2002). Phospholipid and triacylglycerol fatty acid composition of major life stages of sunn pest, *Eurygaster integriceps* (Heteroptera: Scutelleridae). *Comparative Biochemistry and Physiology B* **132**, 375–380.

Bashan, M. and Cakmak, O. (2005). Changes in composition of phospholipid and triacylglycerol fatty acids prepared from prediapausing and diapausing individuals of *Dolycoris baccarum* and *Piezodorus lituratus* (Heteroptera: Pentatomidae). *Annals of Entomological Society of America* **98**, 575–579.

Bayley, M., Petersen, S. O., Knigge, T., Kohler, H.-R., and Holmstrup, M. (2001). Drought acclimation confers cold tolearnce in the soil collembolan *Folsomia candida*. *Journal of Insect Physiology* **47**, 1197–1204.

Behan-Martin, M. K., Jones, G. R., Bowler, K., and Cossins, A. R. (1993). A near perfect temperature adaptation of bilayer order in vertebrate brain membranes. *Biochimica & Biophysica Acta* **1151**, 216–222.

Bennett, V. A., Pruitt, N. L., and Lee Jr., R. E. (1997). Seasonal changes in fatty acid composition associated with cold-hardening in third instar larvae of *Eurosta solidaginis*. *Journal of Comparative Physiology B* **167**, 249–255.

Block, W. (1996). Cold or drought – the lesser of two evils for terrestrial arthropods? *European Journal of Entomology* **93**, 325–339.

Blomquist, G. J., Borgeson, C. E., and Vundla, M. (1991). Polyunsaturated fatty acids and eicosanoids in insects. *Insect Biochemistry* **21**, 99–106.

Brooks, S., Clark, G. T., Wright, S. M., Trueman, R. J., Postle, A. D., Cossins, A. R., and Maclean, N. M. (2002). Electrospray ionisation mass spectrometric analysis of lipid restructuring in the carp (*Cyprinus carpio* L.) during cold acclimation. *Journal of Experimental Biology* **205**, 3989–3997.

Browse, J., Miquel, M., McConn, M., and Wu, J. (1994). Arabidopsis mutants and genetic approaches to the control of lipid composition. In *Temperature Adaptations of Biological Membranes*, ed. A. R. Cossins. London and Chapel Hill: Portland Press, pp. 141–154.

Canavoso, L. E., Jouni, Z. E., Karnas, K. J., Pennington, J. E., and Wells, M. A. (2001). Fat metabolism in insects. *Annual Review of Nutrition* **21**, 23–46.

Chamberlain, P. M. and Black, H. I. J. (2005). Fatty acid compositions of Collembola: unusually high proportions of C-20 polyunsaturated fatty acids in a terrestrial invertebrate. *Comparative Biochemistry and Physiology B* **140**, 299–307.

Chapman, D. (1975). Phase transitions and fluidity characteristics of lipids and cell membranes. *Quarterly Reviews of Biophysics* **8**, 185–235.

Clegg, J. S. (2001). Cryptobiosis – a peculiar state of biological organization. *Comparative Biochemistry and Physiology B* **128**, 613–624.

Colinet, H., Nguyen, T. T. A., Cloutier, C., Michaud, D., and Hance, T. (2007). Proteomic profiling of a parasitic wasp exposed to constant and fluctuating cold exposure. *Insect Biochemistry and Molecular Biology* **37**, 1177–1188.

Cook, H. W. and McMaster C. R. (2002). Fatty acid desaturation and chain elongation in eukaryotes. In *Biochemistry of Lipids, Lipoproteins and Membranes*, ed. D. E. Vance and J. E. Vance. Amsterdam: Elsevier, pp. 181–204.

Cossins, A. R. (1977). Adaptations of biological membranes to temperature – the effect of temperature acclimation of goldfish upon the viscosity of synaptosomal membranes. *Biochimica & Biophysica Acta* **470**, 395–411.

Cossins, A. R. ed. (1994). *Temperature Adaptation of Biological Membranes*. London and Chapel Hill: Portland Press.

Cossins, A. R. and Bowler, K. (1987). *The Temperature Biology of Animals*. London: Chapman and Hall.

Cossins, A. R. and Macdonald, A. G. (1989). The adaptation of biological membranes to temperature and pressure: Fish from the deep and cold. *Journal of Bioenergetics and Biomembranes* **21**, 115–135.

Cossins, A. R., Murray P. A. Gracey, A. Y., Logue, J., Polley, S., Caddick, M., Brooks, S., Postle, T., and Maclean, N. (2002). The role of desaturases in cold-induced lipid restructuring. *Biochemical Society Transactions* **30**, 1082–1086.

Cossins, A. R. and Prosser, C. L. (1978). Evolutionary adaptation of membranes to temperature. *Proceedings of National Academy of Sciences USA* **75**, 2040–2043.

Cossins, A. R. and Sinensky, M. (1984). Adaptations of membranes to temperature, pressure and exogenous lipids. In *Physiology of Membrane Fluidity*, 2nd edn, ed. M. Shinitzky. Boca Raton: CRC Press, pp. 1–20.

Coyne, J. A. and Elwyn, S. (2006). Does the desaturase-2 locus in *Drosophila melanogaster* cause adaptation and sexual isolation? *Evolution* **60**, 279–291.

Crockett, E. L. and Hazel, J. R. (1995). Cholesterol levels explain inverse compensation of membrane order in brush border but not homeoviscous adaptation in basolateral membranes from the intestinal epithelia of rainbow trout. *Journal of Experimental Biology* **198**, 1105–1113.

Dallerac, R., Labeur, C., Jallon, J.-M., Knipple, D. C., Roelofs, W. L., and Wicker-Thomas, C. (2000). A Δ9 desaturase gene with a different substrate specificity is responsible for the cuticular diene hydrocarbon polymorphism in *Drosophila melanogaster*. *Proceedings of National Academy of Sciences USA* **97**, 9449–9454.

Dowhan, W. (1997). Molecular basis for membrane phospholipid diversity: Why are there so many lipids? *Annual Reviews of Biochemistry* **66**, 199–232.

Downer, R. G. H. and Kallapur, V. L. (1981). Temperature-induced changes in lipid composition and transition temperature of flight muscle mitochondria of *Schistocerca gregaria*. *Journal of Thermal Biology* **6**, 189–194.

Egiersdorff, S. and Kacperska, A. (2001). Low temperature effects on growth and actin cytoskeleton organization in suspension cells of winter oilseed rape. *Plant Cell and Tissue Organ Cultures* **40**, 17–25.

Eigenheer, A. L., Young, S., Blomquist, G. J., Borgeson, C. E., Tillman, J. A., and Tittiger, C. (2002). Isolation and molecular characterization of *Musca domestica* delta-9 desaturase sequences. *Insect Molecular Biology* **11**, 533–542.

Garlick, K. M. and Robertson, R. M. (2007). Cytoskeletal stability and heat-shock mediated thermoprotection of central pattern generation in *Locusta migratoria*. *Comparative Biochemistry and Physiology A* **147**, 344–348.

Gonzales, M. S. and Brenner, R. R. (1999). Fatty acid Δ9-desaturation in the *Triatoma infestans* fat body: Response to food and trehalose administration. *Lipids* **34**, 1199–1205.

Greenberg, A. J., Moran, J. R., Coyne, J. A., and Wu, C.-I. (2003). Ecological adaptation during incipient speciation revealed by precise gene replacement. *Science* **302**, 1754–1757.

Haines, T. H. (2001). Do sterols reduce proton and sodium leaks through lipid bilayers? *Progress in Lipid Research* **40**, 299–324.

Hanson, B. J., Cummins, K. W., Cargill, A. S., and Lowry, R. R. (1985). Lipid content, fatty acid composition, and the effect of diet on fats of aquatic insects. *Comparative Biochemistry and Physiology B*, **80**, 257–276.

Harwood, J. L., Jones, A. L., Perry, H. J., Rutter, A. J., Smith, K. L., and Williams, M. (1994). Changes in plant lipids during temperature adaptation. In *Temperature Adaptation of Biological Membranes*. London and Chapel Hill: Portland Press, pp. 107–118.

Hazel, J. R. (1989). Cold adaptation in ectotherms: Regulation of membrane function and cellular metabolism. *Advances in Comparative and Environmental Physiology* **4**, 1–50.

Hazel, J. R. (1995). Thermal adaptation in biological membranes: Is homeoviscous adaptation the explanation? *Annual Reviews of Physiology* **57**, 19–42.

Hazel, J. R. and Williams, E. E. (1990). The role of alterations in membrane lipid composition in enabling physiological adaptation of organisms to their physical environment. *Progress in Lipid Research* **29**, 167–227.

Hayward, S. A. L., Murray, P. A., Gracey, A. Y., and Cossins, A. R. (2007). Beyond the lipid hypothesis: Mechanisms underlying phenotypic plasticity in inducible cold tolerance. In *Molecular Aspects of the Stress Response: Chaperons, Membranes and Networks*, ed. P. Csermely and L. Vigh. Austin: Landes Bioscience, pp. 132–142.

Henriques, V. and Hansen, C. (1901). Vergleichende Untersuchungen über die chemische Zusammenstzung des tierishen Fettes. *Skandinawischen Archive für Physiologie* **11**, 151–165.

Hodková, M., Berková, P., and Zahradníčková, H. (2002). Photoperiodic regulation of the phospholipid molecular species composition in thoracic muscles and fat body of *Pyrrhocoris apterus* (Heteroptera) via an endocrine gland, corpus allatum. *Journal of Insect Physiology* **48**, 1009–1019.

Hodková, M., Šimek, P., Zahradníčková, H., and Nováková, O. (1999). Seasonal changes in the phospholipid composition in thoracic muscles of a heteropteran, *Pyrrhocoris apterus*. *Insect Biochemistry and Molecular Biology* **29**, 367–376.

Holmstrup, M., Hedlund, K., and Boriss, H. (2002). Drought acclimation and lipid composition in *Folsomia candida*: implications for cold shock, heat shock and acute desiccation stress. *Journal of Insect Physiology* **48**, 961–970.

Huang, C.-H., Lin, H., Li, S., and Wang, G. (1997). Influence of the positions of *cis* double bonds in the *sn-2* acyl chain of phosphatidylethanolamine on the bilayer's melting behavior. *Journal of Biological Chemistry* **272**, 21917–21926.

Jurenka, R. A., de Renobales, M., and Blomquist, G. J. (1987). De novo synthesis of polyunsaturated fatty acids in the cockroach *Periplaneta americana*. *Archives of Biochemistry and Biophysics* **255**, 184–193.

Kayukawa, T., Chen, B., Hoshizaki, S., and Ishikawa, Y. (2007). Upregulation of a desaturase is associated with the enhancement of cold hardiness in the onion maggot, *Delia antiqua*. *Insect Biochemistry and Molecular Biology* **37**, 1160–1167.

Kayukawa, T., Chen, B., Miyazaki, S., Itoyama, K., Shinoda, T., and Ishikawa, Y. (2005). Expression of mRNA for the t-complex polypeptide-1, a subunit of chaperonin

CCT, is upregulated in association with increased cold hardiness in *Delia antiqua*. *Cell Stress & Chaperones* **10**, 204–210.

Kim, M., Robich, R. M., Rinehart, J. P., and Denlinger, D. L. (2006). Upregulation of two actin genes and redistribution of actin during diapause and cold stress in the northern house mosquito, *Culex pipiens*. *Journal of Insect Physiology* **52**, 1226–1233.

Kirk, G. L., Gruner, S. M., and Stein, D. L. (1984). A thermodynamic model of the lamellar to inverse hexagonal phase transition of lipid membrane-water systems. *Biochemistry* **23**, 1093–1102.

Knipple, D. C., Rosenfield, C.-L., Nielsen, R., You, K. M., and Jeong, S. E. (2002). Evolution of the integral membrane desaturase gene family in moths and flies. *Genetics* **162**, 1737–1752.

Košťál, V. (2006). Eco-physiological phases of insect diapause. *Journal of Insect Physiology* **52**, 113–127.

Košťál, V., Berková, P., and Šimek, P. (2003). Remodelling of membrane phospholipids during transition to diapause and cold-acclimation in the larvae of *Chymomyza costata* (Drosophilidae). *Comparative Biochemistry and Physiology B* **135**, 407–419.

Košťál, V. and Šimek, P. (1998). Changes in fatty acid composition of phospholipids and triacylglycerols after cold-acclimation of an aestivating insect prepupa. *Journal of Comparative Physiology B* **168**, 453–460.

Košťál, V., Vambera, J., and Bastl, J. (2004). On the nature of pre-freeze mortality in insects: water balance, ion homeostasis and energy charge in the adults of *Pyrrhocoris apterus*. *Journal of Experimental Biology* **207**, 1509–1521.

Kristiansen, E. and Zachariassen, K. E. (2001). Effect of freezing on the transmembrane distribution of ions in freeze-tolerant larvae of the wood fly *Xylophagus cinctus* (Diptera, Xylophagidae). *Journal of Insect Physiology* **47**, 585–592.

Kukal, O., Duman, J. G., and Serianni, A. S. (1989). Cold-induced mitochondrial degradation and cryoprotectant synthesis in freeze-tolerant arctic caterpillars. *Journal of Comparative Physiology B* **158**, 661–671.

Kukal, O. and Kevan, P. G. (1987). The influence of parasitism on the life history of a high arctic insect, *Gynaephora groenlandica* (Wocke) (Lepidoptera: Lymantridae). *Canadian Journal of Zoology* **65**, 156–163.

Lee, K.-Y., Hiremath, S., and Denlinger, D. L. (1998). Expression of actin in the central nervous system is switched off during diapause in the gypsy moth, *Lymantria dispar*. *Journal of Insect Physiology* **44**, 221–226.

Lee, R. E. (1991). Principles of insect low temperature tolerance. In *Insects at Low Temperature*, ed. R. E. Lee and D. L. Denlinger. New York and London: Chapman and Hall, pp. 17–46.

Lee, R. E., Chen, C. P., and Denlinger, D. L. (1987). A rapid cold-hardening process in insects. *Science* **238**, 1415–1417.

Lee, R. E., Damoradan, K., Yi, S.-X., and Lorigan, G. A. (2006). Rapid cold-hardening increases membrane fluidity and cold tolerance of insect cells. *Cryobiology* **52**, 459–463.

Levin, D. B., Danks, H. V., and Barber, S. A. (2003). Variations in mitochondrial DNA and gene transcription in freeze-tolerant larvae of *Eurosta solidaginis* (Diptera:

Tephritidae) and *Gynaephora groenlandica* (Lepidoptera: Lymantriidae). *Insect Molecular Biology* **12**, 281–289.

Lewis, R. N. A. H., Mannock, D. A., McElhaney, R. N., Turner, D. C., and Gruner, S. M. (1989). Effect of fatty-acyl chain length and structure on the lamellar gel to liquid-crystalline and lamellar to reverse hexagonal phase transitions of aqueous phosphatidylethanolamine dispersions. *Biochemistry* **28**, 541–548.

Li, A. Q., Popova-Butler, A., Dean, D. H., and Denlinger, D. L. (2007). Proteomics of the flesh fly brain reveals an abundance of upregulated heat shock proteins during pupal diapause. *Journal of Insect Physiology* **53**, 385–391.

Li, S., Wang, G., Lin, H., and Huang, C.-H. (1998). Calorimetric studies of phosphatidylethanolamines with saturated *sn-1* and dienoic *sn-2* acyl chains. *Journal of Biological Chemistry* **273**, 19009–19018.

Liang, P. and MacRae. T. H. (1997). Molecular chaperones and cytoskeleton. *Journal of Cell Science* **110**, 1431–1440.

Liu, W., Ma, P. W. K., Marsella-Herrick, P., Rosenfield, C. L., Knipple, D. C., and Roelofs, W. (1999). Cloning and functional expression of a cDNA encoding a metabolic acyl-CoA delta 9-destaurase of the cabbage looper moth, *Trichoplusia ni*. *Insect Biochemistry and Molecular Biology* **29**, 435–443.

Macartney, A., Maresca, B., and Cossins, A. R. (1994). Acyl-CoA desaturases and the adaptive regulation of membrane lipid composition. In *The Temperature Biology of Animals*. London: Chapman and Hall, pp. 129–139.

McElhaney, R. N. (1984). The relationship between membrane lipid fluidity and phase state and the ability of bacteria and mycoplasms to grow and survive at various temperatures. *Biomembranes* **12**, 249–276.

McMullen, D. C. and Storey, K. B. (2008). Mitochondria of cold hardy insects: responses to cold and hypoxia assessed at enzymatic, mRNA and DNA levels. *Insect Biochemistry and Molecular Biology* **38**, 367–373.

Michaud, M. R. and Denlinger, D. L. (2006). Oleic acid is elevated in cell membranes during rapid cold-hardening and pupal diapause in the flesh fly, *Sarcophaga crassipalpis*. *Journal of Insect Physiology* **52**, 1073–1082.

Mounier, N. and Arrigo, A. P. (2002). Actin cytoskeleton and small heat shock proteins: how do they interact? *Cell Stress & Chaperones* **7**, 167–176.

Murata, N. and Yamaya, J. (1984). Temperature-dependent phase behavior of phosphatidylglycerols from chilling sensitive and chilling-resistant plants. *Plant Physiology* **74**, 1016–1024.

Murray, P., Hayward, S. A. L., Govan, G. G., Gracey, A. Y., and Cossins, A. R. (2007). An explicit test of the phospholipid saturation hypothesis of acquired cold tolerance in *Caenorhabditis elegans*. *Proceedings of National Academy of Sciences USA* **104**, 5489–5494.

Nozawa, Y., Iida, H., Fukushima, H., Ohki, K., and Ohnishi, S. (1974). Studies on *Tetrahymena* membranes: temperature-induced alterations in fatty acid composition of various membrane fractions in *Tetrahymena pyriformis* and its effect on membrane fluidity as inferred by spin-label study. *Biochemica & Biophysica Acta* **367**, 134–147.

Ohtsu, T., Kimura, M. T., and Katagiri, C. (1998). How *Drosophila* species acquire cold tolerance. Qualitative changes of phospholipids. *European Journal of Biochemistry* **252**, 608–611.

Ohtsu, T., Katagiri, C., and Kimura, M. T. (1999). Biochemical aspects of climatic adaptations in *Drosophila curviceps*, *D. immigrans* and *D. albomicans* (Diptera: Drosophilidae). *Environmental Entomology* **28**, 968–972.

Overgaard, J., Sørensen, J. G., Petersen, S. O., Loeschke, V., and Holmstrup M. (2005). Changes in membrane lipid composition following rapid cold hardening in *Drosophila melanogaster*. *Journal of Insect Physiology* **51**, 1173–1182.

Overgaard, J., Sørensen, J. G., Petersen, S. O., Loeschke, V., and Holmstrup M. (2006). Reorganization of membrane lipids during fast and slow cold hardening in *Drosophila melanogaster*. *Physiological Entomology* **31**, 328–335.

Overgaard, J., Tomčala, A., Sørensen, J. G., Holmstrup, M., Krogh, P. H., Šimek, P., and Koštál, V. (2008). Effects of acclimation temperature on thermal tolerance and membrane phosholipid composition in the fruit fly *Drosophila melanogaster*. *Journal of Insect Physiology* **54**, 619–629.

Pruitt, N. L. and Lu, C. (2008). Seasonal changes in phospholipid class and class-specific fatty acid composition associated with the onset of freeze tolerance in third-instar larvae of *Eurosta solidaginis*. *Physiological and Biochemical Zoology* **81**, 226–234.

Pucciarelli, S, Ballarini, P., and Miceli, C. (1997). Cold-adapted microtubules: characterization of tubulin posttranslational modifications in the Antarctic ciliate *Euplotes focardii*. *Cell Motility and Cytoskeleton* **38**, 329–340.

Qin, W., Neal, S. J., Robertson, R. M., Westwood, J. T., and Walker, V. K. (2005). Cold hardening and transcriptional change in *Drosophila melanogaster*. *Insect Molecular Biology* **14**, 607–613.

Ramesha, C. S. and Thompson, G. A. (1983). Cold stress induces in situ phospholipid molecular species changes in cell surface membranes. *Biochimica & Biophysica Acta* **731**, 251–260.

Riddervold, M. H., Tittiger, C., Blomquist, G. J., and Borgeson, C. E. (2002). Biochemical and molecular characterization of house cricket (*Acheta domesticus*, Orthoptera: Gryllidae) delta 9 desaturase. *Insect Biochemistry and Molecular Biology* **32**, 1731–1740.

Rinehart, J. P., Li, A. Q., Yocum, G. D., Robich, R. M., Hayward, S. A. L., and Denlinger, D. L. (2007). Up-regulation of heat shock proteins is essential for cold survival during insect diapause. *Proceedings of National Academy of Sciences USA* **104**, 11130–11137.

Robich, R. M., Rinehart, J. P., Kitchen, L. J., and Denlinger, D. L. (2007). Diapause-specific gene expression in the northern house mosquito, *Culex pipiens* L., identified by suppressive subtractive hybridization. *Journal of Insect Physiology* **53**, 235–245.

Schunke, M. and Wodtke, E. (1983). Cold-induced increase of delta-nine and delta-six desaturase activities in endoplasmic membranes of carp liver. *Biochimica and Biophysica Acta* **734**, 70–75.

Shreve, S. M., Yi, S.-X., and Lee, R. E. (2007). Increased dietary cholesterol enhances cold tolerance in *Drosophila melanogaster*. *Cryo-Letters* **28**, 33–37.

Sinensky, M. (1974). Homeoviscous adaptation – a homeostatic process that regulates viscosity of membrane lipids in *Escherichia coli*. *Proceedings of National Academy of Sciences USA* **71**, 522–525.

Singer, M. (1981). Permeability of phosphatidylcholine bilayers. *Chemistry and Physics of Lipids* **28**, 253–267.

Šlachta, M., Berková, P., Vambera, J., and Košťál, V. (2002). Physiology of cold-acclimation in non-diapausing adults of *Pyrrhocoris apterus* (Heteroptera). *European Journal of Entomology* **99**, 181–187.

Sørensen, P. G. (1993). Changes of the composition of phospholipids, fatty acids and cholesterol from the erythrocyte plasma membrane from flounders (*Platichthys flesus* L.) which were acclimated to high and low temperatures in aquaria. *Comparative Biochemistry and Physiology B* **106**, 907–912.

Stanley-Samuelson, D. W. and Dadd, R. H. (1983). Long-chain polyunsaturated fatty acids: patterns of occurrence in insects. *Insect Biochemistry* **13**, 549–558.

Stanley-Samuelson, D. W., Jurenka, R. A., Cripps, C., Blomquist, G. J., and de Renobales, M. (1988). Fatty acids in insects: Composition, metabolism and biological significance. *Archives of Insect Biochemistry and Physiology* **9**, 1–33.

Storey, K. B. and Storey, J. M. (2007). Tribute to P. L. Lutz: putting life on "pause" – molecular regulation of hypometabolism. *Journal of Experimental Biology* **210**, 1700–1714.

Suutari, M., Rintamaki, A., and Laakso, S. (1997). Membrane phospholipids in temperature adaptation of *Candida utilis*: alterations in fatty acid chain length and unsaturation. *Journal of Lipid Research* **38**, 790–794.

Tasaka, Y., Gombos, Z., Nishiyama, Y., Mohanty, P., Ohba, T., Ohki, K., and Murata, N. (1996). Targeted mutagenesis of acyl-lipid desaturases in *Synechocystis*: evidence for the important roles of polyunsaturated membrane lipids in growth, respiration and photosynthesis. *EMBO Journal* **15**, 6416–6425.

Thompson, S. N. (1973). A review and comparative characterization of the fatty acid composition of seven insect orders. *Comparative Biochemistry and Physiology* **45**, 467–482.

Tiku, P. E., Gracey, A. Y., Macartney, A. I., Beynon, R. J., and Cossins, A. R. (1996). Cold-induced expression of Δ9-desaturase in carp by transcriptional and posttraslational mechanisms. *Science* **271**, 815–818.

Tomčala, A., Tollarová, M., Overgaard, J., Šimek, P., and Košťál, V. (2006). Seasonal acquisition of chill-tolerance and restructuring of membrane glycerophospholipids in an overwintering insect: triggering by low temperature, desiccation and diapause progression. *Journal of Experimental Biology* **209**, 4102–4114.

Wang, G., Li, S., Lin, H., Brumbaugh, E. E., and Huang, C.-H. (1999). Effects of various numbers and positions of *cis* double bonds in the *sn-2* acyl chain of phosphatidylethanolamine on the chain-melting temperature. *Journal of Biological Chemistry* **274**, 12289–12299.

Wicker-Thomas, C., Henriet, C., and Dallerac, R. (1997). Partial characterization of a fatty acid desaturase gene in *Drosophila melanogaster*. *Insect Biochemistry and Molecular Biology* **27**, 963–972.

Wodtke, E. and Cossins, A. R. (1991). Rapid cold-induced changes of membrane order and Δ9-desaturase actitivy in endoplasmic reticulum of carp liver: A time-course study of thermal acclimation. *Biochimica and Biophysica Acta* **1064**, 343–350.

Yi, S.-X. and Lee, R. E. (2005). Changes in gut and Malpighian tubule transport during seasonal acclimatization and freezing in the gall fly *Eurosta solidaginis*. *Journal of Experimental Biology* **208**, 1895–1904.

Yocum, G. D., Kemp, W. P., Bosch, J., and Knoblett, J. M. (2005). Temporal variation in overwintering gene expression and respiration in the solitary bee *Megachile rotundata*. *Journal of Insect Physiology* **51**, 621–629.

Zachariassen, K. E., Kristiansen, E., and Pedersen, S. A. (2004). Inorganic ions in cold-hardiness. *Cryobiology* **48**, 126–133.

6

Oxygen: Stress and adaptation in cold-hardy insects

KENNETH B. STOREY AND JANET M. STOREY

Oxygen is critical to insect life. Oxygen-based respiration offers a tremendous advantage for cellular energetics by supporting high-efficiency production of ATP from the complete oxidation of substrates to CO_2 and H_2O and is an absolute requirement for powering insect flight muscles. However, once tied to aerobic respiration, organisms have to deal with various consequences of an oxygen-based life, including oxygen limitation (hypoxia, anoxia), oxygen overabundance (hyperoxia) and oxygen in its many reactive forms (e.g. superoxide, hydroperoxides, hydroxyl radical, etc.). The last cause oxidative stress, one of the most pervasive forms of cellular stress, since reactive oxygen species (ROS) can attack and damage most types of cellular macromolecules. This chapter looks at insect cold-hardiness from the point of view of the oxygen-related issues that are important for winter survival. We examine hypoxia/anoxia, freeze-induced ischemia, mitochondria, metabolic-rate depression, oxidative stress and antioxidants as they relate to the low-temperature biology of insects.

6.1 Oxygen challenges in winter

Although there are some unique insect solutions for dealing with winter (e.g. migration of monarch butterflies to Mexico or shivering thermogenesis to provide central heating for honey bee colonies), terrestrial insects generally employ three basic options (sometimes in combination) to survive the cold winter months – underground hibernation below the frostline (in burrows or by digging), freeze-avoidance by supercooling, or freeze-tolerance. Oxygen challenges may be

Low Temperature Biology of Insects, ed. David L. Denlinger and Richard E. Lee Jr. Published by Cambridge University Press.

associated with these. Life underground typically results in reduced oxygen availability (hypoxia) and/or elevated carbon dioxide (hypercapnia), so that respiratory adjustments are needed (e.g. Woodman *et al.*, 2007); fossorial species also typically show enhanced hypoxia-tolerance compared with those that live above ground. One could envision that freezing of the earth above the frostline would also severely limit the possibilities for refreshing the air supply for creatures living below the frostline so it is likely that good hypoxia tolerance is needed for long-term survival. Various species of alpine insects that are encased in ice and snow over the winter also need good hypoxia/anoxia-tolerance and, indeed, several species have been shown to survive many weeks of full anoxia at 0 °C (summarized by Hoback and Stanley, 2001). Insects using the freeze-avoidance strategy of cold-hardiness should have no physical limitation to maintaining normal aerobic metabolism down to their supercooling points and this is probably one of the advantages of this strategy. For example, studies with larvae of the goldenrod gall moth, *Epiblema scudderiana*, show no metabolic indicators of hypoxia stress – i.e. no impairment of tissue ATP levels and no significant accumulation of glycolytic end products (lactate, alanine) over the winter months in nature or during laboratory exposures at −16 °C (Rickards *et al.*, 1987; Churchill and Storey, 1989). However, freeze-tolerant species clearly experience impaired aerobic metabolism when body water freezes. Studies at a high subzero temperature (−5 °C) where it was possible to compare frozen versus unfrozen individuals showed a 69% decrease in O_2 consumption by frozen *Upis ceramboides* (Lundheim and Zachariassen, 1993) and a 47% decrease in CO_2 production by frozen larvae of the goldenrod gall fly, *Eurosta solidaginis* (Irwin and Lee, 2002), compared with unfrozen insects. As temperature falls and more internal ice accumulates, impairment of oxygen-based metabolism would undoubtedly increase; for example, at high ice contents the muscle movements that help to open/shut spiracles and ventilate tracheoles would certainly be halted. Indeed, metabolic indicators confirmed this for frozen *E. solidaginis* at −16 °C; ATP and energy charge decreased over time frozen, and lactate and alanine accumulate (Storey and Storey 1985; Churchill and Storey, 1989).

Aquatic insects also have oxygen issues to deal with in winter (Danks, 2007). Some species probably have quite stable winter situations on the bottom of rivers and lakes where water temperature is near constant at ~4 °C and the high dissolved oxygen content of cold water can easily supply the low metabolic rates of the fauna present. However, ice cover often leads to oxygen depletion because organismal respiration continues, but aeration of the water is halted. Phytoplankton can still produce oxygen via photosynthesis in cold water, but the addition of only a few centimeters of snow cuts off the light needed (Danks, 2007). Hence, some aquatic environments, particularly small ponds, can become hypoxic or even anoxic in winter. Furthermore, if bodies of water freeze to the bottom and/or into the sediment, then strategies for subzero survival are needed just as on land. For example, Danks (1971, 2007) showed that many species of Arctic chironomids are freeze-tolerant. Because they typically live in benthic sediments, chironomid larvae have developed high anoxia tolerance and rather unique (at least among insects) pathways of anaerobic metabolism, with end products including lactate, alanine, succinate and large amounts of ethanol, the latter being excreted (Redecker and Zebe, 1988). Indeed, the pre-existing anoxia-tolerance of this insect group is undoubtedly a major factor in the freezing survival of northern chironomids, as is also the case for other freeze-tolerant species, both vertebrate and invertebrate (Storey and Storey, 1996). Overall, however, relatively little is known about the biochemical responses of aquatic insects in winter so the remainder of this chapter will focus on information about terrestrial model species.

6.2 Hypoxia, ischemia and insect freeze-tolerance

Freeze-tolerant species need well-developed tolerances for hypoxia/anoxia and ischemia for long-term freezing survival (Storey and Storey, 1996). The innate hypoxia- or anoxia-tolerance of different insect species varies widely from hours to weeks with freeze-tolerant alpine species at the high end (Hoback and Stanley, 2001). *E. solidaginis* readily endured 18 days of full anoxia at 13 °C (Storey and Storey, 1990a) and undoubtedly would show even greater tolerance at lower temperatures since larvae can survive constant freezing at −16 °C for at least 12 weeks. There has been no consistent analysis of insect species from the point of view of which features of anoxia-tolerance best support freeze-tolerance, but some factors could be considered for extending viability.

The first is the possibility of sustaining some amount of O_2-based metabolism while frozen. In the past, respiratory proteins were considered unnecessary in most insects because oxygen was delivered directly by the tracheal system. Only a few exceptions were known, such as chironomid midges that have high levels of intracellular and extracellular hemoglobins to support larval life in hypoxic benthic environments. However, intracellular hemoglobins with high oxygen affinity are now known to occur widely in the Holometabola, whereas hemocyanin is found in most Hemimetabola taxa (Burmester and Hankelm, 2007). Nothing is known about these oxygen-binding proteins in relation to cold-hardiness, but it is tempting to speculate that a seasonal adjustment that increases their content in freeze-tolerant species could provide a way to buffer oxygen levels during bouts of freezing.

A second factor is phosphagen buffering of ATP levels. High arginine phosphate levels could potentially provide an energy reserve for cells of frozen insects. There is little data available about this, but a comparison of the goldenrod gall formers does not support this idea; arginine phosphate concentrations were generally higher, 5–10 mM, in the freeze-avoiding *E. scudderiana* than in freeze-tolerant *E. solidaginis* (5–7 mM) (Churchill and Storey, 1989).

Viability while frozen might also be extended by the use of pathways of anaerobic metabolism that provide additional ATP to that available from glycolysis ending in lactate. However, insects and other arthropods rarely use this option, although it is widespread in other invertebrate groups (Hochachka and Somero, 1984). One exception to this is chironomids that have optimized an anaerobic metabolism that is based on ethanol production and excretion. Lactate accumulates during freezing in many species and several studies report the near-equal accumulation of lactate and alanine as glycolytic end products (Storey and Storey, 1990a; Hoback and Stanley, 2001). An interesting link between anoxia exposure and polyol synthesis was revealed in *E. solidaginis.* Larvae placed in an N_2 atmosphere at 13 °C, a temperature that normally stimulates glycerol production, produced much less glycerol compared with aerobic controls, but substituted sorbitol production instead; sorbitol is normally produced only below 5 °C in this species (Storey and Storey, 1990a). The altered pattern under anoxia seems to occur because glycerol biosynthesis requires some ATP consumption, whereas sorbitol does not, so it appears that low-temperature-triggered cryoprotectant synthesis can continue under anoxia if a different pattern of polyols is accumulated. Indeed, the ATP-dependence of glycerol synthesis may be one reason that dual cryoprotectant systems are often seen in freeze-tolerant insects; sorbitol represents a polyol that could still be synthesized if early frosts triggered freezing before full cryoprotection was in place (Storey and Storey, 1991). Note, however, that the primary reason for sorbitol accumulation in freeze-tolerant insects is probably its very low permeability across biological membranes, which keeps it inside cells to help minimize cell water loss during extracellular freezing (Burg, 1995); by contrast, glycerol equilibrates between intra- and extracellular fluids.

Long-term freezing survival could also be facilitated by an enhanced capacity for anaerobic metabolism in the form of elevated activities of glycolytic enzymes. Hypoxia exposure is well known to trigger an increased expression of genes that code for glycolytic enzymes and other proteins that aid low-oxygen survival, a process that is regulated by the hypoxia-inducible transcription factor (HIF-1).

6.3 HIF-1 and freeze-tolerance

A highly conserved response to low oxygen stress across the animal kingdom is the activation of HIF-1 leading to the up-regulation of a variety of genes whose protein products correct or compensate for oxygen deficiency. In general, HIF-1 mediates two basic options: (a) enhancing the capacity for O_2-independent ATP production by up-regulation of glycolytic enzymes, glucose transporters, etc., and (b) increasing oxygen delivery to tissues by proliferating the delivery network (e.g. growth of capillaries or tracheoles) or the oxygen-carrying capacity (e.g. iron uptake, erythropoiesis, etc.) (Semenza, 2003; Gorr *et al.*, 2006). HIF-1 mediated gene expression is key to the growth, proliferation and development of tissues and organs in multicellular organisms and also underlies the ability of organisms to acclimate to variations in environmental oxygen availability. For example, below critical oxygen partial pressure values, HIF-1- and nitric-oxide-mediated pathways produce long-term compensatory changes in tracheal growth, suppressed development and acclimation of ventilation in insects (Harrison *et al.*, 2006). In cold-hardy insects, HIF-1 may be involved in the suppression of growth and development that is part of winter diapause or quiescence and/or the spring arousal from these states, but nothing is yet known about this. Recent work also indicates that HIF-1 has a role in dealing with freeze-induced ischemia (Morin *et al.*, 2005). A HIF-1 mediated increase in glycolytic capacity could enhance anaerobic ATP generation in frozen larvae. Indeed, activities of several glycolytic enzymes are increased in response to cold exposure in freeze-tolerant *E. solidaginis* larvae, including glycogen phosphorylase, hexokinase and phosphofructokinase (Storey and Storey, 1981). This increase in glycolytic capacity near 0 °C would also support cold-induced cryoprotectant synthesis (e.g. sorbitol production is triggered below 5 °C in this species).

Mammalian HIF-1 is a heterodimeric protein that is regulated primarily by the availability of its alpha subunit (the beta subunit is constitutive). The HIF-1α-subunit is stabilized at low oxygen, but rapidly targeted for destruction by oxygen-dependent proline hydroxylation when oxygen levels are high (Semenza, 2003). Only under low oxygen conditions does the alpha subunit persist long enough to allow the heterodimer to form and activate gene transcription. Most aspects of HIF-1 regulation and action are now known to be well-conserved across phylogeny (Gorr *et al.*, 2006). In insects, the HIF-1 dimer was first identified in *Drosophila melanogaster* as two basic helix-loop-helix-PAS proteins, Similar (Sima) and Tango (Tgo), now known as homologues of HIF-1α and HIF-1β, respectively (Lavista-Llanos, 2002).

Morin *et al.* (2005) analysed HIF-1α in *E. solidaginis*. Cloning and sequencing of gall fly larva *hif-1α* revealed strong similarity to *D. melanogaster*, honey bee and shrimp HIF-1α and, like the fruit fly protein (Nambu *et al.*, 1996), *E. solidaginis* HIF-1α proved to be ~1500 amino acids in length, about twice as big as vertebrate HIF-1α. Figure 6.1 shows *E. solidaginis* HIF-1α responses to anoxia and cold exposure, as well as a winter timecourse. Not unexpectedly, oxygen deprivation (under N_2 gas) triggered a strong 2.7-fold increase in *hif-1α* mRNA transcript levels. Acute cold exposure had the same effect; a drop from 15 to 3 °C raised transcript levels by 1.6-fold, a level that was sustained when larvae were subsequently frozen at −16 °C. HIF-1α protein levels also increased sharply by 2.3-fold when larvae were transferred from 15 to 3 °C, but were reduced somewhat at −16 °C. The prominent effect of cold exposure at 3 °C on both transcript and protein levels of HIF-1α

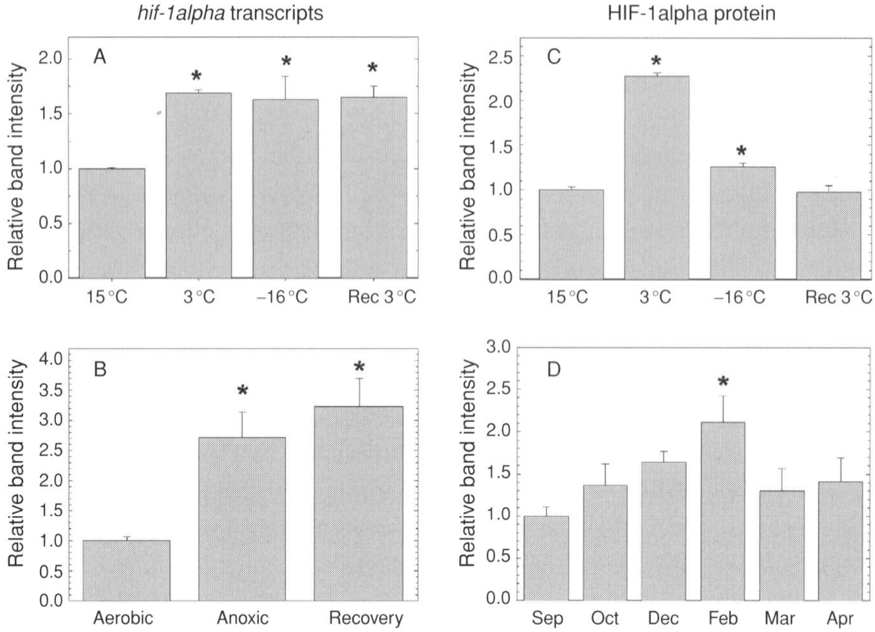

Figure 6.1 Modulation of the hypoxia-inducible transcription factor in *Eurosta solidaginis* (Fitch) (Diptera, Tephritidae). (A) Response to temperature change by *hif-1α* mRNA transcripts in 15 °C controls compared with larvae given serial exposures to 3 °C for 24 h, freezing at −16 °C for 24 h and thawing recovery (Rec) at 3 °C for 24 h. (B) Responses to anoxia by *hif-1α* transcript levels in 15 °C-acclimated larvae comparing aerobic controls, exposure to N_2 gas for 24 h and aerobic recovery for 24 h (Rec). (C) Effects of low temperature and freezing on HIF-1α protein levels, assessed by immunoblotting, in the same experimental groups as in panel A. (D) HIF-1α protein levels over the winter season. Data are means ± SEM, n = 3–5 independent trials. *-Significantly different from the control or September value, $p < 0.01$. (Modified from Morin *et al.*, 2005.)

was intriguing. This implies that cold exposure may lead to hypoxic conditions in the larvae (at least temporarily), triggering HIF-1 mediated metabolic adjustments (such as elevating glycolytic capacity) that could prepare the larvae for survival of subsequent long-term freezing exposures. This would ensure that the energy expenditures on new protein synthesis are paid while the larvae are still unfrozen and able to employ oxygen-dependent ATP production. In larvae held outdoors over the winter, HIF-1α protein levels peaked in February. These larvae would have experienced freeze/thaw episodes multiple times before this time and, therefore, HIF-1 signaling at this time is probably not enhancing ischemia resistance. However, February is the time that *E. solidaginis* in the Ottawa area exit the refractory period of diapause, so HIF-1 signaling at this time may be involved in elevating aerobic capacity to prepare for a resumption of development when permissive temperatures return in the spring.

6.4 Mitochondria

Mitochondria are the center of oxygen-based metabolism in cells. Almost all oxygen consumption in cells is by cytochrome oxidase, the last reaction of the electron transport chain (complex IV). Complex I, the NADH-ubiquinone oxidoreductase, is also the primary endogenous source of oxygen free radicals, superoxide "spilling out" of this complex whenever the respiratory chain is backed-up or stressed (e.g. hypoxia). What are the links between mitochondria and cold-hardiness? In general, aerobic respiration by mitochondria is suppressed during the winter for several reasons: low temperature directly reduces metabolic rate (typically by about one-half for every 10 °C drop in temperature), energy-expensive processes such as digestion, biosynthesis and growth are largely shut down, thereby minimizing the ATP demand, and many species enter winter diapause and/or quiescence, which involves a regulated suppression of metabolic rate. Perhaps not surprisingly, then, the activities of several metabolic enzymes in mitochondria were reduced by 40–70% over the winter months in both freeze-avoiding (*E. scudderiana*) and freeze-tolerant (*E. solidaginis*) species (Joanisse and Storey, 1994, 1996a). However, in *E. scudderiana*, two kinds of evidence show that mitochondrial activity is still important to winter metabolism: (a) products of anaerobic glycolysis do not accumulate, so oxygen-based metabolism clearly still fuels ATP needs (Churchill and Storey, 1989), and (b) activities of the subgroup of mitochondrial enzymes involved in lipid catabolism rise over the autumn months, implying that aerobic lipolysis fuels winter survival (Joanisse and Storey, 1996a). This makes sense for *E. scudderiana*, since the supercooled state presents no physical barrier to continuing to respire at subzero temperatures. Freeze-tolerant species can also use aerobic respiration while unfrozen, but mitochondrial metabolism is compromised when body fluids freeze.

Given this background, do cold-hardy insects adjust their mitochondrial capacity over the winter months to meet reduced demands for oxygen-based metabolism? Studies with freeze-tolerant species indicate that they do. Kukal *et al.* (1989) reported that the numbers of mitochondria were greatly reduced in −15 °C-acclimated individuals of the freeze-tolerant high Arctic caterpillar *Gynaephora groenlandica* as compared with 15 °C controls. Using different methods, including measuring 16S rRNA levels, Levin *et al.* (2003) reached a less dramatic conclusion, but nonetheless found that hibernating caterpillars had about half the amount of mitochondrial DNA (mtDNA) as compared with summer active animals. The authors found a similar ~50% reduction in mtDNA content of winter- versus summer-collected *E. solidaginis*.

McMullen and Storey (2008a) assessed mitochondrial responses in freeze-tolerant and freeze-avoiding species by comparing the two goldenrod gall formers. Figure 6.2 shows transcript levels for two mitochondrially encoded

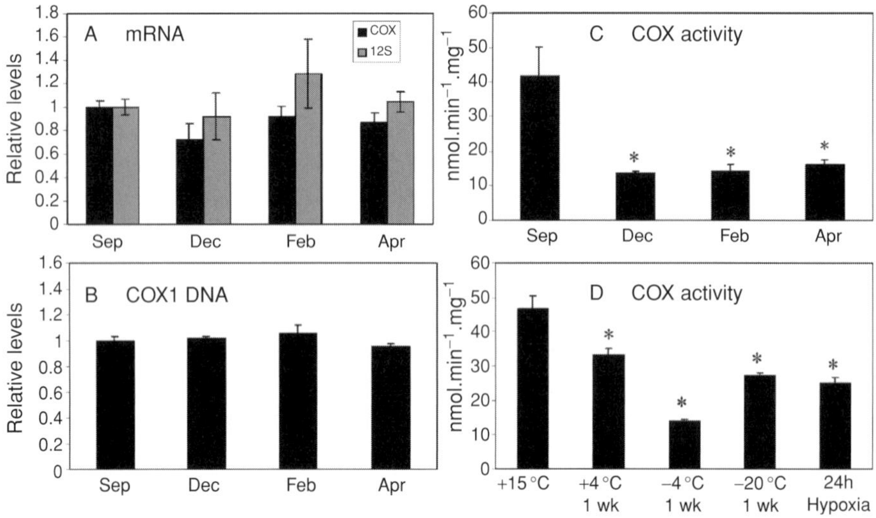

Figure 6.2 Mitochondrial status in larvae of the freeze-avoiding gall moth *Epiblema scudderiana* (Clemens) (Lepidoptera, Olethreutidae). Winter profiles of: (A) transcript levels of two mitochondrially encoded genes, cytochrome oxidase (COX) subunit 1 mRNA and 12 S rRNA, (B) COX 1 genomic DNA content and (C) COX enzymatic activity. (D) COX activity as a function of acclimation temperature or anoxia exposure. Data are means ± SEM, n = 3 independent samples. *-Significantly different from the September or 15 °C control values, $p < 0.05$. Modified from McMullen and Storey (2008a).

genes – cytochrome oxidase (COX) subunit 1 mRNA and 12 S rRNA – along with COX 1 genomic DNA content, and COX maximal enzymatic activity over a winter timecourse for *E. scudderiana*. COX maximal activity decreased in the freeze-avoiding species to a mid-winter low that was about one-third of the September value. This concurs with previous results for some other mitochondrial enzymes (citrate synthase, glutamate dehydrogenase, NAD-isocitrate dehydrogenase, carnitine palmitoyl transferase, malic enzyme) in *E. scudderiana*, which also decreased by ~50% over the winter (Joanisse and Storey, 1994, 1996a). Suppression of mitochondrial enzyme activities could arise due to an overall reduction in mitochondrial numbers, reduced amounts of enzyme proteins per mitochondrion, or suppression of enzyme activity via post-translational regulation (i.e. without a change in protein content). Analysis of other markers of mitochondrial function helps to distinguish between these possibilities. COX1 genomic DNA content did not change over the winter and, in addition, transcript levels of COX 1 mRNA and 12S rRNA were unchanged (these two genes are encoded by the heavy and light chains of the mitochondrial genome, respectively). This indicates that the freeze-avoiding species maintains mitochondrial numbers during the winter and suggests that

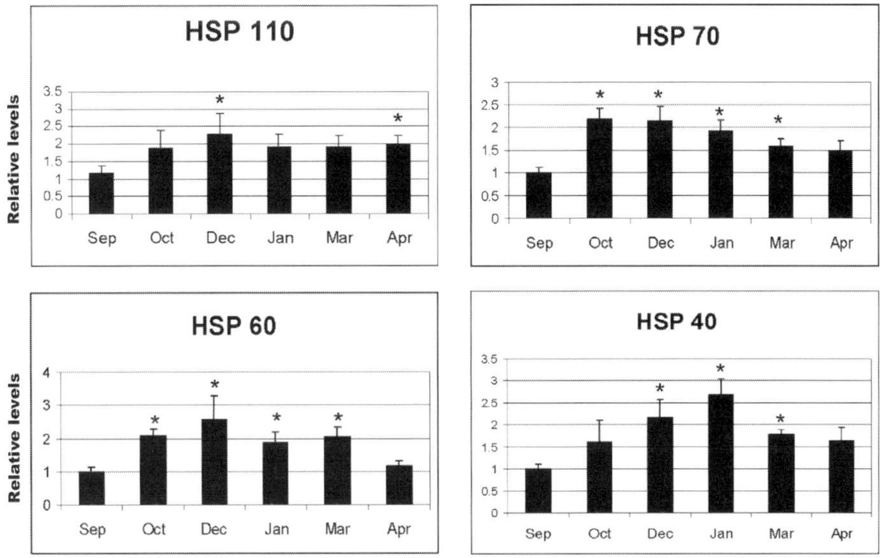

Figure 6.3 Seasonal changes in the levels of four heat-shock proteins in larvae of the freeze-avoiding gall moth *E. scudderiana*. Galls were held outdoors in cloth bags and sampled at selected times over the winter; HSPs were detected via immunoblotting. Data are means ± SEM (n = 4 independent trials) of normalized band intensities for each experimental group. *-Significantly different from the September value, $p < 0.05$. From Zhang (2006).

reduced activities of mitochondrial enzymes (mentioned above) results from selective targeting of individual enzymes, perhaps by a combination of changes in protein levels and post-translational regulation of enzyme activities. This also fits with the responses of the mitochondrial enzymes of fatty-acid oxidation (3-hydroxyacyl-CoA dehydrogenase, thiolase), which showed selective increases of two- to four-fold over the winter in *E. scudderiana* (Joanisse and Storey, 1996a), as well as with an analysis of heat-shock proteins in this species, which found elevated midwinter levels of the mitochondria-specific HSP60 (Fig. 6.3). Hence, freeze-avoiding species demonstrate selective reorganization of mitochondrial metabolism during the winter, but do not appear to undergo mitochondrial degradation.

Freeze-tolerant species do things differently. Analysis of *E. solidaginis* larvae showed that COX maximal activity fell in the winter to 30–50% of the September value (McMullen and Storey, 2008a) as did the activities of many other mitochondrial enzymes, including those involved in fatty-acid oxidation (Joanisse and Storey, 1994, 1996a), as well as levels of the HSP60 chaperone (Zhang, 2006). However, contrary to the freeze-avoiding species, mitochondrial DNA content of *E. solidaginis* larvae was also reduced by about half in frozen, winter-collected larvae, as compared with summer-active animals (Levin *et al.*, 2003). All this evidence suggests that mitochondrial capacity is reduced over the winter months in *E. solidaginis*, probably by a reduction in mitochondrial numbers. For a freeze-tolerant insect this makes sense. While the larvae are frozen for many weeks during the winter, metabolism is anaerobic with lactate and alanine accumulating as end products (Storey and Storey, 1985). Lipid fuels cannot be used when oxygen supply is cut off due to freezing and, furthermore, mitochondria may be vulnerable to structural or metabolic damage arising from osmotic, low oxygen and oxidative stresses associated with freeze/thaw cycles. Therefore, it seems beneficial for *E. solidaginis* larvae to reduce their number of mitochondria over the winter months.

6.5 Metabolic rate depression and environmental stress survival

Entry into a hypometabolic or dormant state is a widespread survival strategy for animals when confronted with environmental stresses, including extremes of heat and cold, oxygen deprivation and water limitation. Strong metabolic rate depression underlies phenomena including hibernation, daily torpor, estivation, diapause, anaerobiosis and anhydrobiosis (Storey and Storey, 1990b, 2004, 2007; Denlinger, 2002). Metabolic rate depression to levels less than 10% of aerobic values is a major contributor to survival by terrestrial insect species with well-developed anoxia tolerance (Hoback and Stanley, 2001). Metabolic rate is clearly suppressed in many species during winter diapause; for example, studies of *E. solidaginis* report that metabolic rate in diapausing larvae in midwinter was just 35–40% of the early autumn value at the same temperature (Irwin *et al.*, 2001; Levin *et al.*, 2003). Many features of hypometabolism are now known to be conserved across phylogeny and the focus in this section is to briefly outline selected features with relevant examples from cold-hardy insects. The biochemical adaptations involved in hypometabolism fit roughly into three broad categories: (a) the signaling and regulatory mechanisms that coordinate and control the suppression of metabolic rate, (b) mechanisms of cell preservation and life extension and (c) specific adaptations that deal with each particular metabolic/environmental situation. For example, synthesis of polyols and antifreeze proteins would be considered examples of the latter category as it applies to cold-hardy insects.

6.5.1 Control of metabolic suppression

Studies on a wide variety of organisms have shown that the regulation of entry into and arousal from hypometabolic states does not involve major reorganization of metabolism, but rather is achieved via reversible controls that change the activity state of key enzymes and functional proteins. The primary mechanism involved is reversible protein phosphorylation, mediated by protein kinases and protein phosphatases, and achieves not just a global suppression of metabolic

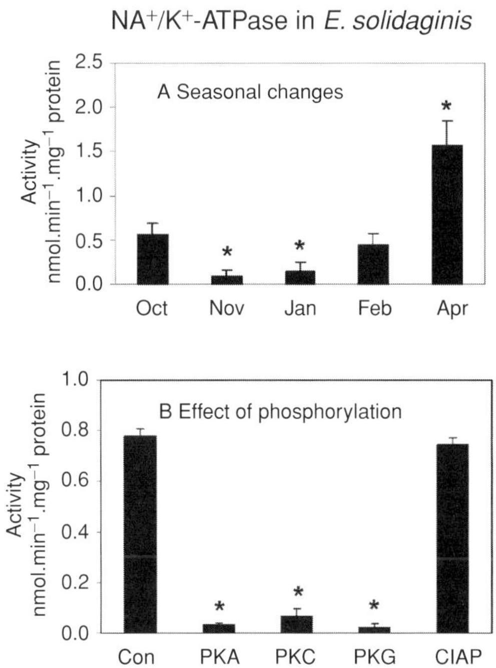

Figure 6.4 Na^+K^+-ATPase in freeze-tolerant *E. solidaginis* larvae. (A) Seasonal changes in Na^+K^+-ATPase activity in larvae sampled in the second week of each month. (B) Effect of *in vitro* incubations to stimulate endogenous protein kinases (PKA, PKG, or PKC) or the addition of exogenous calf-intestinal alkaline phosphatase (CIAP) on Na^+K^+-ATPase activity in extracts from 15 °C-acclimated larvae. Data are means ± SEM, n = 3–5 determinations. *-Significantly different from the corresponding October or control (Con) value, $p < 0.05$. Modified from McMullen and Storey (2008b).

activities, but also a reprioritization of ATP use by different cell functions (Storey and Storey, 2007). For example, transcription and translation are virtually shut off, whereas functions critical to cell survival, such as maintenance of membrane potential difference consume a proportionally larger share of total ATP turnover in the hypometabolic state (Hochachka *et al.*, 1996). Suppression of Na^+K^+-ATPase activity provides a good example of this mechanism in cold-hardy insects. This ion pump is often the greatest single consumer of ATP in cells, so energy savings in hypometabolism must include a net reduction in ATP turnover due to Na^+K^+-ATPase and the oppositely directed actions of membrane ion channels. Figure 6.4A shows the seasonal pattern of Na^+K^+-ATPase activity in freeze-tolerant *E. solidaginis* larvae; activity was high in October and strongly reduced by 80–85% in mid-winter (McMullen and Storey, 2008b). Similar mid-winter suppression of Na^+K^+-ATPase occurs in the freeze-avoiding *E. scudderiana* (a 75% reduction) and Ca^{2+}-ATPase activity also falls by 65–80% in both species (McMullen, 2004). The mechanism of Na^+K^+-ATPase suppression is phosphorylation; *in vitro* studies with extracts from 15 °C acclimated larvae showed that stimulation by protein kinases A, G, or C reduced activity by >90%, whereas incubation with alkaline phosphatase had no effect (Fig. 6.4B). Comparable regulation of Na^+K^+-ATPase by reversible phosphorylation has been demonstrated in hibernating mammals supporting the universality of the mechanism, which is now known to also apply to a wide range of metabolic enzymes, ribosomal proteins and transcription factors (Storey and Storey, 2004, 2007).

6.5.2 Cell preservation and life extension

It has long been recognized that long-term hypometabolism is facilitated by the prior accumulation of large reserves of metabolic fuels and, particularly for anoxia/ischemia resistant species, by adjustments that minimize internal "pollution" by the accumulation of acidic end products (Hochachka and Somero, 1984). More recently, other apparently universal elements of cell preservation during hypometabolism have been recognized. Long, sometimes indeterminate, excursions into hypometabolic states are facilitated by a sharp reduction in ATP turnover, but organisms must then have a way to preserve their macromolecules in a functional state in a situation where they cannot afford to waste ATP on biosynthesis or degradation reactions that normally keep cell constituents in top condition. The solution to this problem is the proliferation of defense mechanisms that protect cells and macromolecules and thereby provide long-term life extension. The two most prominent defense mechanisms identified to date are the proliferation of chaperone proteins and enhancement of antioxidant defenses (Storey and Storey, 2007; Roelofs *et al.*, 2008). Both of these provide cells with enhanced protection against environmental stresses that may be imposed upon them during dormancy.

Chaperones have roles including folding of nascent proteins, refolding of malfolded proteins, preventing or correcting inappropriate aggregation of proteins, and aiding the movement of proteins to their correct subcellular locations. Their up-regulation in response to stresses including heat, cold, heavy metals, low oxygen and UV radiation, among others, provides structural protection for proteins during long-term hypometabolism. The up-regulation of heat-shock protein (HSP) chaperones is a widespread component of diapause in at least four orders of insects (Rinehart *et al.*, 2007). Links with cold-hardiness include a study by Sonoda *et al.* (2006) that reported up-regulation of HSP90 during cold-acclimation in non-diapausing larvae of the rice stem borer, but not in diapausing larvae that already had elevated levels of the chaperone. Similar results were found when expression of the inducible HSP70 was assessed in the onion maggot (Chen *et al.*, 2006). Elevated levels of HSPs are also found over the winter in both goldenrod gall formers.

Levels of four HSPs (110, 70, 60 and 40) rose by 2–2.5 fold in freeze-avoiding *E. scudderiana* larvae over the winter months as compared with September values (Fig. 6.3) (Zhang, 2006). Notably, the inducible ATP-dependent HSP70 and its partner protein HSP40 were both elevated in concert. The mitochondrial ATP-dependent chaperone HSP60 also rose in *E. scudderiana*, but as discussed earlier, HSP60 levels fell in freeze-tolerant *E. solidaginis* larvae. However, HSP110, HSP70 and HSP40 all rose by 40–70% over the mid-winter months in *E. solidaginis*.

The heat-shock transcription factor (HSF1) is active in its trimeric phosphorylated form and translocates to the nucleus, where it binds to the heat-shock element in the promoter regions of heat-shock genes to stimulate their transcription. Analysis of the amount of active HSF1 showed a strong early autumn elevation in both species, rising by about 1.5-fold in *E. solidaginis* in October and 3–4-fold in *E. scudderiana* in October and December, as compared with September values. This implicates enhanced gene expression as the mechanism behind the autumn increase in HSPs in both species. Active HSF1 in *E. solidaginis* also responded strongly to both laboratory anoxia and freezing exposures (Zhang, 2006), which not only supports a role for HSPs in freezing survival, but suggests that enhancement of chaperone protection could be gained via two different routes: (a) as part of seasonal cold-hardening and/or diapause entry, and (b) as a response to freeze-induced ischemia. The latter might be important in cases of very early and deep frosts. Interestingly, recent studies with *Drosophila* have shown that transcription of the *hsf* gene is up-regulated under hypoxia by HIF-1 binding to a response element in an *hsf* intron (Baird *et al.*, 2006). This mechanism ensures the up-regulation of HSPs that are known to be critical for both hypoxia survival and mitigation of oxidative stress during reoxygenation in *Drosophila*. Hence, the seasonal or freeze-responsive acquisition of HSPs in insects might also be regulated in part via HIF-1 regulation of HSF-1 expression.

This link between HIF-1 and HSF-1 shows the interwoven nature of stress responses that ensures that the full complement of cell-preservation mechanism is up-regulated and brings us to our final oxygen-related topic, the mechanisms of cell defense against reactive oxygen species.

6.6 Oxidative stress and antioxidants

Tetravalent reduction of O_2 to H_2O by cytochrome oxidase typically accounts for >95% of all oxygen use in animal cells, but univalent reduction of oxygen also occurs in both enzymatic and non-enzymatic reactions leading to the production of a variety of reactive oxygen species (ROS). Some of these are natural products of various oxidase enzymes and several ROS are now known to function in cellular signaling, including nitric oxide ($NO^\bullet$), superoxide (O_2^{-}), and hydrogen peroxide (H_2O_2) (Rhee *et al.*, 2005). In many cases, however, ROS are overproduced under stress conditions (e.g. ischemia/reperfusion, toxins, heavy metals, UV radiation, etc.) and cause serious damage to cellular macromolecules, including lipid peroxidation, protein oxidation and DNA damage. $NO^\bullet$, O_2^- and H_2O_2 have relatively low reactivity themselves, but they are readily converted into highly reactive forms including hydroxyl radicals ($^\bullet OH$), formed via the Haber–Weiss reaction (see below), and peroxynitrite (ONO_2^-) formed from the reaction of O_2^- with $NO^\bullet$ (Cadenas, 1995; Fridovich, 1998; Hermes-Lima, 2004).

$$H_2O_2 + Fe^{2+} \text{ (or } Cu^+) \rightarrow Fe^{3+} \text{ (or } Cu^{2+}) + OH^- + {}^\bullet OH \qquad \text{[Fenton Reaction]}$$

$$O_2^- + Fe^{3+} \rightarrow Fe^{2+} + O_2$$

$$H_2O_2 + O_2^- \rightarrow O_2 + OH^- + {}^\bullet OH \qquad \text{[net Haber–Weiss Reaction]}$$

For this reason, enzymatic defenses against ROS are critical to the survival of all organisms, and include mechanisms to prevent/limit ROS formation, repair damage and detoxify/dispose of damage products. In the front line of antioxidant defense are metabolites (e.g. ascorbate, vitamin E), peptides (glutathione, thioredoxin) and enzymes that directly react with ROS species. Primary enzymatic defenses include superoxide dismutase (SOD), catalase (CAT), thioredoxin peroxidase and selenium-dependent glutathione peroxidase (Se-GPX), which can catalyse the decomposition of either H_2O_2 or organic hydroperoxides using reduced glutathione (GSH) as a co-substrate (Fridovich, 1998; Hermes-Lima, 2004).

$$2O_2^- + 2H^+ \rightarrow H_2O_2 + O_2 \qquad \text{superoxide dismutase}$$

$$2H_2O_2 \rightarrow H_2O + O_2 \qquad \text{catalase}$$

$$H_2O_2 + 2GSH \rightarrow GSSG + 2H_2O \qquad \text{Se-GPX}$$

$$\text{L-OOH} + 2GSH \rightarrow GSSG + \text{L-OH} + H_2O \qquad \text{Se-GPX}$$

Glutathione S-transferases (GST) are also critical; these catalyse the conjugation of GSH to xenobiotics and display selenium-independent GPX activity toward organic hydroperoxides (Hermes-Lima, 2004). Auxiliary enzymes include glutathione reductase (GR) and thioredoxin reductase that recycle oxidized glutathione (GSSG) or thioredoxin back into reduced forms, as well as the pentose phosphate cycle enzymes glucose-6-phosphate dehydrogenase and 6-phosphogluconate dehydrogenase that supply the NAPDH reducing power needed (note that these two enzymes also have critical roles in polyol synthesis in cold-hardy insects).

Over the last 10–15 years it has become apparent that antioxidant defenses are adaptable to allow animals to cope with the different forms of oxidative stress that are normally associated with their environment or lifestyle. For example, conditions of ischemia/reperfusion cause serious oxidative damage in mammalian tissues due to an inability to cope with a burst of ROS production associated with the reintroduction of oxygen to the highly reduced cytochrome systems of anoxic cells. New studies provide physiological evidence that reperfusion injury also occurs in *Drosophila* (Lighton and Schilman, 2007). However, anoxia-tolerant organisms that naturally experience prolonged periods of oxygen deprivation show little or no ROS-mediated damage when oxygen is reintroduced to their systems. A variety of studies have shown that adaptation of antioxidant defenses generally takes two forms: organisms undergoing frequent bouts of natural stress maintain high constitutive levels of antioxidants, whereas those that experience less frequent or seasonal stress show inducible defenses (Hermes-Lima *et al.*, 2001; Hermes-Lima and Zenteno-Savín, 2002). Furthermore, inducible defenses frequently occur as a preparatory or anticipatory response; for example, anoxia exposure or periods of low oxygen use (e.g. hypometabolic states, such as estivation or hibernation) trigger up-regulation of antioxidant enzymes so that they are in place when oxygen levels and/or consumption rise rapidly again during recovery.

Only a few studies have explored oxidative stress and antioxidants in cold-hardy insects. Three potential sources of increased oxidative stress that seem particularly relevant to winter cold-hardiness of insects can be considered. These are: (a) peroxidation of polyunsaturated fatty acids (PUFAs), (b) metal-ion-induced ROS generation and (c) freeze–thaw transitions that produce conditions of ischemia-reperfusion.

6.6.1 PUFAs

Unsaturated fatty acids, particularly PUFAs, are highly susceptible to free-radical damage because an attack on one lipid residue sets off a chain reaction of lipid peroxidation that can propagate through the PUFA moieties of membrane phospholipids or triglyceride depots until terminated by an antioxidant, commonly α-tocopherol (vitamin E) (Hermes-Lima, 2004). Low temperatures can increase susceptibility to lipid peroxidation because of a strategy called homeoviscous adaptation that retains lipid fluidity at low temperatures by reducing the proportions of saturated fatty acids and increasing monounsaturates and PUFAs as temperature decreases. Homeoviscous adaptation has been reported for a variety of insect species during cold-hardening or diapause along with increased activities of the necessary phospholipid desaturases (summarized in Michaud and Denlinger, 2006). For example, rapid cold-hardening (8 h at 4 °C) elevated the proportion of oleic acid (18:1n–9) in pharate adults of *Sarcophaga crassipalpis* from 30 to 47% of the total fatty acid pool and diapause entry further raised this to 58% (Michaud and Denlinger, 2006). In both *E. solidaginis* and *E. scudderiana* the double-bond index of whole-body lipids was significantly higher in winter due to a decrease in the proportion of saturated fatty acids (16:0, 18:0) and a strong increase in oleic acid (18:1) in *E. solidaginis* and in palmitoleic (16:1) and α-linolenic (18:3) in *E. scudderiana* (Joanisse and Storey, 1996a). Bennett *et al.* (1997) reported a similar steady rise (up to mid-December) in the monoene (18:1n−9) content of both the membrane phospholipids and the triacylglyceride depots of *E. solidaginis* at the expense of saturates. A decrease in saturated and increase in monounsaturated phospholipids also occurred in cold-acclimated and diapausing onion maggots, *Delia antiqua*, along with a 2–10-fold rise in Δ9-acyl-CoA desaturase gene expression (Kayukawa *et al.*, 2007).

6.6.2 Metal ions

The reactive nature of certain metal ions, particularly iron and copper, is critical to their many biological functions, most notably the central role of iron in the redox reactions of the electron-transport system. However, their reactivity also makes them dangerous and both readily react with H_2O_2 to produce $^{\bullet}OH$ (for an extensive review, see Hermes-Lima, 2004). This is the Fenton reaction (see above) and, when coupled with a scheme that uses a reducing agent such as superoxide (O_2^{-}) to regenerate Fe^{2+}, the net Haber–Weiss reaction is perhaps the most toxic chemical reaction in cells. Since this is a non-enzymatic reaction, a critical issue for all organisms is to minimize the availability of free Fe^{2+} (or Cu^{+}) by sequestering these metals in functional proteins (e.g. hemoglobin, hemocyanin, cytochromes, etc.), or binding them to transport proteins (transferrin, ceruloplasmin) or intracellular storage proteins (ferritin, metallothionein). Ferritin, for example, is a huge protein consisting of 24 heavy (H) and light (L) subunits that surrounds a core of up to 4500 iron atoms locked in a low-reactivity ferrihydrite state (Hentze *et al.*, 2004).

6.6.3 Freeze/thaw

Most freeze-tolerant animals can endure the conversion of at least 65% of total body water into ice. Among several consequences of freezing are two that are relevant here: (a) at 65% ice the concentrations of ions and other solutes in remaining body fluids are essentially tripled (this includes the levels of metal catalysts and other potentially toxic trace metals such as cadmium), and (b) freezing is an ischemic event and, even in insects that breathe via trachea, freezing restricts cell oxygenation and causes a build-up of glycolytic end products and a fall in ATP. Ischemia/reperfusion events are well known to trigger oxidative damage caused by a burst of ROS generation when oxygen is reintroduced to the system. Hence, added protection in the form of enhanced levels of antioxidants and metal-binding proteins could be important for long-term viability in the frozen state.

6.6.4 Adaptive strategies for antioxidant defence

Only a little is known to date about the responses of heavy-metal binding proteins during cold-hardening or freezing in insects, but studies with other freeze-tolerant animals suggest the universal importance of this mechanism for freezing survival. Both freezing and anoxia exposures triggered strong up-regulation of metallothionein (MT) in the marine intertidal snail *Littorina littorea*; MT binds a variety of heavy metals including cadmium and copper. MT mRNA levels rose 3.5-fold in muscle and 6-fold in hepatopancreas after 12–24 h of freezing of snails (English and Storey, 2003). Changes in MT gene expression occur within the time-frame of a normal tidal cycle, which supports the idea that variation in MT protein is a natural response to hypoxia and reoxygenation. In addition to metal binding, studies have suggested that MTs also act directly as antioxidants due to their many readily oxidized cysteine residues (typically ~30% of total amino acids) (Viarengo *et al.*, 1999), so there may be a dual purpose to MT up-regulation in freeze-tolerant species. Anoxia exposure also up-regulated ferritin in the snails (freezing was not tested) (Larade and Storey, 2004) and cDNA array screening showed putative up-regulation of both ferritin and MT during freezing in organs of freeze-tolerant wood frogs *Rana sylvatica* (Storey, 2004) and in response to anoxia in freeze-tolerant hatchling turtles *Chrysemys picta marginata* (Storey, 2006).

In recent work, *Drosophila* cDNA arrays have been used to search for freeze- and anoxia-induced gene up-regulation in *E. solidaginis* (K.B. Storey, unpublished data). Larvae acclimated to 15 °C were acutely chilled at 3 °C for 24 h, −4 °C for 4 h, or frozen at −16 °C for 24 h. In all cases, a consistent prominent response was a strong up-regulation (4–5-fold) of ferritin heavy-chain gene expression. A 2-fold increase in ferritin transcript levels also occurred when larvae were given 4 h of anoxia exposure under N_2 gas. A similar ~1.5-fold increase in ferritin heavy-chain mRNA levels occurred in *E. scudderiana* given 4 h of cold or anoxia exposures. Lee *et al.* (2006) also found that expression of the transferrin gene was up-regulated by 2–4-fold by both cold (4 °C) and heat (37 °C) exposures, as well as other stresses (iron overload, H_2O_2 or paraquat exposure) in 25 °C-reared beetles, *Apriona germari*. Furthermore, suppression of the gene expression of this hemolymph iron-transport protein via RNA interference rapidly triggered apoptosis in fat-body cells and this suggested that transferrin acted as an antioxidant by diminishing the Fenton reaction.

Interestingly, another link between cold-hardy insects and metal-binding proteins is known. Zachariassen *et al.* (2004) proposed that the high levels of cysteine in the antifreeze proteins (AFPs) of many freeze-avoiding insects may actually have secondary actions as antioxidants. Studies with *Tenebrio molitor* that have AFPs containing about 20% cysteine showed that metal stress (Cd, Cu, Zn) suppressed the normal developmental increase in AFPs in larvae reared under summer conditions (25 °C, long day) (Pedersen *et al.*, 2006). However, metal exposure did not interfere with AFP production in winter-acclimated larvae (4 °C, short day) suggesting that AFPs take priority in a competition for cysteine under cold conditions. Furthermore, Pedersen *et al.* (2007) recently isolated an inducible 7 kDa protein from *T. molitor* that contained no cysteine and yet appeared to be responsible for most of the Cd-binding in this species. Hence, it appears that species that produce high cysteine AFPs to support winter cold-hardiness, have opted for low cysteine Cd-binding proteins to supplement or replace MT.

Studies with the goldenrod gall formers have profiled antioxidant defenses over the winter months and in response to laboratory cold and anoxia exposures. CAT, SOD, GPX GST and GR were quantified. Overall, activities of all five enzymes were consistently 2–3-fold higher in the lepidopteran larva *E. scudderiana* compared with the dipteran, *E. solidaginis*. In the freeze-tolerant *E. solidaginis* activities were generally high in September and decreased over the mid-winter months (Joanisse and Storey, 1996b). This could be interpreted as a need for high antioxidant defenses while the larvae are still feeding on plant material in the gall, but then a downward adjustment of enzymatic defenses to levels sufficient to support winter survival in diapause. Significantly, CAT, GST and GR rose again by late March when larvae were preparing for pupation. The total glutathione pool in *E. solidaginis* did not change over time, but levels of oxidized GSSG increased significantly from 3% in the autumn to 5% of the total pool in mid-winter before declining to 1.4% in March; this suggests that GSH may be used in the cold months, but perhaps not recycled until temperatures rise again in late winter. The situation was different in the freeze-avoiding *E. scudderiana*. CAT, GPX and GST activities all increased by 2–3-fold in mid-winter as compared with September values, although SOD was largely unchanged and GR activity was reduced (Joanisse and Storey, 1996b). Total glutathione and the % GSSG were stable over the mid-winter months, but reduced as compared with September. These data suggest a greater need for antioxidant defense over the winter months in the freeze-avoiding species than in the freeze-tolerant ones. This might be related to the supercooled state of freeze-avoiding species, which allows unimpeded aerobic metabolism to continue throughout the winter, bringing with it a continuing potential for ROS production and for ROS damage to macromolecules, perhaps particularly damage to PUFAs.

Other research on cold-hardy species has reached similar conclusions about antioxidant defenses. Gulevsky *et al.* (2006) monitored CAT and SOD activities, glutathione (GSH) level and lipid peroxidation (LP) intermediates in larvae, pupae and imagoes of *T. molitor*. They found evidence of oxidative stress at the most cold-sensitive developmental stage (imagoes) with elevated LP products and CAT activity after 2 weeks at 4 °C. Pupae showed a rise in CAT and GSH fell in both larvae and pupae; SOD was unaffected at any stage. Jovanovic-Galovic *et al.* (2007) analysed antioxidant enzyme responses in mitochondria of diapausing and non-diapausing larvae and pupae of the European corn borer (*Ostrinia nubilalis*). The study did not involve cold exposure, but it showed that activities of CAT and GST were significantly lower in diapausing larvae; this correlates with differences in the metabolic rate of diapausing and non-diapausing larvae. In the spruce budworm, *Choristoneura fumiferana*, the gene encoding glutathione S-transferase is highly expressed before the second instar larval diapause, and expression persists throughout diapause and begins to decline after diapause ends (Feng *et al.*, 2001).

In mammalian systems, ischemia/reperfusion events (e.g. stroke, heart attack) cause oxidative damage to tissues primarily during the recovery period, when oxygen is rapidly reintroduced to tissues causing a burst of ROS production. Additional studies with the goldenrod gall formers aimed to determine if such events also occurred in insects in response to experimental anoxia/recovery or freeze/thaw. However, studies with *E. solidaginis* larvae showed that neither 24 h anoxia or freezing exposures, nor aerobic recovery or thawing periods up to 8 h altered two markers of oxidative damage, thiobarbituric acid reactive substances (TBARS) and lipid peroxides (Joanisse and Storey, 1998). Antioxidant enzyme activities were also unaffected, but freezing caused a 6-fold increase in GSSG, showing that GSH consumption increased in frozen larvae, but took >8 h to be corrected after thawing. Anoxia exposure and recovery had no substantial effect on markers of oxidative damage, antioxidant enzymes, or glutathione pools in *E. scudderiana*. Overall, then, both species appear to be well equipped with sufficient antioxidant defenses to deal with oxidative stress arising from low-temperature or low-oxygen stresses associated with winter.

6.7 Future directions

The current chapter has focused on oxygen-related issues with respect to insect cold-hardiness and also on related issues of metabolic rate depression and cell preservation strategies for long-term survival. Where do we go from here? With respect to insect cold-hardiness itself, several physiological issues still need answers. For example, the relationship between the mechanisms of water balance used by different species and their choice of cold-hardiness strategy, newly reported by Zachariassen *et al.* (2008), is very intriguing and needs to be pursued at both physiological and biochemical levels. Various older studies have also revealed what must be a very sophisticated temperature-sensing capacity in cold-hardy insects, for multiple biochemical events are triggered within very narrow temperature ranges. A dramatic example are the diapause eggs of silkmoths that sustain diapause with stable sorbitol levels for >400 days when stored at 1 °C, whereas eggs stored at 5 °C break diapause, catabolize sorbitol and hatch in <100 days (Furusawa *et al.*, 1982). This deserves further exploration. Many issues of the biochemical regulation of cold-hardiness, metabolic rate depression, diapause, freeze-associated ischemia and cell preservation strategies of overwintering insects also remain to be explored. Our studies in recent years have shown major commonalities across phylogeny in the use of reversible phosphorylation to achieve coordinated suppression of ATP use by energy-expensive cell functions (e.g. ion motive ATPases, ribosomal translation) in multiple forms of hypometabolism (Storey and Storey, 2007). Broader aspects of the control of growth/development/reproduction in favorable environments versus survival/preservation/hypometabolism responses in stress environments are now emerging, particularly illuminated by studies of stress-induced dauer formation in nematodes, *Caenorhabditis elegans* (Fielenbach and Antebi, 2008). These studies highlight a central role of insulin/IGF signaling and FOXO transcription factors in regulating hypometabolic states and, indeed, initial studies are confirming their importance in the diapause of overwintering insects (Sim and Denlinger, 2008). This will be a fertile area for study over the next several years to map out the elements of the hypometabolic response to stress, trace the web of metabolic adjustments that are initiated, the signaling mechanisms involved (receptors, protein kinases and phosphatases, transcription factors) and identify both the commonalities and the disparities between different systems that will highlight the underlying principles involved. The mechanisms of stress-responsive gene expression are also intriguing, for they must provide for selective gene expression and protein synthesis against a background of global transcriptional and translational suppression. New techniques and advances provide novel ideas about stress-responsive gene regulation that need to be examined in insect models of cold-hardiness. These include reversible control of transcription factors, methods for storage/preservation of mRNA transcripts during hypometabolism, the role of microRNA species in determining which transcripts are translated, novel methods of reversible protein covalent modification (e.g. sumoylation) and the role that histone modification plays in stress-responsive gene silencing (Morin and Storey, 2008). Much also remains to be learned about the cell-preservation mechanisms that are involved in long-term life extension, for if you save energy by strongly suppressing biosynthetic and degradative pathways, you must then ensure that existing macromolecules are protected and maintain functionality over time spans that may be many-fold longer than the normal turnover times of these molecules. Heat-shock proteins have been well studied in this regard, but broader types of defenses have been recognized in some systems and need to be evaluated in cold-hardy insects as well. For example, the unfolded protein response (UPR) has been documented in a number of systems and provides for a coordinated shut-down of protein synthesis, an enhancement of endoplasmic reticulum chaperones (the glucose-regulated proteins), and selective degradation of damaged proteins in response to cell stress. We are currently evaluating the role of the UPR in the response of cold-hardy insects to cold, freezing and anoxia. Other issues of metal-ion sequestering, antioxidant defense and anti-apoptosis mechanisms also need to be explored as important aspects of long-term life extension that serve winter survival by insects.

References

Baird, N. A., Turnbull, D. W. and Johnson, E. A. (2006). Induction of the heat shock pathway during hypoxia requires regulation of heat shock factor by hypoxia-inducible factor-1. *Journal of Biological Chemistry* **281**, 38675–38681.

Bennett, V. V., Pruitt, N. L. and Lee, R. E. (1997). Seasonal changes in fatty acid composition associated with cold-hardening in third instar larvae of *Eurosta solidaginis*. *Journal of Comparative Physiology B* **167**, 249–255.

Burg, M. B. (1995). Molecular basis of osmotic regulation. *American Journal of Physiology* **268**, F983–F996.

Burmester, T. and Hankeln T. (2007). The respiratory proteins of insects. *Journal of Insect Physiology* **53**, 285–294.

Cadenas, E. (1995). Mechanism of oxygen activation and reactive oxygen species detoxification. In *Oxidative Stress and Antioxidant Defenses in Biology*, ed. S. Ahmad. New York: Chapman & Hall, pp. 1–61.

Chen, B., Kayukawa, T., Monteiro, A. and Ishikawa, Y. (2006). Cloning and characterization of the HSP70 gene and its expression in response to diapauses and thermal stress in the onion maggot, *Delia antiqua*. *Journal of Biochemistry and Molecular Biology* **39**, 749–58.

Churchill, T. A. and Storey, K. B. (1989). Metabolic consequences of rapid cycles of temperature change for freeze avoiding versus freeze tolerant insects. *Journal of Insect Physiology* **35**, 579–586.

Danks, H. V. (1971). Overwintering in some north temperate and arctic Chironomidae. *Canadian Entomologist* **103**, 1875–1901.

Danks, H. V. (2007). How aquatic insects live in cold climates. *Canadian Entomologist* **139**, 443–447.

Denlinger, D. L. (2002). Regulation of diapause. *Annual Review of Entomology* **47**, 93–122.

English, T. E. and Storey, K. B. (2003). Freezing and anoxia stresses induce expression of metallothionein in the foot muscle and hepatopancreas of the marine gastropod, *Littorina littorea*. *Journal of Experimental Biology* **206**, 2517–2524.

Feng, Q. L., Davey, K. G., Pang, A. S. D., Ladd, T. R., Retnakaran, A., Tomkins, B. L., Zheng, S. and Palli, S. R. (2001). Developmental expression and stress induction of glutathione S-transferase in the spruce budworm, *Choristoneura fumiferana*: developmental expression and the induction by various stresses. *Journal of Insect Physiology* **47**, 1–10.

Fielenbach, N. and Antebi, A. (2008). *C. elegans* dauer formation and the molecular basis of plasticity. *Genes and Development* **22**, 2149–2165.

Fridovich, I. (1998). Oxygen toxicity: a radical explanation. *Journal of Experimental Biology* **201**, 1203–1209.

Furusawa, T., Shikata, M. and Yamashita, O. (1982). Temperature dependent sorbitol utilization in diapause eggs of the silkworm, *Bombyx mori*. *Journal of Comparative Physiology B* **147**, 21–26.

Gorr, T. A., Gassmann, M. and Wappner, P. (2006). Sensing and responding to hypoxia via HIF in model invertebrates. *Journal of Insect Physiology* **52**, 349–364.

Gulevsky, A. K., Relina, L. I., and Grishchenkova, Y. A. (2006). Variations of the antioxidant system during development of the cold-tolerant beetle, *Tenebrio molitor*. *Cryo-Letters* **27**, 283–290.

Harrison, J., Frazier, M. R., Henry, J. R., Kaiser, A., Klok, C. J. and Rascón, B. (2006). Responses of terrestrial insects to hypoxia or hyperoxia. *Respiratory Physiology and Neurobiology* **154**, 4–17.

Hentze, M. W., Muckenthaler, M. U. and Andrews, N. C. (2004). Balancing acts: molecular control of mammalian iron metabolism. *Cell* **117**, 285–297.

Hermes-Lima, M. (2004). Oxygen in biology and biochemistry: role of free radicals. In *Functional Metabolism: Regulation and Adaptation*, ed. K. B. Storey. Hoboken, N. J.: Wiley-Liss, pp. 319–368.

Hermes-Lima, M. and Zenteno-Savín, T. (2002). Animal response to drastic changes in oxygen availability and physiological oxidative stress. *Comparative Biochemistry and Physiology* **133**, 537–556.

Hermes-Lima, M., Storey, J. M. and Storey, K. B. (2001). Antioxidant defenses and animal adaptation to oxygen availability during environmental stress. In *Cell and Molecular Responses to Stress*, ed. K. B. Storey and J. M. Storey. vol. 2: *Protein Adaptations and Signal Transduction*. Amsterdam: Elsevier Press, pp. 263–287.

Hoback, W. W. and Stanley, D. W. (2001). Insects in hypoxia. *Journal of Insect Physiology* **47**, 533–542.

Hochachka, P. W. and Somero, G. N. (1984) *Biochemical Adaptation*. Princeton, N. J.: Princeton University Press.

Hochachka, P. W., Buck, L. T., Doll, C. J. and Land, S. C. (1996). Unifying theory of hypoxia tolerance: molecular/metabolic defense and rescue mechanisms for surviving oxygen lack. *Proceedings of the National Academy of Sciences, USA* **93**, 9493–9498.

Irwin, J. T. and Lee, R. E. (2002). Energy and water conservation in frozen vs. supercooled larvae of the goldenrod gall fly, *Eurosta solidaginis* (Fitch) (Diptera: Tephritidae). *Journal of Experimental Zoology* **292**, 345–350.

Irwin, J. T., Bennett, V. A. and Lee, R. E. (2001). Diapause development in frozen larvae of the goldenrod gall fly, *Eurosta solidaginis* Fitch (Diptera: Tephritidae). *Journal of Comparative Physiology B* **171**, 181–188.

Joanisse, D. R. and Storey, K. B. (1994). Mitochondrial enzymes during overwintering in two species of cold-hardy gall insects. *Insect Biochemistry and Molecular Biology* **24**, 145–150.

Joanisse, D. R. and Storey, K. B. (1996a). Fatty acid content and enzymes of fatty acid metabolism in overwintering cold-hardy gall insects. *Physiological Zoology* **69**, 1079–1095.

Joanisse, D. R. and Storey, K. B. (1996b). Oxidative stress and antioxidants during overwintering in larvae of cold-hardy goldenrod gall insects. *Journal of Experimental Biology* **199**, 1483–1491.

Joanisse, D. R. and Storey, K. B. (1998). Oxidative stress and antioxidants in stress and recovery of cold-hardy insects. *Insect Biochemistry and Molecular Biology* **28**, 23–30.

Jovanovic-Galovic, A., Blagojevic, D. P., Grubor-Lajsic, G., Worland M. R. and Spasic, M. B. (2007). Antioxidant defense in mitochondria during diapause and postdiapause development of European corn borer (*Ostrinia nubilalis*, Hubn.). *Archives of Insect Biochemistry and Physiology* **64**, 111–119.

Kayukawa, T., Chen, B., Hoshizaki, S. and Ishikawa Y. (2007). Upregulation of a desaturase is associated with the enhancement of cold hardiness in the onion maggot, *Delia antiqua*. *Insect Biochemistry and Molecular Biology* **37**, 1160–1167.

Kukal, O., Duman, J. G. and Seriani, A. S. (1989). Cold-induced mitochondrial degradation and cryoprotectant synthesis in freeze-tolerant arctic caterpillars. *Journal of Comparative Physiology B* **158**, 661–671.

Larade, K. and Storey, K. B. (2004). Accumulation and translation of ferritin heavy chain transcripts following anoxia exposure in a marine invertebrate. *Journal of Experimental Biology* **207**, 1353–1360

Lavista-Llanos, S., Centanin, L., Irisarri, M., Russo, D. M., Gleadle, J. M. Bocca, S. N., Muzzopappa, M., Ratcliffe, P. J. and Wappner, P. (2002). Control of the hypoxic response in *Drosophila melanogaster* by the basic helix-loop-helix PAS protein similar. *Molecular and Cellular Biology* **22**, 6842–6853.

Lee, K. S., Kim, B. Y., Kim, H. J., Seo, S. J., Yoon, H. J., Choi, Y. S., Kim, I., Han, Y. S., Je, Y. H., Lee, S. M., Kim, D. H., Sohn, H. D. and Jin, B. R. (2006). Transferrin inhibits stress-induced apoptosis in a beetle. *Free Radicals in Biology and Medicine* **41**, 1151–1161.

Levin, D. B., Danks, H. V. and Barber, S. A. (2003). Variations in mitochondrial DNA and gene transcription in freezing-tolerant larvae of *Eurosta solidaginis* (Diptera: Tephritidae) and *Gynaephora groenlandica* (Lepidoptera: Lymantriidae). *Insect Molecular Biology* **12**, 281–289.

Lighton, J. R. and Schilman, P. E. (2007). Oxygen reperfusion damage in an insect. *PLoS ONE* **2**(12):e1267.

Lundheim, R. and Zachariassen, K. E. (1993). Water balance of overwintering beetles in relation to strategies for cold tolerance. *Journal of Comparative Physiology B* **163**, 1–4.

McMullen, D. C. (2004). Molecular and biochemical adaptations conferring cold hardiness in two gall insects. Ph. D. thesis, Carleton University.

McMullen, D. C. and Storey, K. B. (2008a). Mitochondria of cold hardy insects: responses to cold and hypoxia assessed at enzymatic, mRNA and DNA levels. *Insect Biochemistry and Molecular Biology* **38**, 367–373.

McMullen, D. C. and Storey, K. B. (2008b). Suppression of $Na^{+}K^{+}$-ATPase activity by reversible phosphorylation over the winter in a freeze-tolerant insect. *Journal of Insect Physiology* **54**, 1023–1027.

Michaud, R. M. and Denlinger, D. L. (2006). Oleic acid is elevated in cell membranes during rapid cold-hardening and pupal diapause in the flesh fly, *Sarcophaga crassipalpis*. *Journal of Insect Physiology* **52**, 1073–1082.

Morin, P., McMullen, D. C. and Storey, K. B. (2005). HIF-1α involvement in low temperature and anoxia survival by a freeze tolerant insect. *Molecular and Cellular Biochemistry* **280**, 99–106.

Morin, P. and Storey, K. B. (2008). Mammalian hibernation: differential gene expression and novel application of epigenetic controls. *International Journal of Developmental Biology* **53**, 433–442.

Nambu, J. R., Chen, W., Hu, S. and Crews, S. T. (1996). The *Drosophila melanogaster* similar bHLH-PAS gene encodes a protein related to human hypoxia inducible factor 1 alpha and *Drosophila* single-minded. *Gene* **172**, 249–254.

Pedersen, S. A., Kristiansen, E., Andersen, R. A. and Zachariassen, K. E. (2007). Isolation and preliminary characterization of a Cd-binding protein from *Tenebrio molitor* (Coleoptera). *Comparative Biochemistry and Physiology C* **145**, 457–463.

Pedersen, S. A., Kristiansen, E., Hansen, B. H., Andersen, R. A. and Zachariassen, K. E. (2006). Cold hardiness in relation to trace metal stress in the freeze-avoiding beetle *Tenebrio molitor*. *Journal of Insect Physiology* **52**, 846–853.

Redecker, B. and Zebe, E. (1988). Anaerobic metabolism in aquatic insect larvae – studies on *Chironomus thummi* and *Culex pipiens*. *Journal of Comparative Physiology B* **158**, 307–315.

Rhee, S. G., Kang, S. W., Jeong, W., Chang, T. S., Yang, K. S. and Woo, H. A. (2005). Intracellular messenger function of hydrogen peroxide and its regulation by peroxiredoxins. *Current Opinion in Cell Biology* **17**, 183–189.

Rickards, J., Kelleher, M. J. and Storey, K. B. (1987). Strategies of freeze avoidance in larvae of the goldenrod gall moth, *Epiblema scudderiana*: winter profiles of a natural population. *Journal of Insect Physiology* **33**, 443–450.

Rinehart, J. P., Li, A., Yocum, G. D., Robich, R. M., Hayward, S. A. and Denlinger, D. L. (2007). Up-regulation of heat shock proteins is essential for cold survival during insect diapause. *Proceedings of the National Academy of Sciences, USA* **104**, 11130–11137.

Roelofs, D., Aarts, M. G. M., Schat, H. and van Straalen, N. M. (2008). Functional ecological genomics to demonstrate general and specific responses to abiotic stress. *Functional Ecology* **22**, 8–18.

Semenza, G. L. (2003). Targeting HIF-1 for cancer therapy. *Nature Reviews on Cancer* **3**, 721–732.

Sim, C. and Denlinger, D. L. (2008). Insulin signaling and FOXO regulate the overwintering diapause of the mosquito *Culex pipiens*. *Proceedings of the National Academy of Sciences, USA* **105**, 6777–6781.

Sonoda, S., Fukumoto, K., Izumi, Y., Yoshida, H. and Tsumuki, H. (2006). Cloning of heat shock protein genes (hsp90 and hsc70) and their expression during larval diapause and cold tolerance acquisition in the rice stem borer, *Chilo suppressalis* Walker. *Archives of Insect Biochemistry and Physiology* **63**, 36–47.

Storey, K. B. (2004). Strategies for exploration of freeze responsive gene expression: advances in vertebrate freeze tolerance. *Cryobiology* **48**, 134–145.

Storey, K. B. (2006). Reptile freeze tolerance: metabolism and gene expression. *Cryobiology* **52**, 1–16.

Storey, J. M. and Storey, K. B. (1985). Freezing and cellular metabolism in the gall fly larva, *Eurosta solidaginis. Journal of Comparative Physiology B* **155**, 333–337.

Storey, K. B. and Storey, J. M. (1990a). Carbon balance and energetics of cryoprotectant synthesis in a freeze-tolerant insect: responses to perturbation by anoxia. *Journal of Comparative Physiology B* **160**, 77–84.

Storey, K. B. and Storey, J. M. (1990b). Facultative metabolic rate depression: molecular regulation and biochemical adaptation in anaerobiosis, hibernation, and estivation. *Quarterly Review of Biology* **65**, 145–174.

Storey, K. B. and Storey, J. M. (1991). Biochemistry of cryoprotectants. In *Insects at Low Temperature*, ed. R. E. Lee and D. Denlinger. New York: Chapman and Hall, pp. 64–93.

Storey, K. B. and Storey, J. M. (1996). Natural freezing survival in animals. *Annual Review of Ecology and Systematics* **27**, 365–386.

Storey, K. B. and Storey, J. M. (2004). Metabolic rate depression in animals: transcriptional and translational controls. *Biological Reviews of the Cambridge Philosophical Society* **79**, 207–233.

Storey, K. B. and Storey, J. M. (2007). Putting life on "pause" – molecular regulation of hypometabolism. *Journal of Experimental Biology* **210**, 1700–1714.

Storey, K. B. and Storey, J. M. (1981). Biochemical strategies of overwintering in the gall fly larva, *Eurosta solidaginis*: effect of low temperature acclimation on the activities of enzymes of intermediary metabolism. *Journal of Comparative Physiology B* **144**, 191–199.

Viarengo, A., Burlando, B., Cavaletto, M., Marchi, B., Ponzano, E. and Blasco, J. (1999). Role of metallothionein against oxidative stress in the mussel *Mytilus galloprovincialis. American Journal of Physiology* **277**, R1612-R1619.

Woodman, J. D., Cooper, P. D. and Haritos, V. S. (2007). Cyclic gas exchange in the giant burrowing cockroach, *Macropanesthia rhinoceros*: effect of oxygen tension and temperature. *Journal of Insect Physiology* **53**, 497–504.

Zachariassen, K. E., Kristiansen, E. and Pedersen, S. A. (2004). Inorganic ions in cold-hardiness. *Cryobiology* **48**, 126–133.

Zachariassen, K. E., Li, N. G., Laugsand, A. E., Kristiansen, E., and Pedersen, S. A. (2008). Is the strategy for cold hardiness in insects determined by their water balance? A study on two closely related families of beetles: Cerambycidae and Chrysomelidae. *Journal of Comparative Physiology B* **178**, 977–984.

Zhang, G. J. (2006). Protein chaperones and winter cold hardiness in insects: heat shock proteins and glucose regulated proteins in freeze-tolerant and freeze-avoiding species. M.Sc. thesis, Department of Chemistry, Carleton University.

7

Interactions between cold, desiccation and environmental toxins

MARTIN HOLMSTRUP, MARK BAYLEY, SINDRE A. PEDERSEN AND
KARL ERIK ZACHARIASSEN

7.1 Introduction

Climatic stressors are environmental factors of paramount importance for biological systems. Numerous examples from terrestrial plants and animals show that, in particular, cold and drought are factors that dictate the distribution of species (for a review see Hoffmann and Parsons, 1991). Indeed, sediment records of insect remains have been convincingly used to reconstruct past climates (e.g. Atkinson *et al.*, 1987). Other important types of stress that should also be considered are environmental toxins. These may be of anthropogenic origin, but most originate from natural sources. Potentially toxic trace metals are released into ground water by demineralization of rocks and minerals and enter the food chain via plants. A wide variety of toxic organic compounds are produced by aquatic as well as terrestrial plants and animals as protection against herbivory or predation. Insects (and all other animals) have, during evolution, developed adaptive detoxifying mechanisms to cope with toxic stress. These protective systems are further challenged due to increasing levels of anthropogenic chemical pollution (e.g. pesticides and heavy metals) that have been inflicted on the biosphere at a global scale. Environmental toxins per se can have negative effects on the functioning of organisms, but the possibility also exists that some of these chemicals can impact tolerance mechanisms to dominating climatic variables such as cold and drought.

Insects and other organisms have evolved sophisticated physiological and biochemical mechanisms to cope with different environmental stressors. Examples of such mechanisms include various proteins to counteract thermal and other stresses, systems involved in detoxification, such as metallothioneins and the

Low Temperature Biology of Insects, ed. David L. Denlinger and Richard E. Lee Jr. Published by Cambridge University Press.

mixed-function oxidase systems, and the use of compatible solutes to counteract the deleterious effects of osmotic stress, dehydration, or freezing.

In a natural environment, insects and other organisms are seldom exposed to a single stressor. On the contrary, multiple stressors usually act on the organism at the same time. The present chapter aims at providing a summary of the nature of cold and desiccation as environmental stressors and how insects cope with these forms of stress. Further, we then continue to discuss how desiccation may influence cold-tolerance and address the physiological links connecting cold and desiccation in insects. Finally, the role of environmental pollution in insect cold-hardiness is considered.

7.2 Features of cold and dehydration stress

7.2.1 Low-temperature stress

Temperature has a variety of effects on living organisms. In insects and other ectothermic organisms, the body temperature corresponds to ambient temperature and the rates of biochemical processes increase rapidly with increasing temperature. The relationship between temperature and the rate of biological processes is presented as a Q_{10} value, which is the quotient by which the rate of a process increases when the temperature increases by 10 °C (Schmidt-Nielsen, 1997). Temperature also affects the features of lipids, which become more viscous at lower temperatures. This will influence the properties of phospholipids in cellular membranes and alter the permeability of various solutes through the membranes.

Exposure of animals to subzero temperatures above the supercooling point (the temperature where spontaneous freezing occurs) may cause injury not related to ice formation. The nature of these injuries is not clear, but they may involve unfavorably high phospholipid viscosity in cell membranes, changes in protein structure and enzyme inhibition. Since enzymatic processes, including active ionic transport, are affected according to the Q_{10} effects described above, and passive diffusion varies according to absolute temperature (°K), reduced temperature may reduce active transport through cell membranes considerably while diffusion is hardly affected. This may cause an intolerable redistribution of solutes across cell membranes and an osmotic redistribution of water.

Even rapid cooling to a moderately low temperature and re-warming after a short period of time (cold shock) may have injurious effects. The sudden change in temperature may affect protein structures, as well as lipid viscosity. At temperatures sufficiently far below the melting point of the body fluids, the body fluids will freeze. Tiny ice spears may penetrate cell membranes and cause mechanical damage (Mazur, 1977). Ice formation is also associated with a reduction in the amount of solvent water and causes an increase in the concentrations of body fluid solutes. Salts and other solutes may eventually reach toxic levels (Lovelock, 1953).

Tropical insects and temperate summer insects will freeze and die if cooled to about −10 °C. Their supercooling points are far higher than those of similar volumes of physical solution with the same osmolality, indicating that the freezing of their body fluids is triggered by some kind of ice-nucleating structure. These structures seem to be confined to intracellular or intestinal compartments (Zachariassen, 1980; Sømme & Conradi-Larsen, 1977), where ice nucleation and growth of ice will cause osmotic swelling and eventually the rupture of surrounding membrane structures.

7.2.1.1 Freeze-avoidance

Most freeze-avoiding insects seek to avoid the formation of ice in their body fluids by developing a high capacity for supercooling, but some dehydrate, thereby reducing their melting point to the ambient temperature. High supercooling capacity is established by the removal or inactivation of all structures that may trigger ice nucleation (Zachariassen, 1980; Wu and Duman, 1991). This brings the supercooling points down to about −20 °C (Zachariassen, 1980). A further depression of the supercooling point can be achieved by the accumulation of polyols, which depress the supercooling points of these insects by 2–3 times the corresponding melting-point depression (Zachariassen, 1980; Gehrken, 1989). In addition, some freeze-avoiding insects produce highly active antifreeze proteins (AFPs) (see also Chapter 3). AFPs are known for their capacity to prevent the growth of seeding ice crystals upon cooling. There is also evidence that AFPs are involved in the deactivation of ice-nucleating structures in freeze-avoiding insects, where AFPs attach themselves to ice-like structures associated with the nucleators and thus inhibit their ice-nucleating function (Wu *et al.*, 1991; Wang and Duman, 2005). Furthermore, there is experimental data indicating that AFPs prevent inoculation of external ice through the body wall (Olsen *et al.*, 1998; Gehrken, 1992).

7.2.1.2 Freeze-tolerance

In general, freeze-tolerant insects seem to freeze at higher temperatures during winter than summer (Zachariassen, 1980). This high-temperature freezing indicates the presence of highly potent extracellular ice-nucleating agents (INAs), which trigger freezing in the hemolymph before the organism is supercooled to the temperature where intracellular ice-nucleating structures trigger freezing. One important physiological function of this induced extracellular freezing is probably to prevent ice formation in closed compartments, such as cells and the intestine.

As water is transformed into ice in the extracellular fluid, the extracellular solute concentration increases until vapor-pressure equilibrium with the ice is reached. This elevated extracellular solute concentration causes an osmotic efflux of water from the closed compartments and the shrinkage of intestinal and intracellular volumes until osmotic equilibrium is re-established. Thus freezing in these closed compartments is prevented (Kanwisher, 1959; Zachariassen and Hammel, 1976). Eventually, this process can result in the concentration of body fluid solutes to toxic levels, which may in fact define the lethal temperature limit for freeze-tolerant organisms. The organism can lower this temperature by polyol accumulation, which reduces the amount of ice formed colligatively (Zachariassen, 1979; van der Laak, 1982).

This colligative depression of lethal temperature is substantial, imbuing freeze-tolerant insects with greater cold-hardiness than can be achieved by freeze-avoiders, even at moderately high polyol concentrations. With high polyol concentrations, freeze-tolerant insects can tolerate cooling to −90 °C (Miller, 1982), and freeze-tolerance seems to be the preferred strategy of cold-hardiness among insects inhabiting extremely cold areas such as interior Alaska and Siberia (Ring, 1982; Miller, 1982; Zachariassen, 1985).

Not all water in an organism is freezable. A certain fraction of the body water is termed osmotically inactive water (OIW) because it does not act as a solvent for low-molecular-weight solutes in the body fluids. The OIW is probably the water that is bound to cellular (e. g. membranes) and macromolecular (e. g. proteins) components in such a way that it is not readily freezeable, at least not under short-term (ecological) conditions. The role of OIW in freeze-tolerant invertebrates has been much debated (Lee, 1991). However, there does not seem to be any clear relationship between the fraction of OIW and the degree of freeze-tolerance (Zachariassen *et al.*, 1979; Zachariassen, 1991).

7.2.2 Dehydration stress

Water makes up about 70% of the body mass of most animals. The volumes of cells and extracellular fluid are closely related to their water content, and cells and organisms have a limited tolerance to variation in water content. Significant variation in the cellular volume disturbs cellular function, both because it will change the steric organization of the organelles and because it may affect the concentrations of intracellular solutes. Since many solutes interact with intracellular enzymes and other proteins, a change in their concentration will change protein structure and the function of enzymes. Animals using blood to transport dissolved respiratory gases within the organism are particularly sensitive to changes in blood volume, whereas insects, which transport gases via the trachea, appear to be far less sensitive.

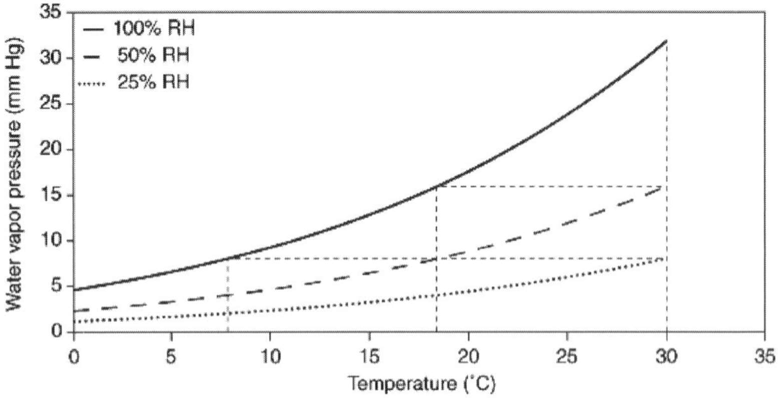

Figure 7.1 Relationship between pressure of water vapor in the atmosphere and temperature for saturated air (100%), and for air at 50% and 25% relative humidity. The percentage of the actual vapor pressure of water in the air (VP) relative to the saturation pressure (VP*) at a given temperature is referred to as the relative humidity (RH), and is calculated according to the formula, $RH(\%) = VP/VP^* \cdot 100$. If the relative humidity at a given temperature is known, the actual vapor pressure can be calculated from the saturation pressure at the same temperature. Dashed lines indicate, for example, that air at 18 °C, 100% RH has the same amount of moisture as air at 30 °C, 50% RH.

When liquid water is brought into contact with a gas phase (air), water molecules will leave the liquid and enter the gas phase. Water molecules will also return to the liquid phase, and the rate of water molecules returning to the liquid phase will eventually become equal to the rate of water molecules leaving the liquid phase, i.e. a steady-state situation is established. The gas phase is then said to be saturated with water. The saturation pressure is strongly influenced by temperature, in that there is a steep increase in saturation pressure when temperature increases (Fig. 7.1).

The difference between the saturation pressure and the actual pressure is referred to as the saturation deficit. The saturation deficit determines the capacity of the gas to absorb vapor, and hence the rate of evaporative water loss from an insect is proportional to the vapor saturation deficit. In terrestrial organisms, this evaporative water loss takes place along two main routes; by evaporation through the cuticle of the outer body wall and through the respiratory spiracles with the exchange of respiratory gases. Because of their small body size, insects have a relatively large body surface compared to their body mass, and they also have a relatively high mass-specific metabolic rate (Schmidt-Nielsen, 1997). Hence, insects and other small invertebrates have a strong disposition for losing body water by evaporation.

7.2.2.1 Mechanisms to counter dehydration stress

Animals have developed a variety of mechanisms to control their water content, both at the cellular level and at the level of the whole organism. Insects have adapted to low humidities by reducing the water permeability of the cuticle. Insects living in arid areas have almost eliminated cuticular water transport, and respiratory water loss is the dominant component of water loss during dry periods (Zachariassen *et al.*, 1987; Zachariassen and Maloiy, 1989). Many insects display discontinuous spiracular opening, where spiracles remain closed most of the time (Lighton, 1994). In some cases this may be considered a water-conserving adaptation, though this has been much debated, and recently it has been proposed that discontinous gas exchange is more an adaptation to counter oxidative stress (Hetz and Bradley, 2005).

Most insects enter diapause during winter, which in combination with low winter temperatures reduces the respiratory water loss to a minimum. Nonetheless, supercooled insects are prone to desiccation, since supercooled water has a higher vapor pressure than ice at the same temperature. In a frozen hibernaculum the air humidity corresponds to the vapor pressure of ice, which may cause water vapor to leave a supercooled insect through the cuticle (see Section 7.3.2 for details). Since hibernating insects can remain supercooled for months, the evaporative loss of water may eventually become intolerable. To reduce transcuticular evaporation many freeze-avoiding insects have reduced cuticular water permeability to a minimum (Lundheim and Zachariassen, 1993). These authors also reported that frozen beetles (*Upis ceramboides* and *Pytho depressus*) lose water at a slower rate during winter than supercooled beetles, suggesting that a freeze-tolerance strategy is advantageous when considering desiccation as a winter mortality factor.

While most vertebrates are intolerant of variations in their water content, some insects can tolerate the loss of up to about 75% of their body water. Here, water loss takes place mainly at the expense of the extracellular fluid, with cellular water loss being minimized. In the desert tenebrionid beetle, *Rhytinota praelonga*, evaporative dehydration is associated with a more rapid drop in hemolymph volume than in cell water. When the hemolymph volume reaches zero, the intracellular water content is reduced by only 50%. These beetles seem to function normally with only a tiny volume of hemolymph remaining, but die when the remaining hemolymph evaporates (Zachariassen and Einarson, 1993). The asymmetric volume reduction of the body fluid compartments requires a redistribution of solutes from the hemolymph to the cellular compartments, but the nature of this solute redistribution is not completely understood. Sodium removed from the hemolymph is not transferred to cellular compartments and is likely to have been excreted (Zachariassen and Einarson, 1993). Similar redistribution of water loss between body fluid compartments in hibernating insects undergoing dehydration has not been studied, but it seems that cold-exposure itself causes the hemolymph volume to drop and the water content of the gut to increase.

7.3 Links between desiccation and cold-tolerance

7.3.1 Role of desiccation in freeze-avoiding insects

Dehydration causes an increase in body-fluid solute concentrations. Since high concentrations of polyols have a cryoprotective effect, dehydration may increase the cold-hardiness of organisms. This type of cryoprotective role of polyols has been reported in several very cold-hardy beetle species from Canada and Alaska. Ring (1982) demonstrated that larvae of the beetle *Pytho deplanatus* could be dehydrated in the laboratory from an initial relative water content of 69% to 30%, or a loss of about two-thirds of their water content. In this highly dehydrated state the larvae had a supercooling point of −54 °C. Due to their extremely high polyol concentrations, they even survived freezing at this low temperature. An interesting example is also provided by Bennett *et al.* (2005), showing that the beetle *Cucujus clavipes* becomes severely dehydrated during winter, increasing glycerol concentrations up to 10 M and elevating thermal hysteresis to almost 13 °C due to increased protein concentrations. This significant cryoprotective effect of dehydration is restricted to animals that can accumulate high concentrations of polyols. As pointed out by Zachariassen (1985), insects with low polyol concentrations display only a moderate depression of their supercooling point as a consequence of dehydration, whereas insects with high polyol concentrations may depress their supercooling point by more than 25 °C.

7.3.2 Cryoprotective dehydration strategy

For permeable insects and other invertebrates that overwinter surrounded by ice (e.g. in rotting wood or in frozen soil), the physical properties of ice and supercooled water are of major importance for both cold-tolerance and the overall water balance. For example, egg capsules ("cocoons") of the earthworm *Dendrobaena octaedra* rapidly dehydrate when subjected to −3 °C in small vials, where the air humidity is defined by ice (Holmstrup and Westh, 1994). As mentioned earlier, the vapor pressure of supercooled water is higher than the vapor pressure of ice (Fig. 7.2). This will cause a transport of water (vapor) from the supercooled body fluids to the ice, where water vapor is condensed, forming ice. The desiccating potential of ice can be expressed as a difference between the "osmotic pressure of ice" and the osmotic pressure of the organisms' body fluids. The difference in osmotic pressure between ice and the supercooled organism increases with

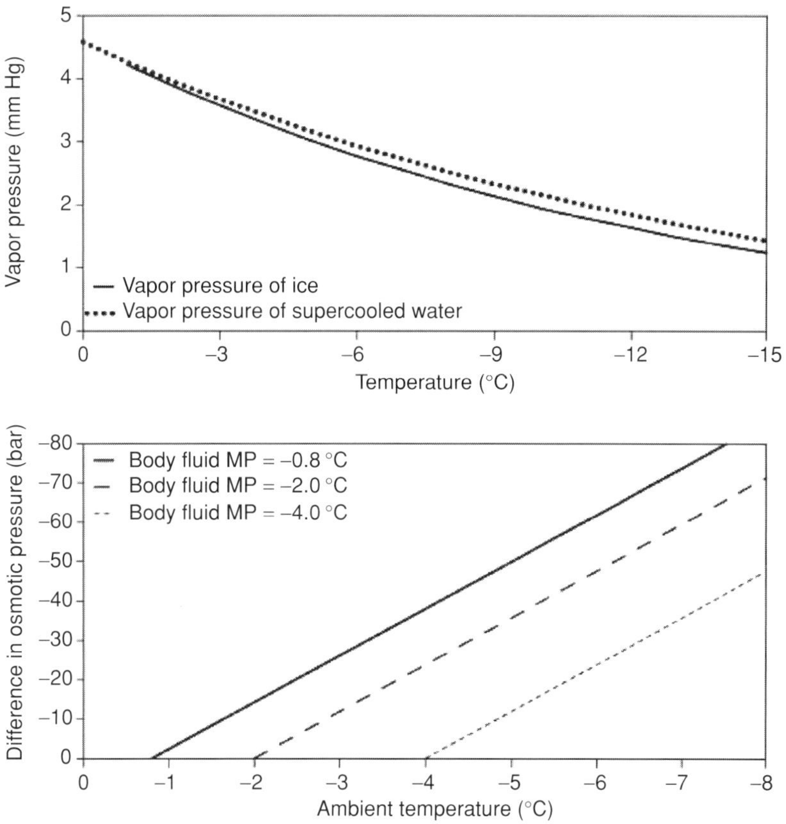

Figure 7.2 The difference in vapor pressure between ice and supercooled water is the driving force for cryoprotective dehydration. Panel A: The vapor pressure of supercooled water is always greater than the vapor pressure of ice. The pressure difference between water and ice increases with decreasing temperature. Data from Weast (1989). Panel B: This figure shows potential differences in "osmotic pressure" (OP) between the body fluids of a supercooled animal and the surrounding ice at varying ambient temperature. Even though we realize that it is not strictly correct to use the term "osmotic pressure differences" in this case where insect hemolymph is separated from the surrounding ice by air, such a comparison is useful to illustrate the magnitude of the desiccating power of ice. The more intense supercooling becomes, the larger becomes the OP difference. A negative value of OP difference means that the organism loses water to the surrounding ice. The osmotic pressure (ψ) of body fluids at a given temperature has been calculated using Van't Hoff's equation: $\psi = Osm{\cdot}R{\cdot}T$, where *Osm* is the osmolality of body fluids, *R* is the gas constant and *T* is absolute temperature (°K). The melting point of body fluids (MP) is calculated by application of the osmolal melting point depression constant (−1.86 °C osmol^{-1} kg). The melting point of a solution is defined by the vapor pressure of ice when there is equilibrium between ice and solution in the system. Thus, the "osmotic pressure of ice" can be calculated by transforming the temperature of ice (ambient temperature) to osmolality, and then use Van't Hoff's equation. Modified from Holmstrup *et al.* (2002).

decreasing temperature, and is substantial, even in situations where there is only a slight difference between the body fluid MP and the temperature of the ice, ranging to 50 bar or more at ecologically relevant temperatures (Fig. 7.2). The dehydration will continue until the vapor pressure of body fluids equals that of the surrounding ice, i.e. until the body-fluid melting point (due to concentration of original solutes and newly synthesized polyols or sugars) equals the temperature of the ice (Holmstrup and Westh, 1994; Holmstrup *et al.*, 2002). At this stage, the risk of tissue ice formation has been eliminated, and subzero survival is ensured.

As shown above, the driving force for dehydration in a frozen environment is indeed substantial. However, for the dehydration mechanism to be effective, the lowering of body-fluid melting point must match the cooling rates encountered in the overwintering habitat in order to avoid extensive supercooling and the associated risk of lethal freezing. Therefore, a prerequisite for frost-induced dehydration is a high integumental permeability for water. Most terrestrial insects have a quite impermeable integument. However, a number of hydrophilic insect larvae and soil-living arthropods such as Collembola are characterized by a high permeability for water through their cuticle. In small permeable invertebrates, the dehydration process may therefore occur rapidly, matching most natural cooling rates (Holmstrup *et al.*, 2002).

The best-known example of an arthropod employing cryoprotective dehydration as a cold-tolerance strategy is found in the Arctic collembolan, *Megaphorura arctica* (formerly known as *Onychiurus arcticus*). These studies show that *M. arctica* becomes severely dehydrated when exposed to low subzero temperatures (Holmstrup and Sømme, 1998; Worland *et al.*, 1998). Thus, as was the case with earthworm cocoons, the water content of this collembolan equilibrates to a level dictated by the temperature (i.e. the "osmotic pressure of ice") of its surrounding frozen environment (Holmstrup *et al.*, 2002). It is clear that these animals must be able to tolerate extensive dehydration, which is indeed the case. *M. arctica* tolerates the loss of practically all osmotically active water (about 90% of the initial body water), by accumulating high trehalose concentrations (Holmstrup and Sømme, 1998; Worland *et al.*, 1998; Bahrndorff *et al.*, 2007), comparable to levels seen in anhydrobiotic organisms (Hawes and Bale, 2007). An outline of the cryoprotective dehydration mechanism is shown in Fig. 7.3.

Cryoprotective dehydration bears some resemblance to freeze-tolerance in which extracellular ice formation induces dehydration of cells, thereby protecting against cellular freezing (Zachariassen, 1985). Parallel to this, organisms using cryoprotective dehydration are protected by the dehydrating effects of environmental ice. As a response to dehydration, both groups of organisms usually accumulate cellular sugars and polyols, which may prevent deleterious effects of extensive water loss, coupled with osmotic pressure equilibration.

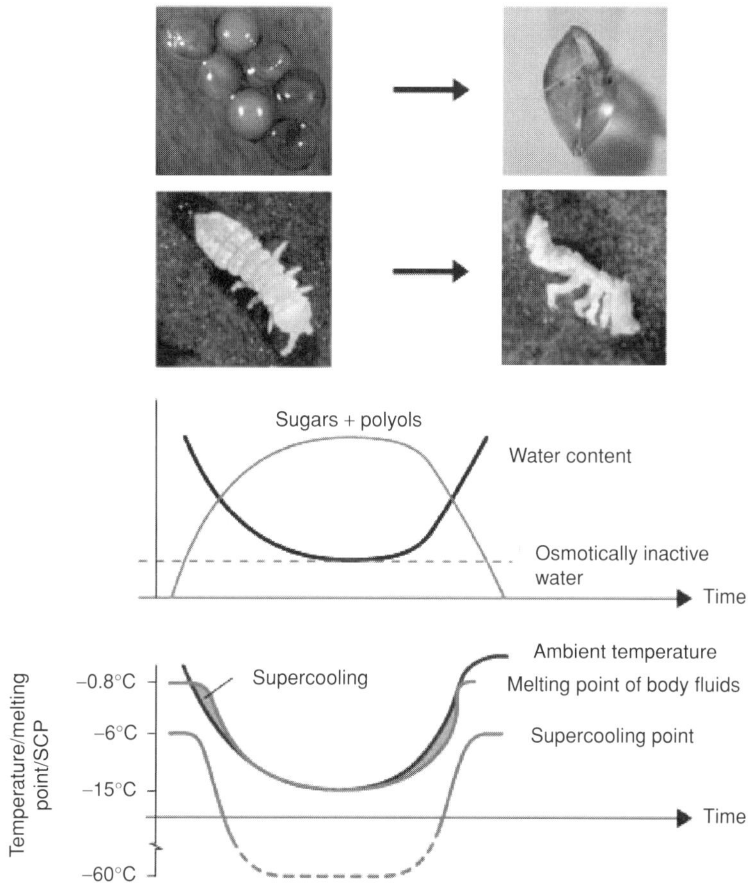

Figure 7.3 A general model of cryoprotective dehydration. Decreasing temperature in frozen soil causes dehydration of the organism, which in turn induces accumulation of sugars and polyols, SPs (upper panel). At low temperature (−15 to −20 °C), practically all osmotically active water is lost. Dehydration and SP accumulation together bring about a lowering of melting point (MP), largely at the same rate as soil temperature decreases. Supercooling is therefore restricted to only a few degrees and only during a short initial period at relatively high subzero temperatures (lower panel). A temperature rise causes the animal to take up water, and increase the body fluid MP. Fully hydrated animals have supercooling points of approximately −6 °C. Dehydration results in a depression of the supercooling point. When dehydration approaches the level of osmotically inactive water, freezing cannot occur. The photos show fully hydrated cocoons of *Dendrobaena octaedra* and collembolan *Megaphorura arctica* (photos at left), and animals dehydrated at −6 °C for 7 days (photos at right).

The cryoprotective dehydration strategy has so far only been studied in a few insect species. However, its effectiveness as an overwintering strategy has been demonstrated in a variety of terrestrial invertebrates whose only relevant common feature in this context is their high permeability for water, suggesting that cryoprotective dehydration is a widespread winter survival strategy in terrestrial invertebrates. Thus, frost-induced dehydration has been shown in nematodes (Wharton *et al.*, 2003), enchytraeids (Sømme and Birkemoe, 1997; Pedersen and Holmstrup, 2003), and in aquatic chironomid larvae (Scholander *et al.*, 1953; Danks, 1971). Recently, cryoprotective dehydration has been shown in the collembolan, *Cryptopygus antarcticus*, and in larvae of the Antarctic midge, *Belgica antarctica* (Elnitsky *et al.*, 2008a,b). These authors also showed that cryoprotective dehydration in *B. antarctica* may be constrained by inoculative freezing if soil moisture is high, whereas at low soil moisture cryoprotective dehydration predominates. *B. antarctica* is at the same time freeze-tolerant and tolerates extensive dehydration and may therefore utilize both mechanisms, depending on the soil moisture content during winters, a dual strategy also known from Arctic enchytraeids (Pedersen and Holmstrup, 2003).

7.3.3 Drought acclimation responses improving cold-tolerance

The early colonization of land by aquatic organisms has necessitated the evolution of desiccation-tolerance mechanisms, which is perhaps the most fundamental feature of terrestrial organisms (Hadley, 1994). Several authors have therefore proposed that many of the adaptations promoting cold-tolerance in insects have in fact originally evolved to tolerate desiccation (e.g. Block, 1996; Ring and Danks, 1994, 1998). One of the most important arguments in favor of this idea is that accumulation of compatible solutes, such as sugars and polyols seems to be central in tolerance to both drought and cold. Thus, sugars and polyols may slow dehydration rates in freeze-avoiding species during winter and have the same effect under summer drought (Ring and Danks, 1994, 1998). Sugars and polyols may reduce cellular dehydration in freeze-tolerant organisms during winter, improve the supercooling capacity of freeze-avoiding species and lower the equilibrium dehydration level in summer drought-exposed organisms (Zachariassen, 1985; Bayley and Holmstrup, 1999). The specific protection of proteins provided by sugars and polyols may prevent denaturation, both due to low winter temperatures and summer dehydration, with anhydrobiotic organisms as the most spectacular example (Crowe *et al.*, 1992). Equally important, a variety of sugars and polyols can stabilize membrane structure, exerting a direct influence on membrane fluidity by lowering the phase-transition temperature (T_m) of cellular phospholipid membranes (Crowe and Crowe, 1986).

Despite these clear links between adaptations to cold and desiccation, there are relatively few insect examples showing that a previous acclimation to desiccation confers increased cold-tolerance, or vice versa. Recently, Hayward *et al.* (2007) showed that *B. antarctica* exposed to moderate dehydration over two days at 98.2% RH and 4 °C (about 30% water loss) resulted in a dramatically increased freeze-tolerance at −10 and −15 °C when compared to non-desiccated midges. Based on measurements of body fluid osmolality the authors suggested that dehydration-induced synthesis of osmolytes could have contributed to increased freeze-tolerance, but other mechanisms such as alterations of membrane phospolipid composition were also proposed as beneficial for freeze-tolerance.

These two physiological modifications, osmolyte production and membrane alterations, probably also play a significant role for the increased cold-tolerance observed in *Folsomia candida* (Collembola) after drought acclimation. In a study of *F. candida*'s responses to drought it was shown that an 8-day acclimation at 98.2% RH induces an increase in the mole percent of monounsaturated phospholipid fatty acids and linoleic acid, resembling typical changes following cold-acclimation in other ectothermic animals (Bayley *et al.*, 2001). In addition to membrane changes during acclimation to 98.2% RH *F. candida* also accumulates high concentrations of myo-inositol and glucose, which together with dehydration greatly decrease cellular water activity (Bayley and Holmstrup, 1999). Hazel and Williams (1990) discuss membrane adaptations to changes in water activity, which seem relevant to the observations made for *F. candida* under drought stress (and organisms that use cryoprotective dehydration). Water activity is important because the interactions between water and phospholipids are the fundamental feature of membrane organization in a bilayer structure. Thus, a decrease in cellular water activity may result in tighter phospholipid packing and therefore lower fluidity, which would also be the result of a sudden drop in temperature. The typical cellular response to this is activation of desaturation enzymes to re-establish "normal" membrane fluidity. If this hypothesis is true, it may explain why acclimation to drought increases cold-tolerance, be it increased freeze-tolerance as in *B. antarctica* or increased chill-tolerance as in *F. candida*. Obviously, if acclimation to drought also initiates production of sugars and polyols these osmolytes will also directly increase the stability of cellular membranes under cold exposure (Crowe *et al.*, 1992). Further study of these cross-tolerance mechanisms for both drought and cold is likely to be rewarding.

Heat-shock and other chaperoning proteins have been suggested to play a significant role in tolerance to both drought and cold (Goto *et al.*, 1998; Tammariello *et al.*, 1999; Bayley *et al.*, 2001; Hayward *et al.*, 2004). However, desiccation-induced up-regulation of heat-shock proteins did not improve cold-tolerance in *Sarcophaga crassipalpis* and the up-regulation was less dramatic than that following heat exposure (Tammariello *et al.*, 1999). Browne *et al.* (2002) showed that specific proteins were up-regulated in drought-exposed nematodes and that some of these proteins showed similarity to LEA proteins (Late Embryogenesis Abundant), until recently only known in plants. Since then, LEA proteins have also been found in dehydrating chironomids (Kikawada *et al.*, 2006) and rotifers (Tunnacliffe *et al.*, 2005). Although the actual functional mechanism of LEA proteins is not known precisely, the hypothesis is that LEAs are "surfactants" capable of inhibiting the coagulation of a range of macromolecules, thereby preserving structural integrity of cell membranes and proteins during desiccation. Although presently un-investigated, these effects of LEAs could play a role in the cross-tolerance with cold observed in Collembola and insects, as reported for several plant species (Kosova *et al.* 2007). Recently, an LEA sequence was obtained from an EST library of *M. arctica*, suggesting the presence of LEA proteins in this collembolan that is protected against cold by cryoprotective dehydration (Clark *et al.*, 2007; Bahrndorff *et al.*, 2009). LEA proteins may therefore have a dual role, offering protection against both cold and desiccation.

7.4 Interactions between cold, drought and toxic stress

The study of interactions between effects of environmental toxins and climatic stress has obvious ecological significance and has greatly increased the understanding of how pollution affects coastal benthic animals where the environment is dominated by variations in salinity and oxygen tension (e.g. Thurberg *et al.*, 1973) and in understanding the effects of acid rain, where reduced cold- and drought-tolerance play an important role in tree death (McLaughlin and Percy, 1999). In spite of the potential rewards of studying such interactions, there are very few examples in the literature and they certainly haven't seen systematic investigation in ecophysiology, ecotoxicology, or climate-change studies. A major difficulty with interaction studies is that the addition of an extra dimension, such as a pollutant in a study of temperature tolerance, dramatically increases the complexity of the study and therefore decreases the likelihood of being able to draw solid conclusions. In addition, when considering temperature or drought effects on ectothermic animals in combination with chemical pollutants, it is important to realize that interactions between the two types of stress factors (environmental and chemical) can occur in several ways. Firstly, as argued above, the pollutant can directly impact the physiological mechanism behind tolerance, or indirectly have an impact, for example by redirecting energy away from a normal acclimation response to detoxification processes. Secondly, environmental variables, such as temperature and water availability, can dramatically affect the bioavailability of the pollutant as a result of changing the binding of the pollutant to environmental matrices, or indeed as a result of changing the organism's activity level in its habitat (Van Gestel, 1997). The vast majority of interaction studies in the literature between temperature and chemical pollutants concern either this second category of bioavailability alone, or involve an experimental design where it is impossible to distinguish between the two overall interaction types, which detracts significantly from the general applicability of the study.

One solution to this dilemma is to temporally separate the two stress forms so that exposure occurs under standard laboratory conditions, where chemical loading of the organism can be assured, after which the tolerance of the organism to the environmental variable of interest is quantified. The extreme of this approach to studying interactions involves the use of full factorial designs, where both stress types are applied over a wide range of intensities and all interactions between the different degrees of stress are assessed. The result is a very large number of treatments, but, if successful, this allows the entire response surface to be mapped. This approach has been used to show synergistic mortality interactions between realistic drought stress and environmentally realistic nonylphenol concentrations in the collembolan *F. candida* (Højer *et al.*, 2001) and by Bindesbøl *et al.* (2005) to show a similar interaction between frost temperatures and environmentally realistic copper concentrations in the earthworm *D. octaedra* (Fig. 7.4.A, C). The approach of mapping the entire survival surface allows the calculation of how the toxicity of the chemical is impacted by subsequent environmental stress (Fig 7.4.B), or the calculation of how the toxin impacts the animal's tolerance of the climatic variable (Fig 7.4.D)

Very little data exist on the effects of environmental toxins on the susceptibility of terrestrial invertebrates to temperatures below their body-fluid melting points, but such adverse effects have been shown with copper (Holmstrup *et al.*, 1998; Bindesbøl *et al.*, 2005), mercury (Holmstrup *et al.*, 2008) and surfactants (Horton *et al.*, 1996). More evidence exists for interactions between drought and a range of toxic compounds (e.g. Skovlund *et al.*, 2006), but this issue will not be discussed further here.

Freeze-tolerance and dehydration during summer both require the rapid movement of water and compatible osmolytes, such as glycerol, across cell membranes, as discussed in previous sections. Environmental toxins can affect membranes by targeting specific membrane molecules such as aquaporins (Izumi *et al.*, 2006; Philip *et al.*, 2008), or by accumulating in the phospholipid core and interacting in a non-specific way with membrane lipids (Chaisuksant *et al.*, 1999). Thus, Izumi *et al.* (2006) showed that frost survival of fat-body cells from the rice stem borer, *Chilo suppressalis*, was severely impacted by mercury, as a result of this chemical's known interaction with aquaporins (Preston *et al.*, 1992). Further, using radiotracer

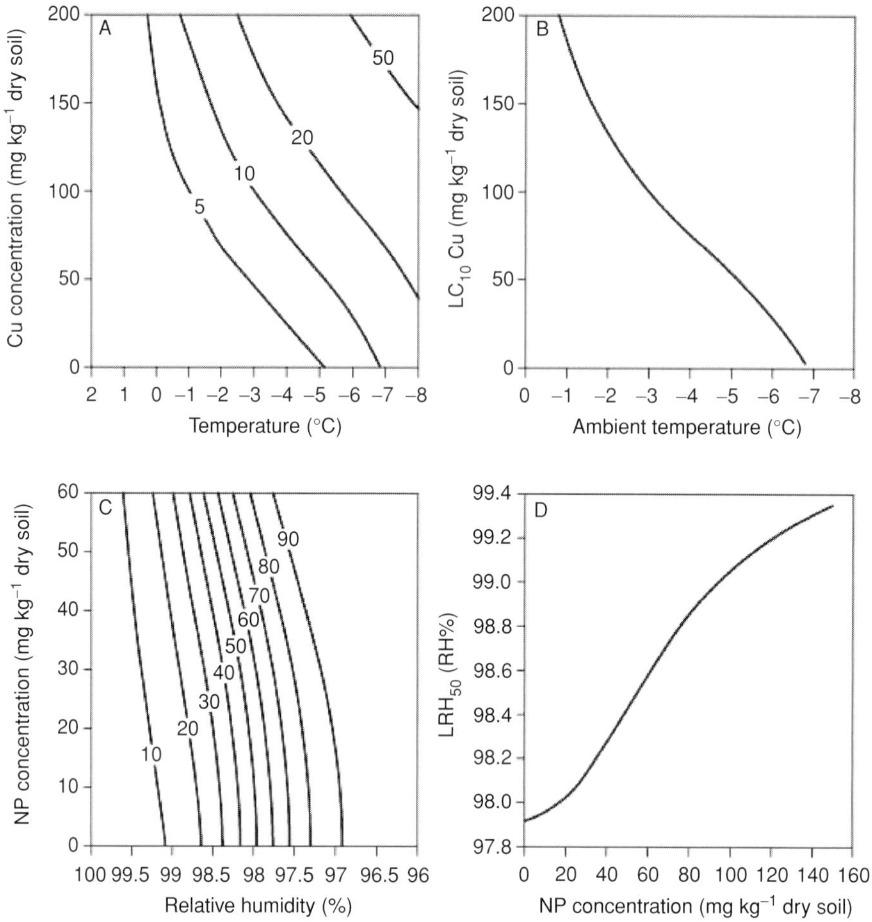

Figure 7.4 Full factorial experiments between environmental toxins at realistic concentrations (Copper (Cu) and Nonylphenol (NP)) and realistic levels of drought and frost. Figures A and C show mortality isoclines where the number shown on the cline indicates the mortality in %. Mapping the entire response surface in this way allows other useful data to be derived. Thus B shows how copper toxicity increases as the animal is exposed to increasingly severe frost. The corollary of this is seen in D, showing how drought survival is increasingly impaired as the concentration of nonylphenol rises. Modified from Bindesbøl *et al.* (2005) and Højer *et al.* (2001).

techniques, Izumi *et al.* (2006) showed that the transport of both water and glycerol was almost eliminated by mercury, suggesting the involvement of aquaporin 3 (AQP3), which allows the passage of both (Ishibashi *et al.*, 1994). Similar effects of mercury were shown in the goldenrod gall fly (*Eurosta solidaginis*) where AQP3 was up-regulated by both desiccation and frost (Philip *et al.*, 2008). Interestingly, in human erythrocytes, both the water and glycerol permeability of AQP3 is rapidly and significantly inhibited by concentrations of copper as low as the 0.5 mM found during several copper toxicity syndromes (Zelenina *et al.*, 2004). Given the ability of copper to impair frost- and drought-tolerance at environmentally realistic concentrations, interactions between this metal and AQP3 may well provide a mechanistic explanation for the response. Metals such as mercury and copper are known to cause protein denaturation as a result of their affinity for sulfhydryl groups and their ability to peroxidize membrane lipids (Valko *et al.*, 2005), both of which may affect membrane integrity. Holmstrup *et al.* (2008) showed that mercury impairs survival of a cold-shock in *F. candida* after rapid cold-hardening, at temperatures between the animal's melting point and supercooling points, where water transport across cell membranes is not an issue. Recent data shows, in addition, that the ability of earthworms to change the lipid composition of their membranes in response to lowered temperatures is also impacted by copper at environmentally realistic concentrations (Bindesbøl *et al.*, 2009). Apart from these mechanisms it should also be considered that both freezing and dehydration will reduce the volume of liquid water in the organism, thereby increasing the concentration of metal ions and the risk of toxic damage.

It has been proposed by Zachariassen *et al.* (2004) that metal stress may affect cryoprotective mechanisms of AFPs, because metal-binding metallothioneins and many types of insect AFPs contain high amounts of cysteine. Exposure to heavy metals could therefore lead to competition for cysteine and a reduced capacity of insects to produce AFP. This idea was explored by Pedersen *et al.* (2006): levels of AFPs, as well as the associated AFP YL-3 mRNA were reduced after exposure to sublethal concentrations of cadmium, zinc and copper in summer-acclimated meal worms (*Tenebrio molitor*). However, cold-acclimation of these animals at 4 °C resulted in an increase in both AFP levels and gene transcription in all groups, such that the negative effects of the metals disappeared.

Chemicals with surfactant properties may also impact survival by more general mechanisms involved in animal water balance. Thus nonylphenol caused an increase in water-loss rate and a reduced compatible osmolyte production (glucose and inositol) in drought-exposed collembolans (Højer *et al.*, 2001). Horton *et al.* (1996) found that spraying pear psylla (*Cacopsylla pyricola*) with a variety of surfactants caused a dramatic increase in frost mortality, which they argued is a result of ice inoculation through the temporarily moistened integument.

So far, very few laboratory studies have been published on interactions between environmentally relevant toxins and tolerance to dominant environmental variables, such as winter frost. It is therefore too early to evaluate the ecological consequences of these interactions, but several studies suggest that interactions can be important in severely polluted areas.

7.5 Future directions

It has been hypothesized that cryoprotective dehydration as a cold-tolerance strategy is most likely to be present in species with a permeable cuticle, because rapid loss of water is a prerequisite for the mechanism to be efficient (Holmstrup *et al.* 2002). However, research is needed to establish whether there is a threshold permeability allowing cryoprotective dehydration and how widespread the mechanism is within insects and other invertebrates. For example, cold-tolerance of the Antarctic collembolan, *C. antarcticus*, has been studied intensively with a main focus on supercooling abilities. A recent study by Elnitsky *et al.* (2008b) demonstrates, though, that cryoprotective dehydration is a mechanism that also applies to this species when it is exposed to environmental conditions simulating the actual conditions experienced in the field. Thus, successful exploration of the interactions between cold and desiccation demands investigations considering and closely resembling environmentally realistic conditions. Although this plea may seem self-evident, such approaches are relatively sparse in the literature.

Many molecular mechanisms underlying cross-tolerance of desiccation- and cold-induced stress are largely unexplored and should receive much more attention in future research. Especially the roles of LEA proteins, heat-shock proteins and other chaperones deserve more attention, since they seem to play an important role of tolerance to both cold and desiccation within insects, and their exploration is likely to increase our understanding of this problem. Modern molecular tools are becoming increasingly available, promoting research in this direction.

As stated earlier in this chapter, much remains to be done in the area of interactions between toxicants and cold stress. A range of ecologically relevant toxicants should be tested to reveal if particular classes of chemicals are more likely to cause synergistic interactions with cold stress than others, which would be of particular interest from an environmental management perspective. For example, environmental pollutants with endocrine-disrupting effects could have an effect on winter survival, since cold-hardening in insects to a great extent is under hormonal control. Also, metal-detoxification mechanisms (metallothioneins) and AFPs seem to have a common ancestry that should be explored further. Although such research is perhaps most interesting for ecotoxicologists it may also be rewarding from an ecophysiological viewpoint.

References

Atkinson, T. C., Briffa, K. R., and Coope, G. R. (1987). Seasonal temperatures in Britain during the past 22 000 years, reconstructed using beetle remains. *Nature* **325**, 587–592.

Bahrndorff, S., Petersen, S. O., Loeschcke, V., Overgaard, J., and Holmstrup, M. (2007). Differences in cold and drought tolerance of high arctic and subarctic populations of *Megaphorura arctica* (Collembola). *Cryobiology* **55**, 315–323.

Bahrndorff, S., Tunnacliffe, A., Wise, M. J., McGee, B., Holmstrup, M., and Loeschcke, V. (2009). Bioinformatics and protein expression analyses implicate LEA proteins in the drought response of Collembola. *Journal of Insect Physiology*, **55**, 210–217.

Bayley, M. and Holmstrup, M. (1999). Water vapor absorption in arthropods by accumulation of myoinositol and glucose. *Science* **285**, 1909–1911.

Bayley, M., Petersen, S. O., Knigge, T., Köhler, H. R., and Holmstrup, M. (2001). Drought acclimation confers cold tolerance in the soil collembolan *Folsomia candida*. *Journal of Insect Physiology* **47**, 1197–1204.

Bennett, V. A., Sformo, T., Walters, K., Toien, O., Jeannet, K., Hochstrasser, R., Pan, Q. F., Serianni, A. S., Barnes, B. M., and Duman, J. G. (2005). Comparative overwintering physiology of Alaska and Indiana populations of the beetle *Cucujus clavipes* (Fabricius): roles of antifreeze proteins, polyols, dehydration and diapause. *Journal of Experimental Biology* **208**, 4467–4477.

Bindesbøl, A. M., Holmstrup, M., Damgaard, C., and Bayley, M. (2005). Stress synergy between environmentally realistic levels of copper and frost in the earthworm *Dendrobaena octaedra*. *Environmental Toxicology and Chemistry* **24**, 1462–1467.

Bindesbøl, A. M., Bayley, M., Damgaard, C., Hedlund, K., and Holmstrup, M. (2009). Changes in membrane phospholipids as a mechanistic explanation for decreased freeze tolerance in earthworms exposed to sublethal copper concentrations. *Environmental Science and Technology*, **43**, 5495–5500.

Block, W. (1996). Cold or drought – the lesser of two evils for terrestrial arthropods. *European Journal of Entomology* **93**, 325–339.

Browne, J., Tunnacliffe, A., and Burnell, A. (2002). Plant desiccation gene found in a nematode. *Nature* **416**, 38.

Chaisuksant, Y., Yu, Q. M., and Connell, D. W. (1999). The internal critical level concept of nonspecific toxicity. *Reviews of Environmental Contamination and Toxicology* **162**, 1–41.

Clark, M. S., Thorne, M. A. S., Purac, J., Grubor-Lasjic, G., Kube, M., Reinhardt, R., and Worland, M. R. (2007). Surviving extreme polar winters by desiccation: clues from Arctic springtail (*Onychiurus arcticus*) EST libraries. *BMC Genomics* **8**, 475.

Crowe, J. H. and Crowe, L. M. (1986). Stabilization of membranes in anhydrobiotic organisms. In *Membranes, Metabolism and Dry Organisms*, ed. C. A. Leopold, London: Comstock Publishing Association, pp. 188–209

Crowe, J. H., Hoekstra, F., and Crowe, L. M. (1992). Anhydrobiosis. *Annual Review of Physiology* **54**, 579–599.

Danks, H. V. (1971). Overwintering of some north temperate and Arctic Chironomidae. II. Chironomid biology. *Canadian Entomologist* **103**, 1875–1910.

Elnitsky, M. A., Hayward, S. A. L., Rinehart, J. P., Denlinger, D. L., and Lee, R. E. (2008a). Cryoprotective dehydration and the resistance to inoculative freezing in the Antarctic midge, *Belgica antarctica*. *Journal of Experimental Biology* **211**, 524–530.

Elnitsky, M. A., Benoit, J. B., Denlinger, D. L., and Lee Jr., R. E. (2008b) Desiccation tolerance and drought acclimation in the Antarctic collembolan *Cryptopygus antarcticus*. *Journal of Insect Physiology* **54**, 1432–1439.

Gehrken, U. (1989). Supercooling and thermal hysteresis in the adult bark beetle, *Ips acuminatus Gyll*. *Journal of Insect Physiology* **35**, 347–352.

Gehrken, U. (1992). Inoculative freezing and thermal hysteresis in the adult beetles *Ips acuminatus* and *Rhagium inquisitor*. *Journal of Insect Physiology* **38**, 519–524.

Goto, S., Yoshida, K., and Kimura, M. (1998). Accumulation of Hsp70 mRNA under environmental stresses in diapausing and nondiapausing adults of *Drosophila triauraria*. *Journal of Insect Physiology* **44**, 1009–1015.

Hadley, N. (1994). *Water Relations of Terrestrial Arthropods*. San Diego, CA: Academic Press.

Hawes, T. C. and Bale, J. S. (2007). Plasticity in arthropod cryotypes. *Journal of Experimental Biology* **210**, 2585–2592.

Hayward, S. A. L., Rinehart, J. P., and Denlinger, D. L. (2004). Desiccation and rehydration elicit distinct heat shock protein transcript responses in flesh fly pupae. *Journal of Experimental Biology* **207**, 963–971.

Hayward, S. A. L., Rinehart, J. P., Sandro, L. H., Lee, R. E., and Denlinger, D. L. (2007). Slow dehydration promotes desiccation and freeze tolerance in the Antarctic midge *Belgica antarctica*. *Journal of Experimental Biology* **210**, 836–844.

Hazel, J. R. and Williams, E. E. (1990). The role of alterations in membrane lipid composition in enabling physiological adaptation of organisms to their physical environment. *Progress in Lipid Research* **29**, 167–227.

Hetz, S. K. and Bradley, T. J. (2005). Insects breathe discontinuously to avoid oxygen toxicity. *Nature* **433**, 516–519.

Hoffmann, A. A. and Parsons, P. A. (1991). *Evolutionary Genetics and Environmental Stress*. Oxford: Oxford University Press.

Højer, R., Bayley, M., Damgaard, C. F., and Holmstrup, M. (2001). Stress synergy between drought and a common environmental contaminant: studies with the collembolan *Folsomia candida*. *Global Change Biology* **7**, 485–494.

Holmstrup, M. and Westh, P. (1994). Dehydration of earthworm cocoons exposed to cold: A novel cold hardiness mechanism. *Journal of Comparative Physiology B* **164**, 312–315.

Holmstrup, M., Petersen, B. F., and Larsen, M. M. (1998). Combined effects of copper, desiccation, and frost on the viability of earthworm cocoons. *Environmental Toxicology and Chemistry* **17**, 897–901.

Holmstrup, M. and Sømme, L. (1998). Dehydration and cold hardiness in the Arctic collembolan *Onychiurus arcticus* Tullberg 1876. *Journal of Comparative Physiology B* **168**, 197–203.

Holmstrup, M., Bayley, M., and Ramløv, H. (2002). Supercool or dehydrate? An experimental analysis of overwintering strategies in small permeable arctic invertebrates. *Proceedings of the National Academy of Sciences, USA* **99**, 5716–5720.

Holmstrup, M., Aubail, A., and Damgaard, C. (2008). Mercury reduces cold tolerance in the springtail *Folsomia candida*. *Comparative Biochemistry and Physiology C* **148**, 172–177.

Horton, D. R., Lewis, T. M., and Neven, L. G. (1996). Reduced cold-hardiness of pear psylla (Homoptera: Psyllidae) caused by exposure to external water and surfactants. *Canadian Entomologist* **128**, 825–830.

Ishibashi, K., Sasaki, S., Fushimi, K., Uchida, S., Kuwahara, M., Saito, H., Furukawa, T., Nakajima, K., Yamaguchi, Y., Gojobori, T., and Marumo, F. (1994). Molecular-cloning and expression of a member of the aquaporin family with permeability to glycerol and urea in addition to water expressed at the basolateral membrane of kidney collecting duct cells. *Proceedings of the National Academy of Sciences, USA* **91**, 6269–6273.

Izumi, Y., Sonoda, S., Yoshida, H., Danks, H. V., and Tsumuki, H. (2006). Role of membrane transport of water and glycerol in the freeze tolerance of the rice stem borer, *Chilo suppressalis* Walker (Lepidoptera: Pyralidae). *Journal of Insect Physiology* **52**, 215–220.

Kanwisher, J. W. (1959). Histology and metabolism of frozen intertidal animals. *Biological Bulletin* **116**, 258–264.

Kikawada, T., Nakahara, Y., Kanamori, Y., Iwata, K. I., Watanabe, M., McGee, B., Tunnacliffe, A., and Okuda, T. (2006). Dehydration-induced expression of LEA proteins in an anhydrobiotic chironomid. *Biochemical and Biophysical Research Communications* **348**, 56–61.

Kosova, K., Vitamvas, P., and Prasil, I. T. (2007). The role of dehydrins in plant response to cold. *Biologia Plantarum* **51**, 601–617.

Lee, R. E. Jr. (1991). Principles of insect low temperature tolerance. In *Insects at Low Temperature*, ed. R. E. Lee Jr. and D. L. Denlinger. New York: Chapman and Hall, pp. 17–46.

Lighton, J. (1994). Discontinous ventilation in terrestrial insects. *Physiological Zoology* **67**, 142–162.

Lovelock, J. E. (1953). The mechanism of the cryoprotective effect of glycerol against freezing and thawing. *Biochimica et Biophysica Acta* **11**, 28–36.

Lundheim, R. and Zachariassen, K. E. (1993). Water balance of over-wintering beetles in relation to strategies for cold tolerance. *Journal of Comparative Physiology B* **163**, 1–4.

Mazur, P. (1977). The role of intracellular freezing in the death of cells cooled at supraoptimal rates. *Cryobiology* **14**, 251–272.

McLaughlin, S. and Percy, K. (1999). Forest health in North America: Some perspectives on actual and potential roles of climate and air pollution. *Water, Air and Soil Pollution* **116**, 151–197.

Miller, K. (1982). Cold-hardiness strategies of some adult and immature insects overwintering in interior Alaska. *Comparative Biochemistry and Physiology* **73A**, 595–604.

Olsen, T. M., Sass, S. J., Li, N., and Duman, J. G. (1998). Factors contributing to seasonal increases in inoculative freezing resistance in overwintering fire-colored beetle larvae *Dendroides canadensis* (Pyrochroidae). *Journal of Experimental Biology* **201**, 1585–1594.

Pedersen, P. G. and Holmstrup, M. (2003). Freeze or dehydrate: only two options for the survival of subzero temperatures in the arctic enchytraeid *Fridericia ratzeli*. *Journal of Comparative Physiology B* **173**, 601–609.

Pedersen, S. A., Kristiansen, E., Hansen, B. H., Andersen, R. A., and Zachariassen, K. E. (2006). Cold hardiness in relation to trace metal stress in the freeze-avoiding beetle *Tenebrio molitor*. *Journal of Insect Physiology* **52**, 846–853.

Philip, B. N., Yi, S. X., Elnitsky, M. A., and Lee, R. E. (2008). Aquaporins play a role in desiccation and freeze tolerance in larvae of the goldenrod gall fly, *Eurosta solidaginis*. *Journal of Experimental Biology* **211**, 1114–1119.

Preston, G. M., Carroll, T. P., Guggino, W. B., and Agre, P. (1992). Appearance of water channels in *Xenopus* oocytes expressing red-cell chip28 protein. *Science* **256**, 385–387.

Ring, R. (1982). Freezing-tolerant insects with low supercooling points. *Comparative Biochemistry and Physiology* **73A**, 605–612.

Ring, R. and Danks, H. (1994). Desiccation and cryoprotection: Overlapping adaptations. *Cryo-Letters* **15**, 181–190.

Ring, R. and Danks, H. (1998). The role of trehalose in cold-hardiness and desiccation. *Cryo-Letters* **19**, 275–282.

Schmidt-Nielsen, K. (1997). *Animal Physiology*. New York: Cambridge University Press.

Scholander, P. F., Flagg, W., Hock, R. J., and Irving, L. (1953). Studies on the physiology of frozen plants and animals in the Arctic. *Journal of Cellular and Comparative Physiology* **42**, 1–56.

Skovlund, G., Damgaard, C., Bayley, M., and Holmstrup, M. (2006). Does lipophilicity of toxic compounds determine effects on drought tolerance of the soil collembolan *Folsomia candida*? *Environmental Pollution* **144**, 808–815.

Sømme, L. and Birkemoe, T. (1997). Cold tolerance and dehydration in Enchytraeidae from Svalbard. *Journal of Comparative Physiology B* **167**, 264–269.

Sømme, L. and Conradi-Larsen, E.-M. (1977). Cold-hardiness of collembolans and oribatid mites from wind-swept mountain ridges. *Oikos* **29**, 118–126.

Tammariello, S., Rinehart, J., and Denlinger, D. L. (1999). Desiccation elicits heat shock protein transcription in the flesh fly, *Sarcophaga crassipalpis*, but does not enhance tolerance to high or low temperatures. *Journal of Insect Physiology* **45**, 933–938.

Thurberg, F. P., Dawson, M. A., and Collier, R. S. (1973). Effects of copper and cadmium on osmoregulation and oxygen-consumption in two species of estuarine crabs. *Marine Biology* **23**, 171–175.

Tunnacliffe, A., Lapinski, J., and McGee, B. (2005). A putative LEA protein, but no trehalose, is present in anhydrobiotic bdelloid rotifers. *Hydrobiologia* **546**, 315–321.

Van Der Laak, S. (1982). Physiological adaptations to low temperature in freezing-tolerant *Phyllodecta laticollis* beetles. *Comparative Biochemistry and Physiology* **73A**, 613–620.

Van Gestel, C. A. M. (1997). Scientific basis for extrapolating results from soil ecotoxicity tests to field conditions and the use of bioassays. In *Ecological Risk Assessment of Contaminants in Soil*, ed. N. M. Van Straalen and H. Løkke. London: Chapman and Hall, pp. 25–50.

Valko, M., Morris, H., and Cronin, M. T. D. (2005). Metals, toxicity and oxidative stress. *Current Medicinal Chemistry* **12**, 1161–1208.

Wang, L. and Duman, J. G. (2005). Antifreeze proteins of the beetle *Dendroides canadensis* enhance one another's activities. *Biochemistry* **44**, 10305–10312.

Weast, R. C. (1989). *Handbook of Chemistry and Physics*. Cleveland: CRC Press.

Wharton, D. A., Goodall, G., and Marshall, C. J. (2003). Freezing survival and cryoprotective dehydration as cold tolerance mechanisms in the Antarctic nematode *Panagrolaimus davidi*. *Journal of Experimental Biology* **206**, 215–221.

Worland, M., Grubor-Lajsic, G., and Montiel, P. (1998). Partial desiccation induced by subzero temperatures as a component of the survival strategy of the Arctic collembolan *Onychiurus arcticus* (Tullberg). *Journal of Insect Physiology* **44**, 211–219.

Wu, D. W. and Duman, J. G. (1991). Activation of antifreeze proteins from larvae of the beetle *Dendroides canadensis*. *Journal of Comparative Physiology* **161B**, 279–281.

Wu, D. W., Duman, J. G., Cheng, C.-H. C., and Castellino, F. J. (1991). Purification and characterization of antifreeze proteins from larvae of the beetle *Dendroides canadensis*. *Journal of Comparative Physiology* **161B**, 271–278.

Zachariassen, K. E. (1979). The mechanism of the cryoprotective effect of glycerol in beetles tolerant to freezing. *Journal of Insect Physiology* **25**, 29–32.

Zachariassen, K. E. (1980). The role of polyols and nucleating agents in cold-hardy beetles. *Journal of Comparative Physiology* **140**, 227–234.

Zachariassen, K. E. (1985). Physiology of cold tolerance of insects. *Physiological Reviews* **65**, 799–832.

Zachariassen, K. E. (1991). The water relations of overwintering insects. In *Insects at Low Temperature*, ed. R. E. Lee Jr., and D. L. Denlinger. New York: Chapman and Hall, pp. 47–63.

Zachariassen, K. E. and Hammel, H. T. (1976). Nucleating agents in the haemolymph of insects tolerant to freezing. *Nature* **262**, 285–287.

Zachariassen, K. E., Hammel, H. T., and Schmidek, W. (1979). Osmotically inactive water in relation to tolerance to freezing in *Eleodes blanchardi* beetles. *Comparative Biochemistry and Physiology* **63A**, 203–206.

Zachariassen, K. E., Andersen, J., Maloiy, G. M. O., and Kamau, J. M. Z. (1987). Transpiratory water loss and metabolism of beetles from arid areas in East Africa. *Comparative Biochemistry and Physiology* **68A**, 403–408.

Zachariassen, K. E. and Maloiy, G. M. O. (1989). Water balance of beetles as an indicator of environmental humidity. *Fauna Norvegica* **36B**, 27–31.

Zachariassen, K. E. and Einarson, S. (1993). Regulation of body fluid compartments during dehydration of the tenebrionid beetle *Rhytinota praelonga*. *Journal of Experimental Biology* **182**, 283–289.

Zachariassen, K. E., Kristiansen, E., and Pedersen, S. A. (2004). Inorganic ions in cold-hardiness. *Cryobiology* **48**, 126–133.

Zelenina, M., Tritto, S., Bondar, A. A., Zelenin, S., and Aperia, A. (2004). Copper inhibits the water and glycerol permeability of aquaporin-3. *Journal of Biological Chemistry* **279**, 51939–51943.

PART II ECOLOGICAL AND EVOLUTIONARY RESPONSES

8

The macrophysiology of insect cold-hardiness

STEVEN L. CHOWN AND BRENT J. SINCLAIR

8.1 Introduction: macrophysiology

During the early part of the twentieth century, comparative physiological studies were as much at home in ecological journals as they were in those devoted to physiology. Indeed, Shelford (1913) considered ecology to be a "branch of general physiology which deals with the organism as a whole . . . and which also considers the organism with particular reference to its usual environment". For reasons that have been discussed elsewhere (e.g. Huey, 1991; Spicer and Gaston, 1999; Chown *et al.*, 2004) ecology and physiology subsequently parted ways with both increasing their focus on smaller-scale questions. Although large-scale ecological and biogeographic work continued, interest in physiological mechanisms waned (see e.g. Myers and Giller, 1988; Lomolino and Heaney, 2004). In much the same way, large-scale comparative physiological ecology dwindled in significance, making studies such as those by Scholander *et al.* (1953) and Brattstrom (1968) milestones along an increasingly deserted road. Clearly, investigations of animal responses to the environment continued (the work of Bartholomew stands out especially (Dawson, 2005) (see also reviews in Prosser, 1986; Angilletta *et al.*, 2002; Hoffmann *et al.*, 2003), and the development of methods to correct for phylogenetic non-independence prompted a resurgence of interest in understanding the evolution of physiological traits and their variation among species and higher taxa (Feder *et al.*, 2000). However, by the late 1980s, the subject of organismal physiological diversity was in several ways thought to be a dead end. In a landmark volume on the subject, Feder (1987) asked: "Having granted that physiological diversity and pattern in physiological diversity are significant findings that have justified

Low Temperature Biology of Insects, ed. David L. Denlinger and Richard E. Lee Jr. Published by Cambridge University Press.

ecophysiological analysis in the past, is either sufficiently inconclusive to merit more substantiation?" and answered that "With the exception of variation within populations [Chapter 11], the answer is obviously no."

As is sometimes the case with such pronouncements, developments elsewhere were about to prove them premature. First, macroecology revitalized the investigation of physiological diversity, confronting the latter, in several aspects, with its own inadequate understanding of both pattern and process (Chown and Gaston, 1999; Gaston *et al.*, 1998; Spicer and Gaston, 1999; Chown *et al.*, 2002, 2004). Second, general appreciation of the reality and consequences of global climate change (IPCC, 1990) prompted a rapid increase in studies of biodiversity responses to such change (see reviews in Parmesan, 2006; Rosenzweig *et al.* 2008). Many of these drew attention to the need for a renewed focus on the relationships between organisms and their abiotic environments, and how these vary through space and over time (Davis and Shaw, 2001; Helmuth *et al.*, 2005; Angilletta *et al.*, 2006; Anonymous, 2007). Later demonstrations that biodiversity loss and the concomitant threats to human wellbeing are the consequence of interactions among five major drivers of global scope (climate change, habitat alteration, biological invasion, over-exploitation and pollution) led to further articulation of the need for large-scale investigations of physiological diversity and the mechanisms underlying it (see Chown and Gaston, 2008).

Of course, arguing that a macrophysiology, or "the investigation of variation in physiological traits over large geographical and temporal scales and the ecological implications of this variation" (Chown *et al.*, 2004) is required does not guarantee that it will prove either useful or novel (given the existence of several different approaches within the general field of environmental physiology, e.g. Feder *et al.*, 2000). However, recent work has started to demonstrate that macrophysiology is providing fresh new insights about fundamental biological patterns, the processes underlying them and human impacts thereon (e.g. Osovitz and Hofmann, 2007). Investigations of insect thermal biology have, in many ways, formed the vanguard of such demonstrations (e.g. Addo-Bediako *et al.*, 2000; Umina *et al.*, 2005; Rako *et al.*, 2007; Deutsch *et al.*, 2008), and those of large-scale variation in insect cold-hardiness are no exception.

That insects surviving subzero temperatures do so typically by adopting one of three major strategies, each with some variation about them, has long been known, and the biochemistry and physiology of these responses much investigated (see reviews elsewhere in this volume). Although the presence of phylogenetic signal in these strategies had been recognized (Sømme, 1982), and some speculation provided about environments that might favor one strategy over another (e.g. Sømme and Zachariassen, 1981; Van Der Laak, 1982; Fields and McNeil, 1986;

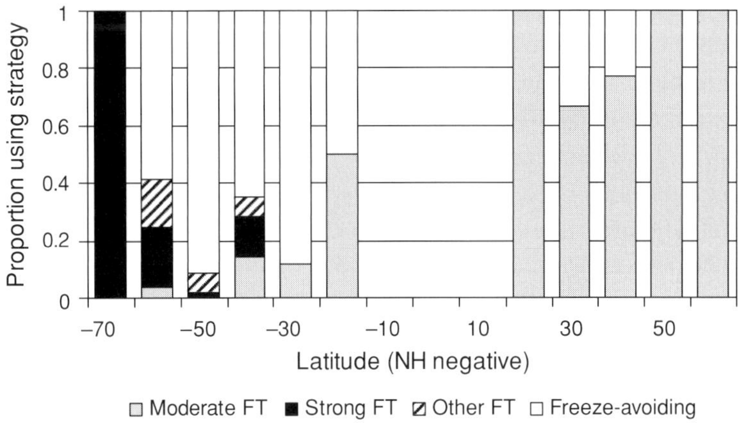

Figure 8.1 Latitudinal distribution of insect cold-tolerance strategies (excluding species from 10° either side of the equator). Moderate FT = moderately freeze-tolerant; Strong FT = strongly freeze-tolerant; Other FT = other freeze-tolerant (includes freeze-tolerant with low supercooling point and species that survive only with external inoculation, also includes species for which data are inadequate to determine class of freeze-tolerance). (Redrawn from Sinclair and Chown, 2005.)

Duman *et al.*, 1991; Gehrken *et al.*, 1991), little thought had been given to whether identifiable, regional spatial patterns in strategies and their variants might be present. Indeed, until at least 2002, most major reviews were silent on the matter (but see Zachariassen, 1985; Duman *et al.*, 1991). However, in an investigation of moderate freeze-tolerance in the caterpillars of a sub-Antarctic moth, Klok and Chown (1997) noted freeze-tolerance was paradoxically common in the southern hemisphere, an observation echoed by Addo-Bediako *et al.* (2000). A comprehensive assessment of spatial variation in strategies revealed that substantial north–south differences in strategies do indeed exist (Sinclair *et al.*, 2003a; Sinclair and Chown, 2005) (Fig. 8.1), and Chown *et al.* (2004) extended this observation to other ecological patterns and processes. Environmental predictability and variability were identified as likely mechanisms underlying these patterns, drawing attention to their importance, not only for the evolution of thermal tolerance (see Levins, 1968; Kingsolver and Huey, 1998; Angilletta *et al.*, 2006; Chown and Terblanche, 2007), but perhaps also for many other patterns at various levels in the biological hierarchy.

Thus, it is clear that a large-scale physiological perspective of insect cold-tolerance strategies can provide significant and fresh insights that might otherwise not have been realized, and has already done so. Although the approach and some of its specific conclusions have been questioned (Hodkinson, 2003; Turnock and Fields, 2005) – about which more later (see also Chown *et al.*,

2003) – such debate reflects the healthy exploration and critical examination of ideas: the essence of good science. In consequence, this chapter provides a review of the macrophysiology of insect cold-hardiness. Specifically, it concerns the measurement of environmental variation at various scales, phylogenetic signal in insect low-temperature responses, the evolution of both fixed and plastic responses to temporally variable environments, and spatial patterns in insect lower lethal limits and the proximate mechanism underlying them.

8.2 Environmental variation

The evolution of responses to the thermal environment is contingent on the characteristics of the environment, including explicit spatial characteristics thereof (such as patch size and isolation), and the generation time, energetics, behaviour, sensing ability and dispersal capability of the focal organism (Levins, 1968; Lynch and Gabriel, 1987; Gilchrist, 1995; Voituron *et al.*, 2002; Angilletta *et al.*, 2006; Makarieva *et al.*, 2006; Chown and Terblanche, 2007). This evolutionary response can take the form of relatively fixed traits, or the shapes of reaction norms (i.e. the extent and form of phenotypic plasticity – see Chown and Terblanche, 2007; Ghalambor *et al.*, 2007). In insects, responses may further be influenced by the presence of stages that differ substantially in their characteristics and in the environments they use (as is true for most holometabolous species), and by the fact that diapause is common in overwintering stages (Denlinger, 2002). Understanding the interactions among these internal and external characteristics is the subject of much of evolutionary physiology as a whole, and so cannot be explored in any inclusive fashion here. However, if the proximate environmental mechanisms underlying large-scale spatial and temporal variation in insect responses to low temperature are to be comprehended, several aspects of environmental variation must be discussed explicitly, if only to highlight their importance and to draw attention to several methodological matters.

8.2.1 Magnitude, duration and variance

The significance of the magnitude and duration of a low-temperature event has long been appreciated in the investigation of insect tolerance of low temperatures, whether or not the event is accompanied by the fundamental physicochemical changes caused by the freezing of water (Salt and James, 1947; Sømme and Zachariassen, 1981; Bale, 1987; Sømme, 1996; Nedvěd, 1998; Ramløv, 2000). The magnitude of an event most clearly determines the outcome (microclimate temperatures below the lower lethal temperature for 100% of the population (LLT_{100}) will remove 100% of a population), and lethal extreme events have often been recorded (e.g. Virtanen *et al.*, 1998; Parmesan *et al.*, 2000; Crozier, 2004).

However, the duration of an event is significant because it may increase mortality (e.g. prolonged low subzero temperatures may by chance result in the freezing and death of supercooled insects that exist in a meta-stable state – Sømme, 1996), or may substantially influence the extent of sublethal injury. The latter might result, not only because a threshold level of damage is overcome via additive effects (Makarieva *et al.*, 2006), but also because temperature effects on performance, at either the biochemical, organ/system, or organismal levels, mean that the fitness of the animal is substantially reduced. Indeed, consideration of these interactions makes clear the somewhat arbitrary distinction between tolerance and performance traits (or resistance and capacity adaptations), as is readily illustrated by the oxygen limitation of thermal tolerance, thought to be widespread in many ectotherms (Pörtner, 2001).

Measurement of temperatures in appropriate microhabitats clearly provides a straightforward means of assessing the magnitude and duration of thermal events, and both hardware to make the measurements and software to facilitate the analysis thereof are now widely available (e.g. Sinclair, 2001a). Moreover, simple statistical assessment of such time-series also enables average conditions and the magnitude of the variation around them, both of which influence the evolution of thermal responses (e.g. Gilchrist, 1995; Angilletta *et al.*, 2006), to be determined in a relatively straightforward fashion at a variety of scales. Nonetheless, where comparisons among sites are to be made, thought must be given to what characteristics of the variation are to be measured and how the comparison is implemented. For example, if considering temperature variation only, absolute variance may be the measure of variation of most interest, but if population responses are also to be investigated, proportional variation may be more relevant (McArdle and Gaston, 1995). In the latter case, standard estimates of variance, such as the standard deviation or coefficient of variation, are unlikely to be suitable for comparing time-series data. Rather, at the minimum, a proportional coefficient of variation should be used. Moreover, data transformations may also bias comparisons, or lead to unsuitable conclusions owing to difficulties such as those associated with Jensen's inequality (Ruel and Ayres, 1999). For example, and most simply, the log of a mean value from a time series may be equal to, but could also be greater than, the mean of the logged values of the series (the arithmetic–geometric means inequality). Similarly, time series of different length are likely to show different variances depending on the color of the noise spectra (Pimm and Redfearn, 1998), and this should be kept in mind, as should the non-independence of time-series information. A variety of techniques for comparing variability, which may also overcome the above problems, and enable unimportant trends in the series to be removed, are now widely available (e.g. Ferguson and Messier, 1996; Denny and Gaines, 2000; Chatfield, 2004).

8.2.2 Frequency or return time

Straightforward microclimate (or macroclimate) measurements can also provide a simple assessment of return times or the frequency of a particular event (Gaines and Denny, 1993; Gutschick and BassiriRad, 2003). However, more sophisticated approaches are required to assess the likelihood that an event of a particular nature may be followed by similar events. Perhaps the most common way of investigating the form and significance of environmental "noise" in ecology is the calculation of spectral densities and the examination of the relationship between spectral density and frequency (e.g. Halley, 1996). White noise contains an equal mix of all frequencies, with a flat spectral density, that is the spectral exponent $\gamma = 0$. If the spectral density is greater at low than at high frequencies, then the spectrum is said to be reddened – low-frequency cycles dominate.

In a broad-ranging assessment of long-term variability (excluding seasonal variation), Vasseur and Yodzis (2004) showed that in the case of mean environmental temperature, noise color varies from brown for sea-surface temperature, to red-brown at coastal locations, to mostly white in terrestrial locations. The difference between mostly white spectra at terrestrial locations and reddened noise in marine systems is probably the consequence of the substantial buffering capacity of the sea (Vasseur and Yodzis, 2004), which also likely explains the predominantly white spectra of minimum temperature between 30 and 60°N, but reddened spectra between 30 and 60°S, owing to the dominance of the vast Southern Ocean (Chown and Terblanche, 2007).

These differences, and indeed those that might be characteristic of different sites in a given area (e.g. above and below the thermal buffering of snow – Irwin and Lee 2003), are especially significant because the color of noise interacts with the nature of density regulation to determine extinction risk in populations (Petchey *et al.*, 1997; Inchausti and Halley, 2003; Ruokolainen and Fowler, 2008). Indeed, models that explore these interactions, especially those that deal explicitly with the presence or absence of catastrophic events (e.g. Schwager *et al.*, 2006), provide a useful way of further assessing the significance of extreme lethal versus repeated non-lethal events for population persistence. The effects of repeated low-temperature exposure on insects are subject to considerable debate, but all of the published work agrees that the physiological effects of a single cold exposure event are different to those of multiple exposures (see e.g. Bale *et al.*, 2001; Kelty and Lee, 2001; Brown *et al.*, 2004; Renault *et al.*, 2004; Sinclair and Chown, 2005). The effects of extreme and sublethal events play out simultaneously through their physiological consequences, through interactions among individuals within a given population, and through interactions among these individuals and those of other species. Nonetheless, their consequences are always realized through demographic

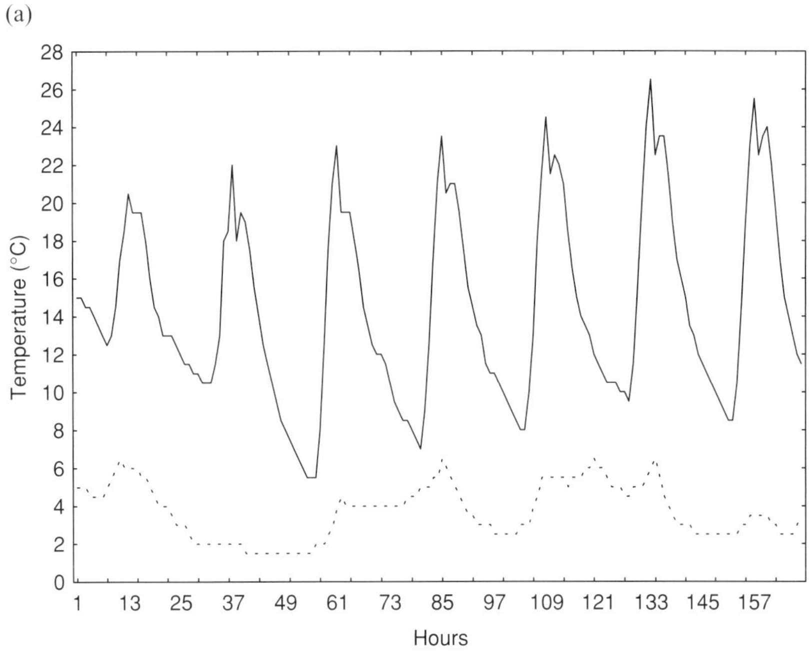

Figure 8.2 (a) Hourly temperatures at the soil surface over a week long period in August 2002 for a sea level site at Lambert's Bay on the west coast of South Africa (solid line) and a sea level site at sub-Antarctic Marion Island (dashed line). (b) Autocorrelation plots for hourly temperatures shown in (a). Left: Lambert's Bay, Right: Marion Island. (c) Spectral density plots for the same sites shown in (b). Note the reduced spectral density values and the lack of any 24 h periodicity at Marion Island by comparison with Lambert's Bay.

rates (Andrewartha and Birch, 1954; Levins, 1968; Feder, 1987; Dillon *et al.*, 2007), which in turn determine abundance and distribution in a given patch (Holt *et al.*, 1997), which is enmeshed in a larger meta-community landscape (Gaston, 2003; Leibold *et al.*, 2004; Ruokolainen and Fowler, 2008).

8.2.3 Predictability

Consideration of noise color reveals that the magnitude of environmental variability is not equivalent to the predictability of future conditions, at least not in any straightforward way. Zero variation means that conditions are always predictable. However, large variation might be perfectly predictable, whilst small changes in conditions might not only be random, but might also in fact show negative temporal autocorrelation. That is, over the time period in question, conditions in the near future may be more different than expected by chance from current ones (see Fig. 8.2). In consequence, providing an estimate of predictability

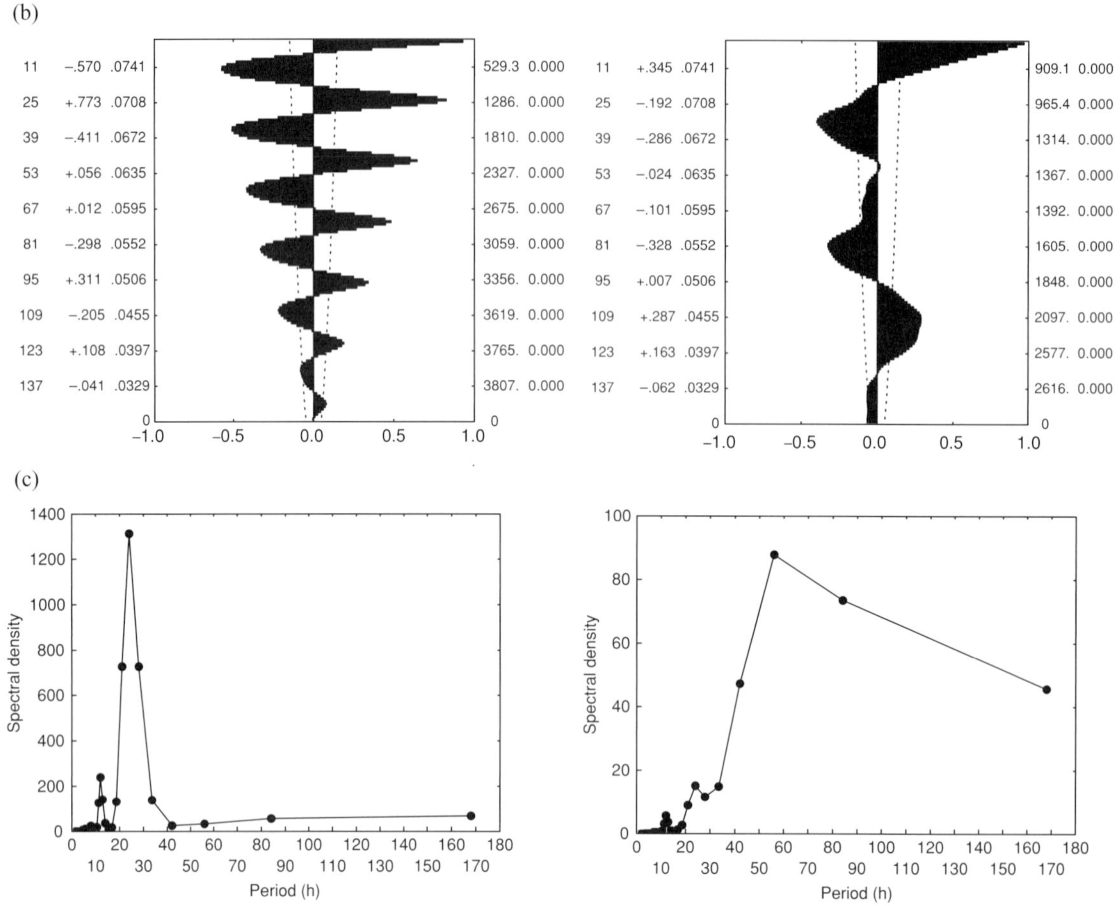

Figure 8.2 *(cont.)*

of events on a given timescale relevant to the organism in question is also important. For cold-hardiness strategies, the significance of so doing has already been demonstrated: it appears that unpredictable, moderate freezing events in moist environments have selected for the preponderance of moderate freeze-tolerance in southern-hemisphere species (Sinclair *et al.*, 2003a, b; Sinclair and Chown, 2005).

Several methods have been proposed to quantify predictability (see e.g. Kingsolver and Huey, 1998), but perhaps the most reliable are the construction of autocorrelation plots (Gaines and Denny, 1993; Chown and Terblanche, 2007), and spectral analysis, which resolves a complex time series with cyclical components into several underlying sinusoidal functions of particular wavelengths (Chatfield, 2004). Spectral densities, readily obtained by fast Fourier transformations, provide an assessment of the periods of the dominant sinusoidal functions, and together with autocorrelation plots offer a clear picture of the extent to which particular conditions may be expected to repeat themselves. For example, at Lambert's Bay on the coast of South Africa, hourly temperature records over a one-week period are variable, but highly predictable (one week after the day of interest one can still expect to experience the same temperature, at, for example, midday), whereas at a coastal site on sub-Antarctic Marion Island the temperature over the same period is less variable, but also much less predictable (Fig. 8.2) (Deere and Chown, 2006).

8.2.4 *Rates of change*

Thermal physiologists have long appreciated the fact that the rate of temperature change used in an experiment is likely to affect the outcome thereof. Although crystallization temperatures are apparently largely unaffected by rate variation (at least over the ranges studied) (Salt, 1966; Slabber *et al.*, 2007), low-temperature survival is strongly influenced by cooling rate (Miller, 1978; Baust and Rojas, 1985; Ramløv, 2000; Sinclair, 2001b), as is critical thermal minimum (CT_{min}) (Kelty and Lee, 1999; Terblanche *et al.*, 2007). However, the direction of the response is unlikely to be consistent among species. For example, in *Drosophila melanogaster*, slow rates of cooling lower the CT_{min}, probably as a consequence of rapid cold-hardening (RCH) (Kelty and Lee, 1999, 2001). By contrast, in *Glossina pallidipes*, slow rates of cooling lead to an elevation of CT_{min}, which may reflect an absence of RCH in this species (Terblanche *et al.*, 2007). For this reason, and because interactions among rates of change, and the extent and direction of acclimation effects are also likely (Chown *et al.*, 2009), rates of change in experiments should not only be reported, but perhaps should also be recorded in the field (especially given that field rates are typically lower than those used in the laboratory). In the case of the tsetse, for example, the slow rates of change brought the critical thermal maximum within the range of temperatures shown to induce mortality, based on mark-recapture studies (Terblanche *et al.*, 2007). Likewise the lower lethal temperature of the New Zealand alpine cockroach, *Celatoblatta quinquemaculata*, appeared to be very close to minimum temperatures recorded in its microhabitat. However, a 100-fold reduction in cooling rate, bringing it within the range observed in the field, resulted in a decrease in lower lethal temperature from −8.9 to −11.1 °C, enough to ensure survival in some microhabitats in all winters for which data are available (Sinclair, 2001b).

These results raise an important question for macrophysiology: should constant rates be used for all species (or populations) across the full spatial extent of a study (thus allowing standardized comparison) or should rates vary according to those found under natural conditions (thus ensuring comparisons with possibly greater ecological relevance)? In many ways this issue is similar to arguments concerning use of mean environmental temperatures used in many macrophysiological studies thus far (see Chown *et al.*, 2003; Hodkinson, 2003). Much of the answer will depend on the nature of the question being posed, the availability of appropriate environmental data, and the relationship between macro- and microclimatic information. What seems likely, though, is that whilst standardized data, or mean macroclimatic rather than exact microclimatic data, will add variance to the relationship being investigated, they are unlikely to introduce a systematic bias favoring one conclusion over another. Nonetheless, it is worth recognizing that substantial differences in opportunities for exploiting microclimatic variation might exist between high and low latitude and altitude sites (Addo-Bediako *et al.*, 2000; Deutsch *et al.*, 2008).

Rates of change refer not only to short-term transitions between high and low temperatures, but also to longer-term changes in temperature means, variances and extremes, and the predictability thereof. Such longer-term change is indeed one of humanity's most pressing current concerns (Easterling *et al.*, 2000; Joss and Spahni, 2008). Sophisticated ways exist to detect such change, especially in the predictability of events (such as wavelet analysis – e.g. Torrence and Campo, 1998), although they are typically not applied to data on insects, largely because the biotic variables are not measured for sufficiently long time periods (though see Convey *et al.*, 2003; Stige *et al.*, 2007 for exceptions).

8.3 Low-temperature responses through time

8.3.1 The long term

Many insect orders are represented in low-temperature habitats (see tables in Sinclair *et al.*, 2003a and Turnock and Fields, 2005 for partial summaries), and a variety of cold-tolerance strategies exist across these orders. Although information on the phylogenetic distribution of insect responses to low temperature is increasing, how low-temperature survival strategies evolved nonetheless remains largely conjectural (Sinclair *et al.*, 2003a). Vernon and Vannier (2002) suggested that freeze-avoidance is the basal state, presumably derived from an improvement of chilling-tolerance in species that are otherwise chill-susceptible, and indeed this strategy is most common at the higher taxonomic levels in the insects (Fig. 8.3). Nonetheless, several orders are phylogenetically predisposed to freeze-tolerance (e.g. cockroaches and Orthoptera, Sinclair *et al.*, 2003a), although the evolutionary trajectories that have resulted in this pattern remain poorly known. Preponderance of freeze-tolerance in a particular group might be the consequence of its radiation subsequent to, but not as a result of, the acquisition of this suite of traits. This could be termed phylogenetic constraint, though the term should be used cautiously (Roff and Fairbairn, 2007). Alternatively, the suite of traits may

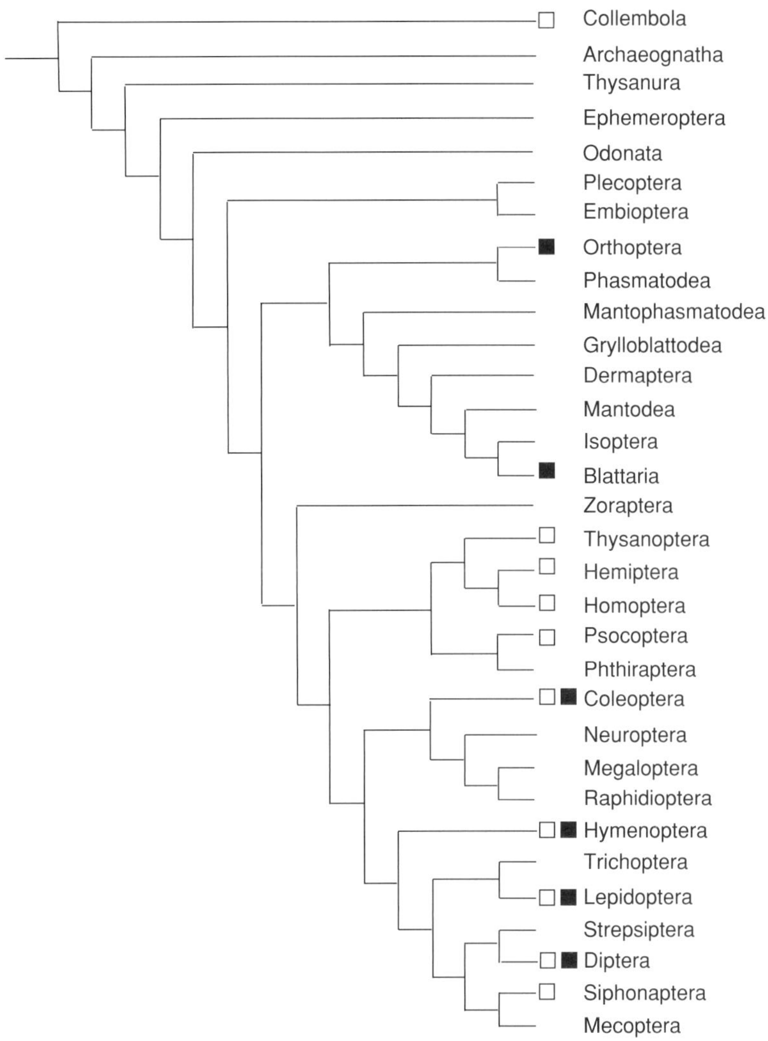

Figure 8.3 Phylogeny of the insect Orders (Collembola as an outgroup) indicating freeze-avoidance (open squares) and freeze-tolerance (closed squares) where they have been investigated. (Modified from Chown and Nicolson, 2004.)

well have enabled the group to diversify in a given set of habitats (i.e. a key innovation – see Schluter, 2000), as may well have been the case for southern-hemisphere mountain cockroaches (Sinclair and Chown, 2005). At present, the extent of information on strategies across the arthropod and insect phylogenies is insufficient to develop fully hypotheses concerning the evolution of different cold-hardiness strategies, irrespective of whether these involve avoidance of injuries associated with freezing or those associated with chilling (Sinclair *et al.*, 2003a; Chown and Nicolson, 2004; Turnock and Fields, 2005). Nonetheless, freeze-tolerance probably evolved more than once (Fig. 8.3), and the characteristics of the environment, especially the predictability and regularity of freeze–thaw events, the duration and intensity of low-temperature events, and the likelihood of inoculation, have been significant in promoting selection for this suite of traits. Low supercooling points, but considerable pre-freeze mortality (see Bale, 1987), and exceptionally low, lower lethal temperatures plainly demonstrate that not all variation in lower thermal limit traits can be considered adaptive or the product of selection. For example, larvae of *Helicomyza borealis* survive to −60 °C (Worland *et al.*, 2000), and diapausing larvae of *Chymomyza costata* are able to survive immersion in liquid nitrogen (Moon *et al.*, 1996). Unfortunately, data on responses such as RCH are too sparse to tempt speculation about their origins, though it is clear that some species lack the ability to alter rapidly their lower lethal limits in response to a low-temperature event (e.g. Sinclair and Chown, 2003; Terblanche *et al.*, 2008).

Additional insights into the evolution of cold-tolerance may come from the relationship between tolerance to cold and to other environmental stresses. Many of the biochemical responses to low temperature and to desiccation are shared, presumably because the damage caused by these stressors is similar (Ring and Danks, 1994; Koštál and Šimcek, 1996; Pullin, 1996; Storey and Storey, 1996; Holmstrup *et al.*, 2002a; Schliess and Haüssinger, 2002; Hayward *et al.*, 2004; Williams *et al.*, 2004; Yoder *et al.*, 2006). Moreover, dehydration provides one way to avoid injury associated with low temperature (Worland *et al.*, 1998; Holmstrup *et al.*, 2002b), and may also influence the strategy that evolves: a supercooled insect is likely to dehydrate much faster in the presence of ice than is a frozen one (Lundheim and Zachariassen, 1993). In consequence, and given the climates during the periods when many insect higher taxa first evolved (late Paleozoic to mid-Mesozoic) (Grimaldi and Engel, 2005), it seems plausible to presume that responses to low temperature were derived or modified from those that initially evolved to enable insects to cope with desiccation (Chown and Terblanche, 2007). However, artificial selection experiments with *Drosophila* suggest that selection for desiccation-tolerance does not always result in an increase in cold-tolerance (Bubliy and Loeschcke, 2005; Sinclair *et al.*, 2007), and that trade-offs with other components of the physiological phenotype may be complex (Hoffmann *et al.*, 2005).

8.3.2 The short term

Over shorter timescales, physiological responses to seasonal environmental variation are certainly the most widely known and indeed were the first to be recognized. That winter and summer physiological states differ is virtually axiomatic, at least for continental northern-hemisphere species. Moreover, much of the field has been, and rightly continues to be, concerned with the biochemistry, physiology and ecological implications of these seasonal, or programmed, responses to low temperature (Bale, 2002; Voituron *et al.*, 2002; Sinclair *et al.*, 2003b; Makarieva *et al.*, 2006). That responses may also be more rapid, developing sometimes within a few hours, was only recognized much later (Lee *et al.*, 1987; Worland and Convey, 2001). Now, the mechanisms and consequences of these responses are also being carefully scrutinized (Michaud and Denlinger, 2005; Overgaard *et al.*, 2005, 2007; Hawes *et al.*, 2006; Yoder *et al.*, 2006).

At least from the perspective of evolutionary physiology, the continuum of temporal responses (from the seasonal to the hourly) represents, in several respects, phenotypic plasticity, or "the ability of an organism to react to an environmental input with a change in form, state, movement, or rate of activity" (West-Eberhard, 2003). However, care must be taken not to imagine that the full range of evolved strategies to low temperature, and particularly the cold-hardiness strategies recognized so widely in the literature (Bale, 2002), can routinely be labelled plasticity, such as the cryotypic plasticity suggested by Hawes and Bale (2007). Much variation in these strategies has been documented (Bale, 1993; Sinclair, 1999), and some responses have arisen from selection for a fixed strategy under all environmental conditions, while others from selection for marked phenotypic plasticity (Chown *et al.*, 2008). For example, the year-round moderate freeze-tolerance shown by several southern-hemisphere insect species may represent a fixed strategy developed in response to an unpredictably variable environment (Sinclair and Chown, 2005), rather than phenotypic plasticity. Thus, even following substantial acclimation (e.g. 0 °C and 9L:15D photoperiod versus 15 °C and 14L:10D) indigenous springtail species on Marion Island show limited variation in crystallization temperature (= lower lethal temperature in these species) (Slabber *et al.*, 2007). Lack of response to acclimation and to hardening is characteristic also of crystallization temperatures and lower lethal temperatures in adults and larvae of the sub-Antarctic kelp fly, *Paractora dreuxi* (Marais *et al.*, 2009). A similar set of circumstances may be responsible for year-round maintenance of freeze-tolerance in the cockroach, *Celatoblatta quinquemaculata* (Sinclair, 1997), although its lower lethal temperatures vary through time.

This distinction between fixed and plastic responses is important, but is nowhere as clear-cut between freeze-tolerant and freeze-avoiding species as Hawes and Bale (2007) maintain. They argued that while increasing cold-hardiness (which they assume to indicate a more evolutionarily derived state) is associated with increased plasticity in freeze-avoiders, the reverse is true for species that are freeze-tolerant. If plasticity is equated with mean seasonal variation in cold-hardiness (measured as lower lethal temperature change among seasons), the freeze-tolerant and freeze-avoiding species show little difference (Fig. 8.4), demonstrating that slim grounds exist to suppose that all freeze-tolerant species are either less phenotypically plastic or more specialized than freeze-avoiders (Chown *et al.*, 2008). Rather,

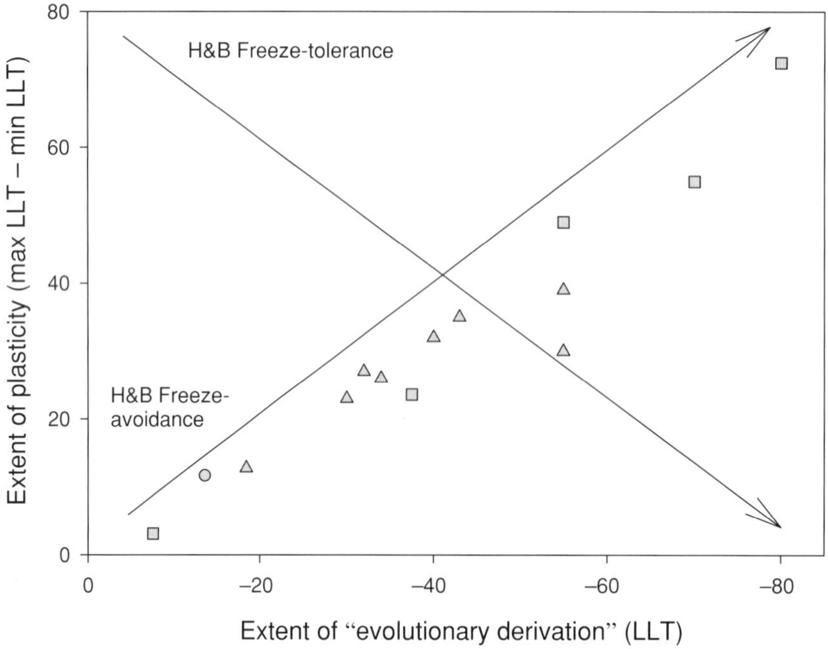

Figure 8.4 The extent of plasticity (maximum–minimum lower lethal temperature across seasons) plotted against winter lower lethal temperature as a test of the hypothesis (indicated with the dark arrows) established by Hawes and Bale (2007) that freeze-tolerant (squares) and freeze-avoiding (triangles) species differ in these relationships. The empirical data falsify this hypothesis. (Redrawn from Chown *et al.*, 2008.)

the situation probably reflects the interactions between spatial and temporal environmental variation, and organismal characteristics that result in a continuum from fixed to plastic traits.

In this context, theoretical and empirical studies have demonstrated that phenotypic plasticity is favored in populations when environmental variability is substantial, the costs of plasticity are low relative to their benefits, migration among patches is not uncommon, and cue reliability and the accuracy of the response are high. The latter depend on environmental lability and predictability, as well as on the lag time of the response (reviewed in Chown and Terblanche, 2007). Where environments are unpredictably variable on a scale relevant to the organism and the rate at which a response can be mobilized, a fixed strategy tends to evolve and no plasticity is present (see also Deere and Chown, 2006). Thus, it is clear that physiological responses to seasonal change can be considered plasticity, especially in those regions where the duration of the stressor is long relative to the time taken to develop the full response (such as in temperate, continental North America). However, maintenance of a given strategy irrespective of season might not be.

Of course, just as in the case of the problems that have arisen with the terms 'r- and K-selection' in ecology (Roff, 2002), so too must qualification be adopted here because different components of the response may vary, even though the response as a whole may be stable through time.

Indeed, it is clear that the extent of plasticity can differ substantially among traits, and within traits among stages, of the same species (Deere *et al.*, 2006; Terblanche and Chown, 2006; Slabber *et al.*, 2007). Such variation no doubt reflects interactions among the traits themselves (Sinclair and Chown, 2003), the rate at which a response can be effected relative to the duration of the event, the longevity of the stage concerned (Lee *et al.*, 2006) and the history of the organisms involved (e.g. recent colonist versus evolved *in situ*) (Slabber *et al.*, 2007; Chown *et al.*, 2007). For example, Lee *et al.* (2006) suggested that the absence of RCH in adults of the Antarctic midge, *Belgica antarctica*, might be due in part to their short lifespan, by contrast with the larvae. They also argued that the lack of an RCH response could be due to mobility in the adults, which would allow them to seek out thermal refuges, so avoiding low-temperature events (see also Hawes *et al.*, 2006). Although it is widely appreciated that behavioral regulation can inhibit evolutionary change by buffering the effects of the environment (known as the Bogert effect – see Huey *et al.*, 2003), its role in mediating the extent to which phenotypic plasticity evolves has not been extensively examined. However, a recent study has made clear that the Bogert effect may indeed substantially affect the extent of plasticity. In the sub-Antarctic kelp fly, *Paractora dreuxi*, chill-coma recovery is characterized by beneficial acclimation in the less mobile larvae that have broader thermal preferences. By contrast, the more mobile adults that have narrower thermal preferences perform best under all conditions, so long as they previously experienced low temperatures (Marais and Chown, 2008) (Fig. 8.5). In other words, the adults simply select their preferred temperatures, whereas the less mobile larvae must alter performance to cope with the changed environment to which they are exposed.

The rapid response to changing thermal circumstances, originally described as rapid cold-hardening and involving a change in the extent of chilling injury without freezing (Lee *et al.*, 1987, 2006), is a form of phenotypic plasticity that has been documented in a wide variety of species (Chown and Nicolson, 2004). Rapid changes in the crystallization temperature, or supercooling point, documented in some (Worland and Convey, 2001; Hawes *et al.*, 2007a), but not all (Hawes *et al.*, 2006; Sinclair *et al.*, 2003c), polar arthropods have also been termed RCH, although the term RCH was originally not intended to describe such changes (Lee *et al.*, 2006). Recently, it has been argued that all forms of RCH should be considered examples of "superplasticity" (Hawes and Bale, 2007; Hawes *et al.*, 2007a). The term was coined for cases of high levels of plasticity that should be

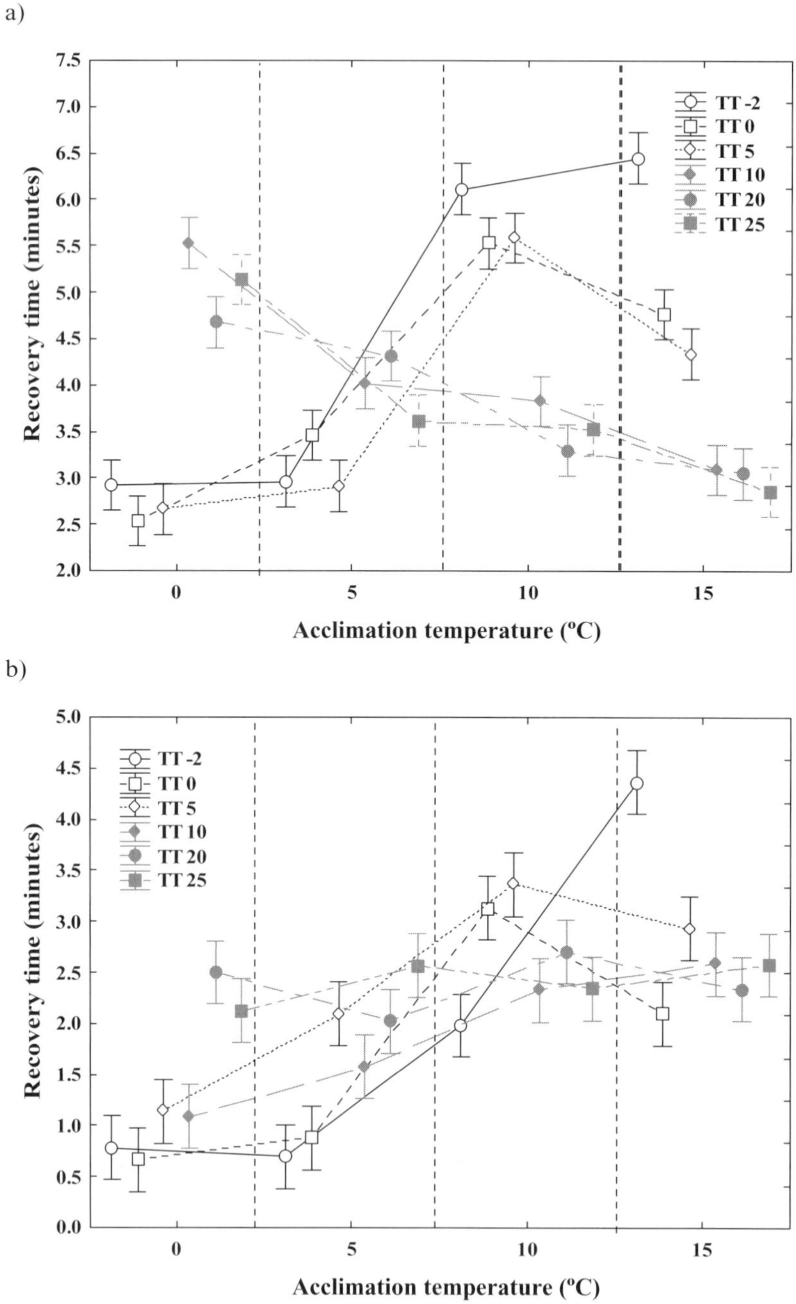

Figure 8.5 Mean (± SE) chill-coma recovery times (minutes) in *Paractora dreuxi* larvae (a) and adults (b) after a seven day acclimation at 0, 5, 10 and 15 °C with 2-hour treatments (−2, 0, 5, 10, 20 and 25 °C). The acclimation temperatures are provided along the *x*-axis, and recovery times at each of the test temperatures (TT) indicated using a different symbol. Lower test temperature responses are indicated with open symbols and higher test temperature responses with closed symbols. (Redrawn from Marais and Chown, 2008.)

"distinguished from standard 'labile' responses" because they "operate at temporal and/or physiological scales in excess of environmental variation" (Hawes and Bale, 2007: 2590).

In our view, caution should be exercised in inventing and using yet another term for acclimation, acclimatization, or plastic responses (see also West-Eberhard, 2003; Loeschcke and Sørensen, 2005). Here, in particular, little evidence exists that the rapid and sometimes large responses described by Hawes and Bale (2007) as "superplasticity" really do exceed environmental variation. This is the justification for the use of the term, because their "standard labile responses" refer to Scheiner's (1993) definition of a labile trait as one where "the phenotype of the individual can change at least as fast as the environment" Hawes and Bale (2007) use two examples further to justify the use of this term. The first, by Worland and Convey (2001), includes microclimate data indicating concurrent rapid change in temperature and physiology. The second example from their own work (Hawes *et al.*, 2007a), includes no relevant data on short-term variation in temperature, with the exception of reference to an earlier paper, which does not include such explicit data either. Moreover, Scheiner's (1993) definition of labile traits suggests that the change is "at least as fast as the environment" and therefore includes responses that are faster. In consequence, it is our view that the existing terminology in both the phenotypic plasticity and rapid cold-hardening (RCH) literature is adequate and that the term "superplasticity" is redundant (Chown *et al.*, 2008). Similarly, the hypotheses proposed in the context of temperature transients (or short-term extreme temperatures) (Dillon *et al.*, 2007) require further integration into the broader field of plasticity and thermal responses.

8.4 Low-temperature responses over space

8.4.1 Spatial patterns in lower limits and cold-hardiness strategies

Because clear geographic variation in low temperatures exists, spatial variation in insect lower lethal limits has been examined at multiple large geographic scales, ranging from within continents (e.g. Bahrndorff *et al.*, 2006) to biome-wide studies (Klok and Chown, 2003). Although *Drosophila* species have been a useful model in these studies, owing to their tractability (Hoffmann *et al.*, 2003), many taxa have been investigated. It is not surprising that the general conclusion from these studies is that insects living in colder places tend to be more cold-tolerant (Addo-Bediako *et al.*, 2000; Turnock and Fields, 2005). However, significant among-species variation exists in the relationship between lower lethal temperature and latitude (especially when this is corrected for altitude, Addo-Bediako *et al.*, 2000) (Fig. 8.6), and it seems most likely that the variation is the

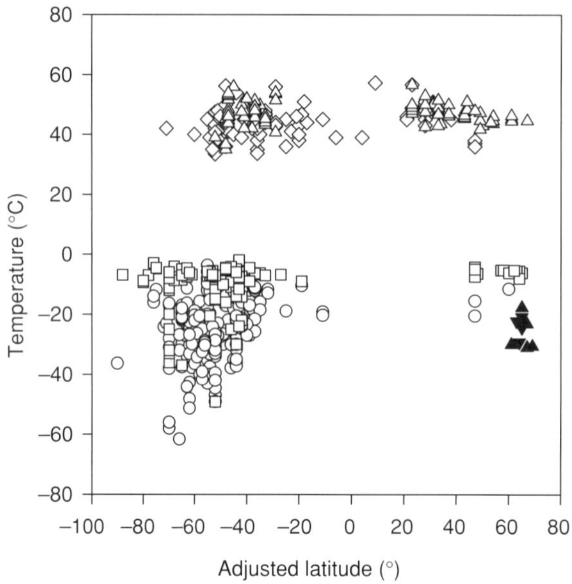

Figure 8.6 Latitudinal variation (northern latitudes are negative) in supercooling points (SCP) (freeze-avoiding (○) and freeze-tolerant insect species (□)) and upper thermal limits (critical thermal maxima (Δ) and upper lethal temperatures (◇)). Data for some freeze-avoiding Antarctic Collembola (▼) and Acari (▲) are shown for comparative purposes. The data are corrected for altitude. (Redrawn from Addo-Bediako *et al.*, 2000.)

consequence of microhabitat selection. However, many cases exist where species tolerate temperatures well in excess of those experienced in the microhabitat (e.g. the goldenrod gall fly can survive as low as −80 °C, Baust and Nishino, 1991).

Several reasons exist why a global-scale fingerprint of insect cold-tolerance strategies might be expected, if only because the nature of temperature mean and variability varies substantially over space (Bonan, 2002). As noted in the Introduction, a broad-scale geographic signal in insect cold-hardiness strategies is apparent (Fig. 8.1), consisting of a preponderance of freeze-avoiding species in northern temperate habitats, dominance of strong freeze-tolerance in high arctic areas and frequent, moderate freeze-tolerance in high-latitude southern-hemisphere insects. Previously, it has been argued that freeze-tolerance provides two different forms of advantage (Sinclair *et al.*, 2003a). The first form is that once frozen, an insect is no longer at risk of freezing, and freeze-tolerance is thus advantageous for insects that are exposed to very low subzero temperatures (particularly those below −50 °C), or to low subzero temperatures for long periods (see also Bale, 1996; Turnock and Fields, 2005). However, in a situation where an insect may be at high risk of freezing (e.g. on moist sub-Antarctic islands, Klok and Chown, 1997), or where there is a risk of unexpected freezing events that would not allow an insect to "prepare" by producing cryoprotectants or managing ice nucleation, freeze-tolerance provides a means to reduce the risk of mortality from freezing. Mild, variable microclimates are more common in the temperate latitudes of the southern hemisphere, where the mean annual temperature is much closer to 0 °C than in continental climates, and this likely explains some of the apparent interhemispheric differences in cold-tolerance strategies adopted by insects (Sinclair *et al.* 2003a; Sinclair and Chown, 2005). By contrast, Turnock and Fields (2005) suggested that the availability of water in the environment (which at subzero temperatures may increase inoculation probability) might be most important in promoting selection for moderate freeze-tolerance (in keeping with remarks by Klok and Chown, 1997, and by previous researchers – e.g. Zachariassen, 1985). Although Turnock and Fields (2005) failed to distinguish northern- from southern-hemisphere data, and therefore missed the significance of unpredictable environmental conditions, it does seem plausible that both unpredictability and inoculation probability are significant factors influencing the evolution of freeze-tolerance. In many instances the two will be inseparable because of the oceanic nature of many unpredictable environments, and further exploration of more detailed information on microclimates and strategies will reveal the relative significance of each of these proposed proximate environmental explanations.

One advantage of Sinclair *et al.*'s (2003a) hypothesis about climate variability and cold-tolerance strategy is that it leads to several further predictions about the relationship between cold-tolerance strategy and the environment: (a) freeze-avoiding species should have a mix of high and low supercooling points at the population level; (b) most freeze-avoiding and chill-susceptible species in variable environments should show a rapid cold-hardening response; (c) cool, aseasonal unpredictable environments in the northern hemisphere should also have a preponderance of freeze-tolerant species.

Although numerous instances of bimodal supercooling point distributions have been documented, particularly in Collembola (e.g. Cannon and Block, 1988), Worland and colleagues (Worland, 2005, Worland *et al.*, 2006) demonstrated that many field observations of bimodal supercooling points are a consequence of patterns of molting behavior (whereby freshly molted individuals have low SCPs) (see also Hawes *et al.*, 2007b). Moreover, most non-freeze-tolerant insects (aside from Collembola) in mild environments tend towards chill-susceptibility (Turnock and Fields, 2005), so few data are available to test prediction (a) in higher insects (though see Klok and Chown, 1998 for some support for this idea). The number of species demonstrating RCH responses is ever-increasing (see the Chapter 2 in this volume), often from habitats that support prediction (b). Although several studies have not found evidence for RCH, this may be a consequence of behavioral avoidance or the rate of change in the environment relative to the rate of development of the response, its cost or the longevity of the life stage concerned, none of which have been thoroughly explored across a range of environments. Few data are available to test prediction (c), although Turnock and Fields (2005) argue that what data are available do not support the prediction.

8.4.2 Spatial covariation in lethal thermal limits

Surprisingly, across a range of geographic scales, little evidence exists of broad covariation between high- and low-temperature tolerances. Among species along elevational and latitudinal gradients, low-temperature tolerances vary much more than do high-temperature tolerances (Gaston and Chown, 1999; Addo-Bediako *et al.* 2000; Klok and Chown, 2003; Ayrinhac *et al.*, 2004; Calosi *et al.*, 2008), and indeed this pattern appears to be common for many insect groups, extending also to responses to acclimation and laboratory selection (Chown, 2001; Chown and Terblanche, 2007). However, this is not true of all groups, with some indication of a trade-off among (or covariation in) high- and low-temperature tolerances in *Drosophila* (Hoffmann *et al.*, 2002).

The actual mechanism of trade-off (or lack thereof) between high- and low-temperature tolerances remains to be determined, and may vary among species. Membrane phospholipid composition appears to be a key component of both high- and low-temperature tolerances, and might be expected to result in a trade-off, if basal characteristics that enhance high-temperature function (e.g. decreased fluidity) also hamper performance in cold (Hochachka & Somero, 2002). In marine invertebrates, a high–low-temperature tradeoff has been ascribed to mismatch between oxygen supply at the organismal level and oxygen demand at the subcellular level (Pörtner, 2001). However, Klok *et al.* (2004) showed that high-temperature tolerance of a terrestrial beetle is largely oxygen-independent. If low-temperature tolerances remain oxygen-dependent, this could explain why upper and lower thermal tolerances appear decoupled in at least some insect taxa. Current data suggest that ATP, and therefore probably O_2, supply does not decline with chilling mortality of insects (Košťál *et al.*, 2004), but O_2 supply and demand in insects at low temperature has not yet received substantial attention.

8.5 Conclusions and future directions

As the severity of humanity's self-induced climate and habitat alteration predicament becomes more obvious, so too will calls for further and more rapid development of the understanding required to mitigate and adapt to these conditions become more vocal (Anonymous, 2007). Most practically, the ecological implications of organismal physiological responses will have to be explored more explicitly at multiple spatial and temporal scales. How environmental variation results in the responses organisms show to the environment, how rapidly these responses might change, and how these changes interact with other aspects of the environment, such as the responses of other organisms, are key elements of this understanding (Soberón, 2007). Although the challenge at times seems unmanageable, considerable progress has been made in meeting it. For example, bottom-up physiological models such as those developed by Porter and his colleagues (Porter *et al.*, 2002, see also Crozier, 2004) are helping to understand better the problems with top-down bioclimatic approaches (Botkin *et al.*, 2007), and drawing attention to the value of both (Kearney, 2006). Likewise, physiologists can draw attention to subtle, though important, variation in the environment that may confound predictions of species or assemblage responses. For example, at Marion Island, ongoing warming is being accompanied by increasing frequency of freeze–thaw events because of an increase in clear-sky evenings, especially in winter. Although the former will likely favor insect species given positive growth-rate–temperature relationships, repeated freeze–thaw events could counter all of that gain and indeed substantially prolong the immature period (Sinclair and Chown, 2005).

Meeting these challenges certainly does not mean that all insect physiologists should turn their attention to a single level in the biological hierarchy. Mechanistic understanding at all levels is just as important, or perhaps even more so, than it has ever been. For example, while some insects can sense temperature changes of $< 0.1^{\circ}C$, acclimation responses typically show temperature thresholds. How sensing and physiological responses interact to result in the phenotypic adjustments typically seen remains opaque (Chown and Terblanche, 2007; Marais and Chown, 2008). Likewise, the mechanisms underlying cross-tolerance and the inter-relationships between resistance and performance traits require further exploration. In particular, overcoming the experimentalist's notion that only one variable should be altered at a time, when it is clear that this does not happen outside the laboratory, is important. However, as the number of variables to be altered increases, so too does the number of interactions, eventually precluding a sensible design. Field and common-garden experiments (or reciprocal transplants), a regular feature of botanical studies, represent one way of overcoming these problems that might readily be applied to insect models (e.g. Gilbert, 1980; Kristensen *et al.*, 2008). Nonetheless, as we have demonstrated here, much may also be learned about the way insects respond to the environment by adopting an explicitly macrophysiological approach. We are encouraged by the increasing number of studies that do so, and look forward to the field benefiting increasingly, in the way that ecology has, from this perspective.

Acknowledgements

We thank Elrike Marais and Katie Marshall for their insightful comments on a previous draft of the work. Anel Garthwaite assisted with manuscript curation. This work was partially supported by NRF Grant SNA2007042400003 and BJS by an NSERC Discovery Grant.

References

Addo-Bediako, A., Chown, S. L. and Gaston, K. J. (2000). Thermal tolerance, climatic variability and latitude. *Proceedings of the Royal Society, Series B* **267**, 739–745.

Andrewartha, H. G. and Birch, L. C. (1954). *The Distribution and Abundance of Animals.* Chicago: University of Chicago Press.

Angilletta, M. J., Bennett, A. F., Guderley, H., Navas, C. A., Seebacher, F. and Wilson, R. S. (2006). Coadaptation: a unifying principle in evolutionary thermal biology. *Physiological and Biochemical Zoology* **79**, 282–294.

Angilletta, M. J., Niewiarowski, P. H. and Navas, C. A. (2002). The evolution of thermal physiology in ectotherms. *Journal of Thermal Biology* **27**, 249–268.

Anonymous. (2007). *Biodiversity-Climate Interactions: Adaptation, Mitigation and Human Livelihoods. Summary of an International Meeting held at the Royal Society 12–13 June 2007.* UNEP/CBD/SBSTTA/12/INF/19, Paris.

Ayrinhac, A., Debat, V., Gibert, P., Kister, A. G., Legout, H., Moreteau, B., Vergilino, R. and David, J. R. (2004). Cold adaptation in geographical populations of *Drosophila melanogaster*: phenotypic plasticity is more important than genetic variability. *Functional Ecology* **18**, 700–706.

Bahrndorff, S., Holmstrup, M., Petersen, H. and Loeschcke, V. (2006). Geographic variation for climatic stress resistance traits in the springtail *Orchesella cincta*. *Journal of Insect Physiology* **52**, 951–959.

Bale, J. S. (1987). Insect cold hardiness: freezing and supercooling – an ecophysiological perspective. *Journal of Insect Physiology* **33**, 899–908.

Bale, J. S. (1993). Classes of insect cold hardiness. *Functional Ecology* **7**, 751–753.

Bale, J. S. (1996). Insect cold hardiness: A matter of life and death. *European Journal of Entomology* **93**, 369–382

Bale, J. S. (2002). Insects and low temperatures: from molecular biology to distributions and abundance. *Philosophical Transactions of the Royal Society of London B* **357**, 849–861.

Bale, J. S., Worland, M. R. and Block, W. (2001). Effects of summer frost exposures on the cold tolerance strategy of a sub-Antarctic beetle. *Journal of Insect Physiology* **47**, 1161–1167.

Baust, J. G. and Nishino, M. (1991). Freezing tolerance in the goldenrod gall fly *(Eurosta solidaginis)*. In *Insects at Low Temperatures*, ed. R. E. Lee and D. L. Denlinger, New York: Chapman and Hall, pp. 260–275.

Baust, J. G. and Rojas, R. R. (1985). Insect cold hardiness: facts and fancy. *Journal of Insect Physiology* **31**, 755–759.

Bonan, G. B. (2002). *Ecological Climatology. Concepts and Applications.* Cambridge: Cambridge University Press.

Botkin, D. B., Saxe, H., Araújo, M. B., Betts, R., Bradshaw, R. H. W., Cedhagen, T., Chesson, P., Dawson, T. P., Etterson, J. R., Faith, D. P., Ferrier, S., Guisan, A., Hansen, A. S., Hilbert, D. W., Loehle, C., Margules, C., New, M., Sobel, M. J. and Stockwell, D. R. B. (2007). Forecasting the effects of global warming on biodiversity. *BioScience* **57**, 227–236.

Brattstrom, B. H. (1968). Thermal acclimation in anuran amphibians as a function of latitude and altitude. *Comparative Biochemistry and Physiology* **24**, 93–111.

Brown, C. L., Bale, J. S. and Walters, K. F. A. (2004). Freezing induces a loss of freeze tolerance in an overwintering insect. *Proceedings of the Royal Society, Series B* **271**, 1507–1511.

Bubliy, O. A. and Loeschcke, V. (2005). Correlated responses to selection for stress resistance and longevity in a laboratory population of *Drosophila melanogaster*. *Journal of Evolutionary Biology* **18**, 789–803.

Calosi, P., Bilton, D. J. Spicer, J. I. and Atfield, A. (2008). Thermal tolerance and geographical range size in the *Agrabus brunneus* group of European diving beetles (Coleoptera: Dytiscidae). *Journal of Biogeography* **35**, 295–305.

Cannon, R. J. C. and Block, W. (1988). Cold tolerance of microarthropods. *Biological Reviews* **63**, 23–77.

Chatfield, C. (2004). *The Analysis of Time Series. An Introduction.* 6th edn. Boca Raton: Chapman and Hall/CRC.

Chown, S. L. (2001). Physiological variation in insects: hierarchical levels and implications. *Journal of Insect Physiology* **47**, 649–660.

Chown, S. L., Addo-Bediako, A. and Gaston, K. J. (2002). Physiological variation in insects: large-scale patterns and their implications. *Comparative Biochemistry and Physiology B* **131**, 587–602.

Chown, S. L., Addo-Bediako, A. and Gaston, K. J. (2003). Physiological diversity: listening to the large-scale signal. *Functional Ecology* **17**, 568–572.

Chown, S. L. and Gaston, K. J. (1999). Exploring links between physiology and ecology at macro-scales: the role of respiratory metabolism in insects. *Biological Reviews* **74**, 87–120.

Chown, S. L. and Gaston, K. J. (2008) Macrophysiology for a changing world. *Proceedings of the Royal Society, Series B* **275**, 1469–1478.

Chown, S. L., Gaston, K. J. and Robinson, D. (2004). Macrophysiology: large-scale patterns in physiological traits and their ecological implications. *Functional Ecology* **18**, 159–167.

Chown, S. L., Jumbam, K. R., Sørensen, J. G. and Terblanche, J. S. (2009). Phenotypic variance, plasticity and heritability estimates of critical thermal limits depend on methodological context. *Functional Ecology* **23**, 133–140.

Chown, S. L. and Nicolson, S. W. (2004). *Insect Physiological Ecology. Mechanisms and Patterns.* Oxford: Oxford University Press.

Chown, S. L., Slabber, S., McGeoch, M. A., Janion, C. and Leinaas, H. P. (2007). Phenotypic plasticity mediates climate change responses among invasive and indigenous arthropods. *Proceedings of the Royal Society, Series B* **274**, 2661–2667.

Chown, S. L., Sørensen, J. G. and Sinclair, B. J. (2008). Physiological variation and phenotypic plasticity – A response to 'Plasticity in arthropod cryotypes' by Hawes and Bale. *Journal of Experimental Biology* **211**, 3353–3357.

Chown, S. L. and Terblanche, J. S. (2007). Physiological diversity in insects: ecological and evolutionary contexts. *Advances in Insect Physiology* **33**, 50–152.

Convey, P., Block, W. and Peat, H. J. (2003). Soil arthropods as indicators of water stress in Antarctic terrestrial habitats? *Global Change Biology* **9**, 1718–1730.

Crozier, L. (2004). Warmer winters drive butterfly range expansion by increasing survivorship. *Ecology* **85**, 231–241.

Davis, M. B. and Shaw, R. G. (2001). Range shifts and adaptive responses to Quaternary climate change. *Science* **292**, 673–679.

Dawson, W. R. (2005). George A. Bartholomew's contributions to integrative and comparative biology. *Integrative and Comparative Biology* **45**, 219–230.

Deere, J. A. and Chown, S. L. (2006). Testing the beneficial acclimation hypothesis and its alternatives for locomotor performance. *American Naturalist* **168**, 630–644.

Deere, J. A., Sinclair, B. J., Marshall, D. J. and Chown, S. L. (2006). Phenotypic plasticity of thermal tolerances in five oribatid mite species from sub-Antarctic Marion Island. *Journal of Insect Physiology* **52**, 693–700.

Denlinger, D. L. (2002). Regulation of diapause. *Annual Review of Entomology* **47**, 93–122.

Denny, M. and Gaines, S. (2000). *Chance in Biology. Using Probability to Explore Nature.* Princeton: Princeton University Press.

Deutsch, C. A., Tewskbury, J. A., Huey, R. B., Sheldon, K. S., Ghalambor, C. K., Haak, D. C. and Martin, P. R. (2008). Impacts of climate warming on terrestrial ectotherms across latitude. *Proceedings of the National Academy of Sciences, USA* **105**, 6668–6672.

Dillon, M. E., Cahn, L. R. Y. and Huey, R. B. (2007). Life history consequences of temperature transients in *Drosophila melanogaster*. *Journal of Experimental Biology* **210**, 2897–2904.

Duman, J. G., Wu, D. W., Xu, L., Tursman, D. and Olsen, T. M. (1991). Adaptations of insects to subzero temperatures. *Quarterly Review of Biology* **66**, 387–410.

Easterling, D. R., Meehl, G. A., Parmesan, C., Changnon, S. A., Karl, T. R. and Mearns, L. O. (2000). Climate extremes: observations, modeling, and impacts. *Science* **289**, 2068–2074.

Feder, M. E. (1987). The analysis of physiological diversity: the prospects for pattern documentation and general questions in ecological physiology. In *New Directions in Ecological Physiology*, ed. M. E. Feder, A. F. Bennett, W. W. Burggren and R. B. Huey, Cambridge: Cambridge University Press, pp. 38–75.

Feder, M. E., Bennett, A. F. and Huey, R. B. (2000). Evolutionary physiology. *Annual Review of Ecology and Systematics* **31**, 315–341.

Ferguson, S. H. and Messier, F. (1996). Ecological implications of a latitudinal gradient in inter-annual climatic varibility: a test using fractal and chaos theories. *Ecography* **19**, 382–392.

Fields, P. G. and McNeil, J. N. (1986). Possible dual cold-hardiness strategies in *Cisseps fulvicollis* (Lepidoptera: Arctiidae). *Canadian Entomologist* **118**, 1309–1311.

Gaines, S. D. and Denny, M. W. (1993). The largest, smallest, highest, lowest, longest and shortest: extremes in ecology. *Ecology* **74**, 1677–1692.

Gaston, K. J. (2003). *The Structure and Dynamics of Geographic Ranges*. Oxford: Oxford University Press.

Gaston, K. J., Blackburn, T. M. and Spicer, J. I. (1998). Rapoport's rule: time for an epitaph? *Trends in Ecology and Evolution* **13**, 70–74.

Gaston, K. J. and Chown, S. L. (1999). Elevation and climatic tolerance: a test using dung beetles. *Oikos* **86**, 584–590.

Gehrken, U., Strømme, A., Lundheim, R. and Zachariassen, K. E. (1991). Inoculative freezing in overwintering tenebrionid beetle, *Bolitophagus reticulatus* Panz. *Journal of Insect Physiology* **37**, 683–687.

Ghalambor, C. K., McKay, J. K., Carroll, S. P. and Reznick, D. N. (2007). Adaptive versus non-adaptive phenotypic plasticity and the potential for contemporary adaptation in new environments. *Functional Ecology* **21**, 394–407.

Gilbert, N. (1980). Comparative dynamics of a single-host aphid. I. The evidence. *Journal of Animal Ecology* **49**, 351–369.

Gilchrist, G. W. (1995). Specialists and generalists in changing environments. I. Fitness landscapes of thermal sensitivity. *American Naturalist* **146**, 252–270.

Grimaldi, D. and Engel, M. S. (2005). *Evolution of the Insects*. Cambridge: Cambridge University Press.

Gutschick, V. P. and BassiriRad, H. (2003). Extreme events as shaping physiology, ecology, and evolution of plants: toward a unified definition and evaluation of their consequences. *New Phytologist* **160**, 21–42.

Halley, J. M. (1996). Ecology, evolution and 1/f-noise. *Trends in Ecology and Evolution* **11**, 33–37.

Hawes, T. C. and Bale, J. S. (2007). Plasticity in arthropod cryotypes. *Journal of Experimental Biology* **210**, 2585–2592.

Hawes, T. C., Bale, J. S., Worland, M. R. and Convey, P. (2007a). Moulting reduces freeze susceptibility in the Antarctic mite *Alaskozetes antarcticus* (Michael). *Physiological Entomology* **32**, 301–304.

Hawes, T. C., Bale, J. S., Worland, M. R. and Convey, P. (2007b). Plasticity and superplasticity in the acclimation potential of the Antarctic mite *Halozetes belgicae* (Michael). *Journal of Experimental Biology* **210**, 593–601.

Hawes, T. C., Couldridge, C. E., Bale, J. S., Worland, M. R. and Convey, P. (2006). Habitat temperature and the temporal scaling of cold hardening in the high Arctic collembolan, *Hypogastrura tullbergi* (Schäffer). *Ecological Entomology* **31**, 450–459.

Hayward, S. A. L., Rinehart, J. P. and Denlinger, D. L. (2004). Desiccation and rehydration elicit distinct heat shock protein transcript responses in flesh fly pupae. *Journal of Experimental Biology* **207**, 963–971.

Helmuth, B., Kingsolver, J. G. and Carrington, E. (2005). Biophysics, physiological ecology, and climate change: does mechanism matter? *Annual Review of Physiology* **67**, 177–201.

Hochachka, P. W. and Somero, G. N. (2002). *Biochemical Adaptation*. New York: Oxford University Press.

Hodkinson, I. D. (2003). Metabolic cold adaptation in arthropods: a smaller-scale perspective. *Functional Ecology* **17**, 562–567.

Hoffmann, A. A., Anderson, A. and Hallas, R. (2002). Opposing clines for high and low temperature resistance in *Drosophila melanogaster*. *Ecology Letters* **5**, 614–618.

Hoffmann, A. A., Hallas, R., Anderson, A. R. and Telonis-Scott, M. (2005). Evidence for a robust sex-specific trade-off between cold resistance and starvation resistance in *Drosophila melanogaster*. *Journal of Evolutionary Biology* **18**, 804–810.

Hoffmann, A. A., Sørensen, J. G. and Loeschcke, V. (2003). Adaptation of *Drosophila* to temperature extremes: bringing together quantitative and molecular approaches. *Journal of Thermal Biology* **28**, 175–216.

Holmstrup, M., Bayley, M. and Ramløv, H. (2002b). Supercool or dehydrate? An experimental analysis of overwintering strategies in small permeable Arctic invertebrates. *Proceedings of the National Academy of Sciences, USA* **99**, 5716–5720.

Holmstrup, M., Hedlund, K. and Boriss, H. (2002a). Drought acclimation and lipid composition in *Folsomia candida*: implications for cold shock, heat shock and acute desiccation stress. *Journal of Insect Physiology* **48**, 961–970.

Holt, R. D., Lawton, J. H., Gaston, K. J. and Blackburn, T. M. (1997). On the relationship between range size and local abundance: back to the basics. *Oikos* **78**, 183–190.

Huey, R. B. (1991). Physiological consequences of habitat selection. *American Naturalist Supplement* **137**, 91–115.

Huey, R. B., Hertz, P. E. and Sinervo, B. (2003). Behavioral drive versus behavioral inertia in evolution: a null model approach. *American Naturalist* **161**, 357–366.

Inchausti, P. and Halley, J. (2003). On the relation between temporal variability and persistence time in animal populations. *Journal of Animal Ecology* **72**, 899–908.

IPCC (1990). *Climate Change: The IPCC Scientific Assessment*. Cambridge: Cambridge University Press.

Irwin, J. T. and Lee, R. E. (2003). Cold winter microenvironments conserve energy and improve overwintering survival and potential fecundity of the goldenrod gall fly, *Eurosta solidaginis*. *Oikos* **100**, 71–78.

Joss, F. and Spahni, R. (2008). Rates of change in natural and anthropogenic radiative forcing over the past 20 000 years. *Proceedings of the National Academy of Sciences, USA* **105**, 1425–1430.

Kearney, M. (2006). Habitat, environment and niche: what are we modelling? *Oikos* **115**, 186–191.

Kelty, J. D. and Lee, R. E. (1999). Induction of rapid cold hardening by cooling at ecologically relevant rates in *Drosophila melanogaster*. *Journal of Insect Physiology* **45**, 719–726.

Kelty, J. D. and Lee, R. E. (2001). Rapid cold-hardening of *Drosophila melanogaster* (Diptera: Drosophilidae) during ecologically based thermoperiodic cycles. *Journal of Experimental Biology* **204**, 1659–1666.

Kingsolver, J. G. and Huey, R. B. (1998). Evolutionary analyses of morphological and physiological plasticity in thermally variable environments. *American Zoologist* **38**, 545–560.

Klok, C. J. and Chown, S. L. (1997). Critical thermal limits, temperature tolerance and water balance of a sub-Antarctic caterpillar, *Pringleophaga marioni* Viette (Lepidoptera: Tineidae). *Journal of Insect Physiology* **43**, 685–694.

Klok, C. J. and Chown, S. L. (1998). Interactions between desiccation resistance, host-plant contact and the thermal biology of a leaf-dwelling sub-antarctic caterpillar, *Embryonopsis halticella* (Lepidoptera: Yponomeutidae). *Journal of Insect Physiology* **44**, 615–628.

Klok, C. J. and Chown, S. L. (2003). Resistance to temperature extremes in sub-Antarctic weevils: interspecific variation, population differentiation and acclimation. *Biological Journal of the Linnean Society* **78**, 401–414.

Klok, C. J., Sinclair, B. J. and Chown, S. L. (2004). Upper thermal tolerance and oxygen limitation in terrestrial arthropods. *Journal of Experimental Biology* **207**, 2361–2370.

Koštál, V. and Šimcek, P. (1996). Biochemistry and physiology of aestivo-hibernation in the adult apple blossom weevil, *Anthonomus pomorum* (Coleoptera: Curculionidae). *Journal of Insect Physiology* **42**, 727–733.

Koštál, V., Vambera, J. and Bastl, J. (2004). On the nature of pre-freeze mortality in insects: water balance, ion homeostasis and energy charge in the adults of *Pyrrhocoris apterus*. *Journal of Experimental Biology* **207**, 1509–1521.

Kristensen, T. N., Hoffmann, A. A., Overgaard, J., Sorensen, J. G., Hallas, R. and Loeschcke, V. (2008). Costs and benefits of cold acclimation in field-released *Drosophila*. *Proceedings of the National Academy of Sciences, USA* **105**, 216–221.

Lee, R. E., Chen, C.-P. and Denlinger, D. L. (1987). A rapid cold-hardening process in insects. *Science* **238**, 1415–1417.

Lee, R. E., Elnitsky, M. A., Rinehart, J. P., Hayward, S. A. L., Sandro, L. H. and Denlinger, D. L. (2006). Rapid cold-hardening increases the freezing tolerance of the Antarctic midge *Belgica antarctica*. *Journal of Experimental Biology* **209**, 399–406.

Leibold, M. A., Holyoak, M., Mouquet, N., Amarasekare, P., Chase, J. M., Hoopes, M. F., Holt, R. D., Shurin, J. B., Law, R., Tilman, D., Loreau, M. and Gonzalez, A. (2004). The metacommunity concept: a framework for multi-scale community ecology. *Ecology Letters* **7**, 601–613.

Levins, R. (1968). *Evolution in Changing Environments. Some Theoretical Explorations.* Princeton: Princeton University Press.

Loeschcke, V. and Sørensen, J. G. (2005). Acclimation, heat shock and hardening – a response from evolutionary biology. *Journal of Thermal Biology* **30**, 255–257.

Lomolino, M. V. and Heaney, L. R. (2004). *Frontiers of Biogeography. New Directions in the Geography of Nature.* Sunderland: Sinauer Associates.

Lynch, M. and Gabriel, W. (1987). Environmental tolerance. *American Naturalist* **129**, 283–303.

Lundheim, R. and Zachariassen, K. E. (1993). Water balance of over-wintering beetles in relation to strategies for cold tolerance. *Journal of Comparative Physiology B* **163**, 1–4.

Makarieva, A. M., Gorshkov, V. G., Li, B.-L. and Chown, S. L. (2006). Size- and temperature-independence of minimum life-supporting metabolic rates. *Functional Ecology* **20**, 83–96.

Marais, E. and Chown, S. L. (2008). Beneficial acclimation and the Bogert effect. *Ecology Letters* **11**, 1027–1036.

Marais, E., Terblanche, J. S. and Chown, S. L. (2009) Life stage-related differences in hardening and acclimation of thermal tolerance traits in the kelp fly, *Paractora dreuxi* (Diptera, Helcomyzidae). *Journal of Insect Physiology* **55**, 336–343.

McArdle, B. H. and Gaston, K. J. (1995). The temporal variability of densities: back to basics. *Oikos* **74**, 165–171.

Michaud, M. R. and Denlinger, D. L. (2005). Molecular modalities of insect cold survival: current understanding and future trends. In *Animals and Environments*, ed. S. Morris and A. Vosloo. Amsterdam: Elsevier International Congress Series 1275, pp. 32–46.

Miller, L. K. (1978). Freezing tolerance in relation to cooling rate in an adult insect. *Cryobiology* **15**, 345–349.

Moon, I., Fujikawa, S. and Shimada, K. (1996). Cryopreservation of *Chymomyza* larvae (Diptera: Drosophilidae) at −196 °C with extracellular freezing. *Cryo-Letters* **17**, 105–110.

Myers, A. A. and Giller, P. S. (1988). *Analytical Biogeography. An Integrated Approach to the Study of Animal and Plant Distributions*. London: Chapman and Hall.

Nedvěd, O. (1998). Modelling the relationship between cold injury and accumulated degree days in terrestrial arthropods. *CryoLetters* **19**, 267–274.

Osovitz, C. J. and Hofmann, G. (2007). Marine macrophysiology: studying physiological variation across large spatial scales in marine systems. *Comparative Biochemistry and Physiology A* **147**, 821–827.

Overgaard, J., Malmendal, A., Sørensen, J. G., Bundy, J. G., Loeschcke, V., Nielsen, N. C. and Holmstrup, M. (2007). Metabolomic profiling of rapid cold hardening and cold shock in *Drosophila melanogaster*. *Journal of Insect Physiology* **53**, 1218–1232.

Overgaard, J., Sørensen, J. G., Petersen, S. O., Loeschcke, V. and Holmstrup, M. (2005). Changes in membrane lipid composition following rapid cold hardening in *Drosophila melanogaster*. *Journal of Insect Physiology* **51**, 1173–1182.

Parmesan, C. (2006). Ecological and evolutionary responses to recent climate change. *Annual Review of Ecology Evolution and Systematics* **37**, 637–669.

Parmesan, C., Root, T. L. and Willig, M. R. (2000). Impacts of extreme weather and climate on terrestrial biota. *Bulletin of the American Meteorological Society* **81**, 443–450.

Petchey, O. L., Gonzales, A. and Wilson, H. B. (1997). Effects on population persistence: the interaction between environmental noise colour, intraspecific competition and space. *Proceedings of the Royal Society, Series* B **264**, 1841–1847.

Pimm, S. L. and Redfearn, A. (1988). The variability of animal populations. *Nature* **334**, 613–614.

Porter, W. P., Sabo, J. L., Tracy, C. R. Reichman, O. J. and Ramankutty, N. (2002). Physiology on a landscape scale: plant-animal interactions. *Integrative and Comparative Biology* **42**, 431–453.

Pörtner, H. O. (2001). Climate change and temperature-dependent biogeography: oxygen limitation of thermal tolerance in animals. *Naturwissenschaften* **88**, 137–146.

Prosser, C. L. (1986). *Adaptational Biology. Molecules to Organisms*. New York: John Wiley & Sons.

Pullin, A. S. (1996). Physiological relationships between insect diapause and cold tolerance: coevolution or coincidence? *European Journal of Entomology* **93**, 121–129.

Rako, L., Blacket, M. J., McKechnie, S. W. and Hoffmann, A. A. (2007). Candidate genes and thermal phenotypes: identifying ecologically important genetic variation for thermotolerance in the Australian *Drosophila melanogaster* cline. *Molecular Ecology* **16**, 2948–2957.

Ramløv, H. (2000). Aspects of natural cold tolerance in ectothermic animals. *Human Reproduction* **15**, 26–46.

Renault, D., Nedved, O., Hervant, F. and Vernon, P. (2004). The importance of fluctuating thermal regimes for repairing chill injuries in the tropical beetle *Alphitobius diaperinus* (Coleoptera: Tenebrionidae) during exposure to low temperature. *Physiological Entomology* **29**, 139–145.

Ring, R. A. and Danks, H. V. (1994). Desiccation and cryoprotection: overlapping adaptations. *Cryoletters* **15**, 181–190.

Roff, D. A. (2002). *Life History Evolution*. Sunderland: Sinauer Associates.

Roff, D. A. and Fairbairn, D. J. (2007). The evolution of trade-offs: where are we? *Journal of Evolutionary Biology* **20**, 433–447.

Rosenzweig, C., Karoly, D., Vicarelli, M., Neofotis, P., Wu, Q., Casassa, G., Menzel, A., Root, T. L., Estrella, N., Seguin, B., Tryjanowski, P., Liu, C., Rawlins, S. and Imeson, A. (2008). Attributing physical and biological impacts to anthropogenic climate change. *Nature* **453**, 353–357.

Ruel, J. J. and Ayres, M. P. (1999). Jensen's inequality predicts effects of environmental variation. *Trends in Ecology and Evolution* **14**, 361–366.

Ruokolainen, L. and Fowler, M. S. (2008). Community extinction in coloured environments. *Proceedings of the Royal Society, Series* B **275**, 1775–1783.

Salt, R. W. (1966). Effect of cooling rate on the freezing temperatures of supercooled insects. *Canadian Journal of Zoology* **44**, 655–659.

Salt, R. W. and James, H. G. (1947). Low temperature as a factor in the mortality of eggs of *Mantis religiosa* L. *Canadian Entomologist* **79**, 33–36.

Scheiner, S. M. (1993). Genetics and evolution of phenotypic plasticity. *Annual Review of Ecology and Systematics* **24**, 35–68.

Schliess, F. and Haüssinger, D. (2002). The cellular hydration state: a critical determinant for cell death and survival. *Biological Chemistry* **383**, 577–583.

Schluter, D. (2000). *The Ecology of Adaptive Radiation*. Oxford: Oxford University Press.

Scholander, P. F., Flagg, W., Walters, V. and Irving, L. (1953). Climatic adaptation in arctic and tropical poikilotherms. *Physiological Zoology* **26**, 67–92.

Schwager, M., Johst, K. and Jeltsch, F. (2006). Does red noise increase or decrease extinction risk? Single extreme events versus series of unfavorable conditions. *American Naturalist* **167**, 879–888.

Shelford, V. E. (1913). *Animal Communities in Temperate America*. Chicago: Chicago University Press.

Sinclair, B. J. (1997). Seasonal variation in freezing tolerance of the New Zealand alpine cockroach *Celatoblatta quinquemaculata*. *Ecological Entomology* **22**, 462–467.

Sinclair, B. J. (1999). Insect cold tolerance: how many kinds of frozen? *European Journal of Entomology* **96**, 157–164.

Sinclair, B. J. (2001a). Biologically relevant environmental data: macros to make the most of microclimate recordings. *Cryoletters* **22**, 125–134.

Sinclair, B. J. (2001b). Field ecology of freeze tolerance: interannual variation in cooling rates, freeze-thaw and thermal stress in the microhabitat of the alpine cockroach *Celatoblatta quinquemaculata*. *Oikos* **93**, 286–293.

Sinclair, B. J., Addo-Bediako, A. and Chown, S. L. (2003a). Climatic variability and the evolution of insect freeze tolerance. *Biological Reviews* **78**, 181–195.

Sinclair, B. J. and Chown, S. L. (2003). Rapid responses to high temperature and desiccation but not to low temperature in the freeze tolerant sub-Antarctic caterpillar *Pringleophaga marioni* (Lepidoptera, Tineidae). *Journal of Insect Physiology* **49**, 45–52.

Sinclair, B. J. and Chown, S. L. (2005). Deleterious effects of repeated cold exposure in a freeze-tolerant sub-Antarctic caterpillar. *Journal of Experimental Biology* **208**, 969–879.

Sinclair, B. J., Klok, C. J., Scott, M. B., Terblanche, J. S. and Chown, S. L. (2003b). Diurnal variation in supercooling points of three species of Collembola from Cape Hallett, Antarctica. *Journal of Insect Physiology* **49**, 1049–1061.

Sinclair, B. J., Nelson, S., Nilson, T. L., Roberts, S. P. and Gibbs, A. G. (2007). The effect of selection for desiccation resistance on cold tolerance of *Drosophila melanogaster*. *Physiological Entomology* **32**, 322–327.

Sinclair, B. J., Vernon, P., Klok, C. J. and Chown, S. L. (2003c). Insects at low temperatures: an ecological perspective. *Trends in Ecology and Evolution* **18**, 257–262.

Slabber, S., Worland, M. R., Leinaas, H. P. and Chown, S. L. (2007). Acclimation effects on thermal tolerances of springtails from sub-Antarctic Marion Island: indigenous and invasive species. *Journal of Insect Physiology* **53**, 113–125.

Soberón, J. (2007). Grinnellian and Eltonian niches and geographic distributions of species. *Ecology Letters* **10**, 1115–1123.

Sømme, L. (1982). Supercooling and winter survival in terrestrial arthropods. *Comparative Biochemistry and Physiology A* **73**, 519–543.

Sømme, L. (1996). The effect of prolonged exposures at low temperatures in insects. *CryoLetters* **17**, 341–346.

Sømme, L. and Zachariassen, K. E. (1981). Adaptations to low temperature in high altitude insects from Mount Kenya. *Ecological Entomology* **6**, 119–204.

Spicer, J. I. and Gaston, K. J. (1999). *Physiological Diversity and its Ecological Implications*. Oxford: Blackwell Science.

Stige, L. C., Chan, K.-S., Zhang, Z. Frank, D. and Stenseth, N. C. (2007). Thousand-year-long Chinese time series reveals climatic forcing of decadal locust dynamics. *Proceedings of the National Academy of Sciences, USA* **104**, 16188–16193.

Storey, K. B. and Storey, J. M. (1996). Natural freezing survival in animals. *Annual Review of Ecology and Systematics* **27**, 365–386.

Terblanche, J. S. and Chown, S. L. (2006). The relative contributions of developmental plasticity and adult acclimation to physiological variation in the tsetse fly, *Glossina pallidipes* (Diptera, Glossinidae). *Journal of Experimental Biology* **209**, 1064–1073.

Terblanche, J. S., Deere, J. A., Clusella Trullas, S., Janion, C. and Chown, S. L. (2007). Critical thermal limits depend on methodological context. *Proceedings of the Royal Society, Series B* **274**, 2935–2942.

Terblanche, J. S., Clusella-Trullas, S., Deere, J. A. and Chown, S. L. (2008). Thermal tolerance in a south-east African population of the tsetse fly *Glossina pallidipes* (Diptera, Glossinidae): implications for forecasting climate change impacts. *Journal of Insect Physiology* **54**, 114–127.

Torrence, C. and Campo, G. P. (1998). A practical guide to wavelet analysis. *Bulletin of the American Meteorological Society* **79**, 61–78.

Turnock, W. J. and Fields, P. G. (2005). Winter climates and cold hardiness in terrestrial insects. *European Journal of Entomology* **102**, 561–576.

Umina, P. A., Weeks, A. R., Kearney, M. R., McKechnie, S. W. and Hoffmann, A. A. (2005). A rapid shift in a clinal pattern in *Drosophila* reflecting climate change. *Science* **308**, 691–693.

Van Der Laak, S. (1982). Physiological adaptations to low temperature in freezing-tolerant *Phyllodecta laticollis* beetles. *Comparative Biochemistry and Physiology A* **73**, 613–620.

Vasseur, D. A. and Yodzis, P. (2004). The color of environmental noise. *Ecology* **85**, 1146–1152.

Vernon, P. and Vannier, G. (2002). Evolution of freezing susceptibility and freezing tolerance in terrestrial arthropods. *Comptes Rendus Biologies* **325**, 1185–1190.

Virtanen, T., Neuvonen, S. and Nikula, A. (1998). Modelling topoclimatic patterns of egg mortality of *Epirrita autumnata* (Lepidoptera: Geometridae) with a geographical information system: predictions for current climate and warmer climate scenarios. *Journal of Applied Ecology* **35**, 311–322.

Voituron, Y., Mouquet, N., de Mazancourt, C. and Clobert, J. (2002). To freeze or not to freeze? An evolutionary perspective on the cold-hardiness strategies of overwintering ectotherms. *American Naturalist* **160**, 255–270.

West-Eberhard, M. J. (2003). *Developmental Plasticity and Evolution*. New York: Oxford University Press.

Williams, J. B., Ruehl, N. C. and Lee, R. E. (2004). Partial link between the seasonal acquisition of cold-tolerance and desiccation resistance in the goldenrod gall fly *Eurosta solidaginis* (Diptera: Tephritidae). *Journal of Experimental Biology* **207**, 4407–4414.

Worland, M. R. (2005). Factors that influence freezing in the sub-Antarctic springtail *Tullbergia antarctica*. *Journal of Insect Physiology* **51**, 881–894.

Worland, M. R., Block, W. and Grubor-Lajsic, G. (2000). Survival of *Heleomyza borealis* (Diptera, Heleomyzidae) larvae down to −60 °C. *Physiological Entomology* **25**, 1–5.

Worland, M. R. and Convey, P. (2001). Rapid cold hardening in Antarctic microarthropods. *Functional Ecology* **15**, 515–524.

Worland, M. R., Grubor-Lajsic, G. and Montiel, P. O. (1998). Partial desiccation induced by sub-zero temperatures as a component of the survival strategy of the Arctic collembolan *Onychiurus arcticus* (Tullberg). *Journal of Insect Physiology* **44**, 211–219.

Worland, M. R., Leinaas, H. P. and Chown, S. L. (2006). Supercooling point frequency distributions in Collembola are affected by moulting. *Functional Ecology* **20**, 323–329.

Yoder, J. A., Benoit, J. B., Denlinger, D. L. and Rivers, D. B. (2006). Stress-induced accumulation of glycerol in the flesh fly, *Sarcophaga bullata*: evidence indicating anti-desiccant and cryoprotectant functions of this polyol and a role for the brain in coordinating the response. *Journal of Insect Physiology* **52**, 202–214.

Zachariassen, K. E. (1985). Physiology of cold tolerance in insects. *Physiological Reviews* **65**, 799–837.

9

Evolutionary physiology of insect thermal adaptation to cold environments

RAYMOND B. HUEY

9.1 Introduction

Body temperature influences all aspects of the physiology and ecology of insects and indeed of all other ectotherms (Cossins and Bowler, 1987; Chown and Nicolson, 2004; Angilletta, 2009). Extreme low or high temperatures are physiologically damaging or even lethal, but temperatures well within those critical limits have profound effects on performance and fitness. Not surprisingly, an insect's thermal sensitivity plays a key role in its behavior, ecology and fitness.

Here I review several consequences of insect adaptation to low temperature. My focus is largely on organismal and population-level consequences. I take a macrophysiological approach (Chown *et al.*, 2004) and focus on four key issues: some are classical, but others have only recently received attention: (a) Do optimal and critical temperatures correlate inversely with latitude, indicating that high-latitude species have adapted evolutionarily to low temperatures? (b) Are high-latitude species, which live in seasonal environments, physiologically adapted to a broader range of temperatures than are low-latitude species? (c) Do species with low optimal temperatures have the same maximal rates of population growth as do species with high optimal temperatures, as might occur if biochemical adaptation compensates fully for the thermodynamically depressing effects of low temperature? (d) Are cold-adapted, high-latitude species at greater risk from climate warming than are warm-adapted, low-latitude species, as might be expected given the faster rate of climate warming at high latitudes?

The vast majority of insects live in and probably evolved in the lowland tropics, where they encounter chronically warm environments with little daily and

Low Temperature Biology of Insects, ed. David L. Denlinger and Richard E. Lee Jr. Published by Cambridge University Press.

seasonal variation in temperature (Janzen, 1967). Nevertheless, many insect lineages have independently invaded and diversified in colder environments found at higher altitudes or latitudes (Convey, Chapter 12, this volume; Chown and Sinclair, Chapter 8, this volume). Such insects encounter multiple physiological challenges: in particular, they must deal with ambient temperatures that are seasonally cold and that have exaggerated daily and seasonal variation (Janzen, 1967; MacArthur, 1984; Ghalambor *et al.*, 2006; Fig. 10.1 in Bradshaw and Holzapfel, Chapter 10). Moreover, they must persist despite shortened growing seasons (Ragland and Kingsolver, 2008; Fig. 10.1 in Bradshaw and Holzapfel, Chapter 10, this volume) and despite attendant disruption of photoperiodic timing (Bradshaw and Holzapfel, Chapter 10, this volume). Colder and more variable ambient temperatures may reduce overall physiological performance and population growth rates, and shortened growing seasons may restrict rates of population growth: together these factors may increase insect vulnerability to local extinction.

Insects can and often do adjust to these challenges in various ways. They can use thermoregulatory behaviors (e.g., habitat selection, adjusting daily time of activity, Heinrich, 1981; Huey *et al.*, 2003; Angilletta, 2009) and phenotypic plasticity (Levins, 1968; Hoffmann and Watson, 1993; Kingsolver and Huey, 1998) to buffer daily and seasonal variation in body temperature. They can adjust their genetic response to photoperiodic cues so that key phenological events (e.g., seasonal activity, reproduction, diapause) are resynchronized with suitable thermal and biotic environments (Beck, 1980; Bradshaw and Holzapfel, 2001; Denlinger *et al.*, 2001; Bradshaw and Holzapfel, Chapter 10, this volume). Given sufficient time and genetic variation (Hoffmann *et al.*, 2003; Bradshaw and Holzapfel, Chapter 10, this volume; Overgaard *et al.*, Chapter 11, this volume), their physiological and biochemical sensitivity can adapt to novel thermal regimes such that performance is enhanced at lower and more variable temperatures (Hochachka and Somero, 2002; Huey and Bennett, 1987; Angilletta *et al.*, 2002; Angilletta, 2009). This could be achieved either by enhanced acclimation capacity (Levins, 1968; Hoffmann and Watson, 1993; Kingsolver and Huey, 1998), or by adaptive shifts in thermal sensitivity (Huey, 1987; Overgaard *et al.*, Chapter 11, this volume), or both.

I focus on some consequences of evolution of insects in cold environments. My analyses require quantitative information on how body temperature influences some aspect of absolute (not merely relative) performance. Rather than treat thermal sensitivity as a continuous reaction norm (see Kingsolver and Gomulkiewicz, 2004; Izem and Kingsolver, 2005), I extract descriptive statistics that characterize key aspects of performance or fitness curves (T_o, the optimal temperature at which some trait is maximized and CT_{max} and CT_{min}, the upper and lower critical temperatures at which performance is zero, Huey and Stevenson, 1979).

In principle, one could review the above questions by analysing the thermal sensitivity of any organismal (e.g., assimilation efficiency, locomotor speed, visual acuity), or lower-level (e.g., enzyme activity, metabolic flux) traits. However, I choose here to focus on a population-level trait, namely, the intrinsic rate of population growth (*r*). These rates are equivalent to compound-interest rates in financial investments and are often used as a measure of Darwinian fitness of insects (Carey, 1993; Charlesworth, 1994; Bradshaw and Holzapfel, 2001; but see Huey and Berrigan, 2001): they represent an integrated measure of the physiological and ecological impact of temperature on an ectotherm (Bennett *et al.*, 1992). Note that a population-level approach necessarily ignores the actual physiological and biochemical bases of enhanced performance at low temperatures, which is often relevant (Helmuth *et al.*, 2005; Pörtner *et al.*, 2007): other chapters in this volume address these subjects.

Although a focus on the thermal sensitivity of Darwinian fitness (*r*) has appeal, several caveats must be recognized, simply because intrinsic rates of population growth are almost always calculated from life-table data (i.e., age-specific schedules of survival and reproduction) that were obtained for insects held in the laboratory at several fixed body temperatures with unlimited food. These measurements are laborious and time-consuming, and so are usually done on insects of agricultural concern: consequently, the available data are for a biased subset of insect diversity. Some assays use lab-adapted stocks, which may respond to temperature very differently than do wild stocks (Kingsolver and Nagle, 2007; Huey and Rosenzweig, 2009); and variation in experimental protocol no doubt adds noise (Chown *et al.*, 2008). Moreover, few, if any, insects naturally live under fixed thermal regimes with unlimited food, so whether the resultant laboratory data are informative of patterns for field populations is a reasonable concern. I share these concerns, but hope that any artifacts they induce will be random with respect to whether species are warm-adapted or cold-adapted.

9.2 Conceptual background

The theme of this book is adaptation of insects to low temperature, which implies that low-temperature insects are derived from insects adapted from warm-adapted lineages. For present purposes, I assume only that the ancestral state of insects (whether warm- versus cold-adapted) doesn't matter for evolutionary trajectories. In other words, the thermal sensitivity of an animal adapting from a high to a low optimal temperature will involve along essentially the same trajectory as an animal adapting from a low to a high optimal temperature. Thus any biophysical and thermodynamic constraints on adaptation temperature are assumed symmetrical and thus independent of the direction of temperature adaptation (so there is no thermal hysteresis of evolutionary trajectories). Whether this is actually true is an unresolved question in thermal biology (Huey, 1987)!

9.2.1 A Panglossian view of adaptation to cold and variable temperatures

Let's imagine that some ancestral insect group is perfectly adapted to a warm environment: consequently, it is routinely active at body temperatures (T_b) that are close to optimal physiological levels (T_o). Imagine that this lineage invades a much cooler environment (high mountain, high latitude) or experiences prolonged climate cooling, as occurred during the Pleistocene. If thermal sensitivity does not shift, and if the insect's behavioral and acclimation adjustments are incapable of compensating for lower operative (equilibrium) temperatures (Bakken, 1992), then it will now experience T_b colder than in the past and its average performance and its fitness will decline (see Fig. 1 in Huey and Bennett, 1987).

If our lineage has standing genetic variation for performance and fitness at suboptimal temperatures (Gilchrist, 1995; Gilchrist, 1996; Kingsolver *et al.*, 2004; Overgaard *et al.*, Chapter 11, this volume), then selection should favor individuals that perform relatively well at low temperature (Huey and Bennett, 1987; Bennett *et al.*, 1992; Hoffmann and Blows, 1993; Kingsolver and Gomulkiewicz, 2004; Angilletta *et al.*, 2006). If the genetic variation is sufficient, and if selection is strong, then (all else equal and given no genetic or thermodynamic constraints, see below) our lineage could evolve a sufficiently "left-shifted" thermal fitness curve, so that its maximum performance and fitness now matches that prior to climate warming (Fig. 9.1a). This clearly is a Panglossian view (Gould and Lewontin, 1979): it holds that adaptation to low temperature is rapid and perfectly compensatory, and has no deleterious side effects and no limiting constraints.

Similarly, imagine a lineage that encounters environments with greater daily and seasonal fluctuations in operative temperatures, but with no shift in the mean temperatures. Selection might favor increased thermal performance breadth or in effect decreased sensitivity to temperature (Levins, 1968; Huey and Slatkin, 1976; Gilchrist, 1995). However, increased breadth might come at a cost to maximal performance if a "jack-of-all-temperatures is a master of none" (Fig. 9.1b, Levins, 1968; Huey and Hertz, 1984; Gilchrist, 1996; Pörtner, 2002; Izem and Kingsolver, 2005). Moreover, maximal performance might be reduced even further in a species adapting to low temperatures because of thermodynamic constraints (Fig. 9.1c, Frazier *et al.*, 2006; Savage *et al.*, 2004, see below).

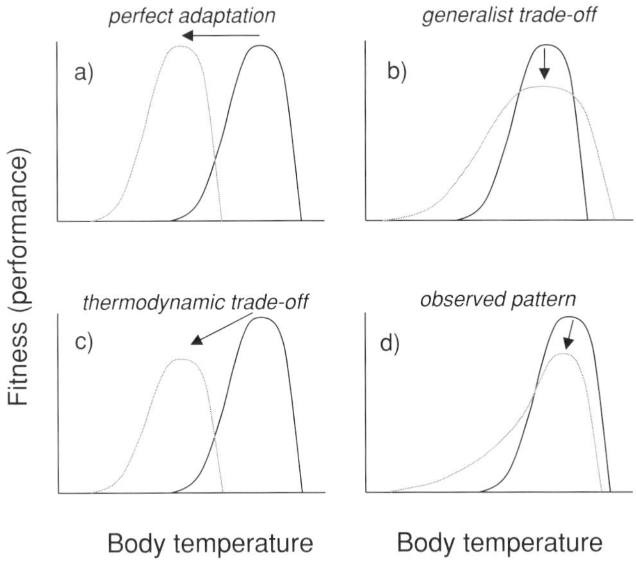

Figure 9.1 Schematic patterns of the evolution of thermal sensitivity in cold or fluctuating environments (black lines = ancestral state, gray lines = derived response). (a) If adaptation to a low-temperature regime is perfectly compensatory (a "Panglossian" view), then the fitness of the derived, low-temperature species at its optimal temperature is the same as that of the ancestral, warm-temperature species at its optimal temperature. No trade-off is evident. (b) If thermal variability increases without a shift in mean temperature, selection might favor a generalist physiology, with an attendant trade-off in maximal performance ("a jack-of-all-temperatures-is-a master of none"). (c) If thermodynamic constraints dominate the evolution of thermal sensitivity, a lineage undergoing selection in a cold environment for a low optimal temperature will have relatively reduced fitness at its optimal temperature (thus "colder is slower"). (d) A schematic view of the "average" thermal sensitivity of low-latitude (black) and of high-latitude (gray line) insects.

9.3 Empirical patterns

9.3.1 Latitudinal patterns of heat and cold tolerance

As one moves away from the lowland tropics, ambient temperatures become colder and also more variable on average (Janzen, 1967), such that climatic "seasonality" increases with latitude (MacArthur, 1984; Ghalambor *et al.*, 2006), especially in the northern hemisphere (Addo-Bediako *et al.*, 2000; Chown *et al.*, 2004). However, the lowest temperatures in winter decline much faster with latitude than do the warmest temperatures in summer (Fig. 10.1 in Bradshaw and Holzapfel, Chapter 10, this volume, Fig. 2 in Ghalambor *et al.*, 2006) and in fact maximum daytime temperatures in summer can be highest at mid-latitudes (e.g.,

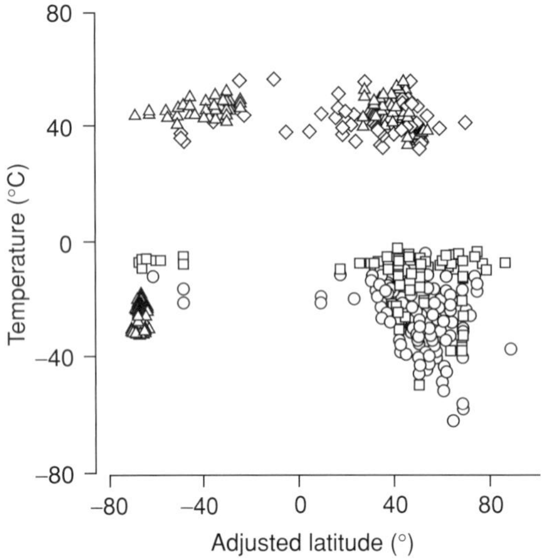

Figure 9.2 Latitudinal variation in heat tolerance and cold-tolerance of insects (redrawn from Addo-Bediako *et al.*, 2000), with latitude adjusted for source altitude (see Price *et al.*, 1998). Heat tolerance shows little systematic shift with latitude, but cold-tolerance is lower and more variable at high latitude, especially in the southern hemisphere (negative latitudes). For upper thermal limits open triangles represent CT_{max}, and open diamonds are upper lethal temperatures. For supercooling points, open circles are for freeze-intolerant insects and open squares are for freeze-tolerant species. Antarctic Collembola are filled inverted triangles, and Acari are filled triangles. Redrawn from Fig. 2c in Addo-Bediako *et al.* (2000): see that reference for details.

30° to 40°N or S) where the world's deserts are concentrated (MacArthur, 1984; Ghalambor *et al.*, 2006).

Given these global patterns of ambient temperature, one might expect that the critical thermal minimum temperature (CT_{min}) would drop faster with latitude than does the critical thermal maximum (CT_{max}). Addo-Bediako *et al.* (2000) exhaustively reviewed the large insect literature and compiled a huge data set. Their patterns (Fig. 9.2) strongly support the expectations (above): insects living at high latitudes have a relatively large tolerance range ($CT_{max} - CT_{min}$). Similar patterns occur in specific insect taxa (Goto and Kimura, 1988; Gibert *et al.*, 2001; Hoffmann *et al.*, 2003; Zani *et al.*, 2005a) and in other terrestrial ectotherms (van Berkum, 1988; Ghalambor *et al.*, 2006; Deutsch *et al.*, 2008).

9.3.2 Latitudinal pattern of optimal body temperatures

How do optimal temperatures (T_o) change with latitude? Here I use a data set compiled by Huey and Berrigan (2001) and Frazier *et al.* (2006), where intrinsic rates of population growth (*r*) were estimated for diverse insects at multiple

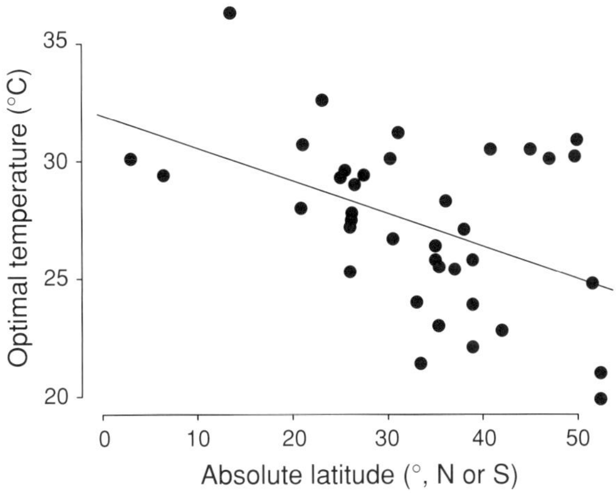

Figure 9.3 The optimal temperature (T_o) that maximizes r (the intrinsic rate of population growth) decreases with absolute latitude. Data from Frazier *et al.* (2006) and Deutsch *et al.* (2008).

temperatures. We (Deutsch *et al.*, 2008) fit a curve to these data for 46 species and estimated the T_b at which r was maximal (T_o). Analyses done in Frazier *et al.* (2006) used both phylogenetic and traditional regression methods: but because the patterns were parallel, I present traditional methods here for simplicity.

Optimal temperature declined significantly with absolute latitude (Fig. 9.2, $r = -0.458$, $p = 0.004$), consistent with expectations based on climate. Note, however, that the interquartile range is surprisingly small (25.3 to 30.1 °C) and that some high latitude species seemingly have T_o as high as some tropical species. By implication, T_o has *not* diversified greatly in most insects – at least those in this sample. Temperature-fitness data on very high-latitude species (Lee and Denlinger, 2006; Klok and Chown, 1997) would definitely be of interest.

Is the decline in T_o with latitude somewhat shallow (Fig. 9.3) because T_o tends to track mean maximum summer temperatures more than it tracks minimum winter temperatures, which decline relatively rapidly with latitude? T_o is in fact strongly correlated with T_{max} ($p = 0.007$, slope $= 0.38 \pm 0.133$) and not at all with T_{min} ($p = 0.711$). This pattern shouldn't be surprising: most insects are active primarily during warm seasons, not in cold ones (see below).

9.3.4 Latitudinal pattern of thermal performance breadth

Because seasonality increases with latitude, one might expect that high-latitude insects would have relatively large performance breadths (e.g., the range of temperature over which fitness is $\geq 80\%$ of r_{max}, Huey and Stevenson, 1979) and thus be able to achieve high fitness over a broad range of temperatures. Performance breadth does increase slightly with absolute latitude, but the relationship is far from significant ($r = 0.212$, $p = 0.202$).

The lack of a significant correlation between performance breadth and absolute latitude may seem surprising, given the strong relationship between tolerance range and absolute latitude (Fig. 9.2, Addo-Bediako *et al.*, 2000). Note, however, that a similar pattern has been documented for lizards (van Berkum, 1988): only their tolerance range, but not performance breadth (at least at high levels of performance), increases with latitude.

9.3.5 Conclusions on latitudinal patterns of thermal sensitivity

Available data suggest that insects moving into higher latitudes have evolved much lower CT_{min} (or related indices, see Addo-Bediako *et al.*, 2000), but only slightly lower CT_{max} and T_o than those of their (presumed) tropical ancestors (Zani *et al.*, 2005, Bradshaw and Holzapfel, Chapter 10, this volume). Moreover, their performance breadths (at high levels of performance, such as 80%) are not shifting conspicuously. The clear implication here is that the primary physiological adaptation to cold and variable temperature regimens is an increased tolerance of very low temperature, with little if any loss in heat tolerance (Fig. 9.1d) or in lowering of optimal temperatures.

These patterns may make sense when related to patterns of climate. The lowest temperatures occur in winter, when most insects are inactive or in diapause and thus have limited capacity to use behavior to evade extreme temperature events. Hence a low CT_{min} may be critical for overwinter survival and perhaps for performance at other times (Lee and Denlinger, 1991; Chown and Nicolson, 2004; Chown and Sinclair, Chapter 8, this volume). The highest ambient temperatures of course occur in summer, when many insects are active, mobile and reproducing: thus T_o and CT_{max} may well be tuned to high-activity temperatures during summer more than to mean annual temperature. The small shifts in CT_{max} (Fig. 9.2) and T_o (Fig. 9.3) with latitude may thus reflect the fact that T_{max} drops much less with latitude than does T_{min} (Fig. 9.4 in Addo-Bediako *et al.* 2000; Bradshaw and Holzapfel, Chapter 10, this volume), that many insects can use thermoregulatory behavior during activity and thus buffer selection on T_o and CT_{max} (Bogert, 1949; Heinrich, 1981; Huey *et al.*, 2003), and perhaps because of thermodynamic disadvantages of low T_o (see below).

9.4 Is warmer better, and is colder slower?

Earlier I described a Panglossian caricature of adaptation of thermal sensitivity to cold; namely, that adaptation to a low optimal temperature is perfect and does not come at any cost to performance or fitness (Fig. 9.1a). Thus, physiological adaption to low temperature (or conversely to high temperature) is complete and perfectly compensatory.

That view was challenged decades ago by W. J. Hamilton III, via his work with coloration of desert insects and lizards. Hamilton (1973) argued that low temperature inevitably slowed reaction rates, such that ectotherms with low optimal temperatures would have reduced performance and fitness at the T_o than would species with higher T_o (Fig. 9.1c). Bennett (1987) later referred to this hypothesis as "warmer is better," but in the context of this volume it could also be called "colder is slower."

A few papers addressed this issue in subsequent years (Garland, 1994; Huey and Kingsolver, 1989). Then in 2004, Savage *et al.* (2004) developed a formal thermodynamic model of the concept. They noted that lower temperatures will slow the velocity of molecules, and that this in turn will reduce the rate of molecular collision, as well as reduce the kinetic energy of any collisions (via a Boltzmann–Arrhenius effect). Consequently, low temperatures must inevitably slow physiological reaction rates. The thermodynamic model of Savage *et al.* (2004) predicts that the log of fitness ($\ln r$) will be linearly related to "inverse" temperature (i.e., $1/KT$, where $K =$ the Boltzmann constant and $T =$ absolute temperature). This model formalizes "warmer is better" in a thermodynamic context, and it implicitly assumes that biochemical compensation is nil (see Clarke, 2006; Frazier *et al.*, 2006; Kingsolver and Huey, 2008).

Savage *et al.* (2004) tested their model by examining comparative data (including data on six species of insects) on $\ln r$ versus $1/KT$. They found the expected inverse relationship that supports "warmer is better," but their comparative methods are problematic (see p. 518, Frazier *et al.*, 2006). In any case, their finding seemed in conflict with well-documented evidence of biochemical and physiological adaptation to low temperature (Lee and Denlinger, 1991; Hochachka and Somero, 2002).

Frazier *et al.* (2006) readdressed this debate by analysing data on r versus T_b from a much larger sample of insects than used by Savage *et al.* (2004). They reasoned that if "warmer is better," then the maximal rate of population growth ($\ln r_{max}$), which, by definition, occurs at T_o, would be inversely related to inverse optimal temperature ($1/KT_o$). Consequently, they compiled data on intrinsic rates of population growth versus temperature from the literature, fitted a curve to the data and estimated the maximum rate of population growth (r_{max}) and T_o. Because r_{max} is strongly influenced by body mass, Frazier and colleagues (2006) corrected for dry body mass, as did Savage *et al.* (2004).

If hotter is better, r_{max} should be positively correlated with T_o (or inversely correlated with $1/KT$). This pattern was strongly evident in both traditional and in phylogenetic analyses (Fig. 9.4). Importantly, the slope of $\ln r_{max}$ on $1/kT$ was

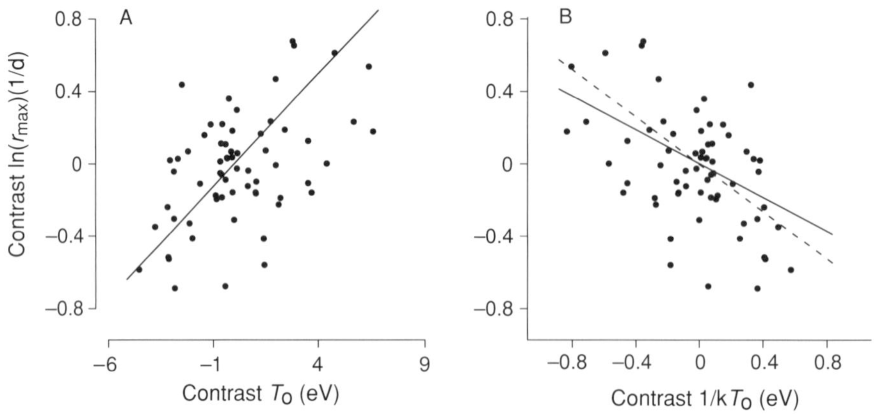

Figure 9.4 Standardized independent contrasts for: (A) ln r_{max} (mass independent) versus T_o and B) ln r_{max} (body mass independent) versus $1/KT_o$ for insects (from Frazier *et al.*, 2006). Both panels support "warmer is better" hypothesis, such that warm-adapted insects have higher r_{max} than do cold-adapted insects. In (B) the actual slope (solid line) is steeper than that predicted (dashed line) from a thermodynamic model (Savage *et al.*, 2004).

even steeper than predicted (Fig. 9.4B) by the thermodynamic model (Savage *et al.*, 2004). If significant evolutionary compensation had occurred, the slope should have been shallower, not steeper.

Although the observed pattern (Fig. 9.4) is qualitatively consistent with thermodynamics (Savage *et al.*, 2004), that pattern could have evolved for non-thermodynamic reasons. For example, insects evolving in cold (high-latitude) environments are also experiencing more variable thermal environments (Addo-Bediako *et al.*, 2000; Ghalambor *et al.*, 2006). This could select for increased thermal performance breadth (i.e., increased thermal generalization), with a correlated trade-off in maximal performance (Levins, 1968; Huey and Slatkin, 1976; Gilchrist, 1995; Palaima and Spitze, 2004; Izem and Kingsolver, 2005). Similarly, cold environments have reduced primary productivity, and so selection might favor down-regulation of performance and rates of population growth (Frazier *et al.*, 2006).

To test all three hypotheses simultaneously, Frazier *et al.* (2006) used a structural-equations (path-analysis) approach and incorporated data on the magnitude of seasonal variation in ambient temperature and on net primary productivity. Analyses (with either raw data or with independent contrasts) supported only "warmer is better": no significant support was detected for a trade-off induced by seasonality or from an impact of net primary productivity (Fig. 4 in Frazier *et al.*, 2006).

These comparative patterns strongly support "warm is better" and imply that insects adapting to cold environments sacrifice maximal rates of population growth as a correlated response: the magnitude of the sacrifice is non-trivial. For example, the data suggest that a decline in T_o of 1 °C causes a ~ 10% reduction in r_{max} (Frazier *et al.* 2006). Perhaps this is one reason why T_o of insects shows little interspecific variation (Fig. 9.3). In any case, colder appears slower for insects.

The lack of evidence for significant physiological compensation of r_{max} for low temperature (Frazier *et al.*, 2006; Savage *et al.*, 2004) is superficially inconsistent with a vast number of studies demonstrating biochemical and physiological adaptations of diverse ectotherms to low temperature (Lee and Denlinger, 1991; Cossins and Bowler, 1987; Hochachka and Somero, 2002; Addo-Bediako *et al.*, 2000; Chen *et al.*, 1990; Clarke, 2006; Pörtner *et al.*, 2007). This conflict may be more apparent than real (Frazier *et al.*, 2006). Obviously, there is overwhelming evidence that physiological and biochemical adaptation to low temperature does occur (e.g., chapters in this volume), and it is reasonable to expect that these adaptations have enabled ectotherms to shift their thermal sensitivity and thus to invade cold environments. Nonetheless, those ectotherms can't evade thermodynamic constraints. As Frazier *et al.* (2006, p. 518), noted: "although physiological adaptation to cold allows organisms to invade cold environments, it is seemingly incapable of compensating for reduced rates of maximal population growth."

9.5 Are cold-adapted insects especially vulnerable to climate warming?

Ample evidence documents that global temperatures are warming, and that the rate of warming is greatest at higher latitudes, especially in the northern hemisphere (IPPC, 2007). Because insects living in such high latitudes have lower optimal temperatures (Fig. 9.3) and sometimes lower heat tolerance (Addo-Bediako *et al.*, 2000) than do insects living in tropical climes, some have predicted that high-latitude species should be especially vulnerable to climate warming (Root *et al.*, 2003).

To be sure, warming at high latitude is having demonstrable impact on insects. Many species are extending their ranges to higher latitudes (Bradshaw *et al.*, 2004; Crozier and Dwyer, 2006; Parmesan, 2006; Bradshaw and Holzapfel, Chapter 10, this volume), and growing seasons are extended as well. In some high-latitude insects, the observed impact of warming on population growth is clearly negative, but primarily because warming is disrupting the synchrony of insects and their host plants (van Asch and Visser, 2007), not because warmer temperatures are physiologically deleterious to the insects themselves.

Nevertheless, Deutsch *et al.* (2008) and Tewksbury *et al.* (2008) recently argued that climate warming will actually have its most negative impact on tropical (thus

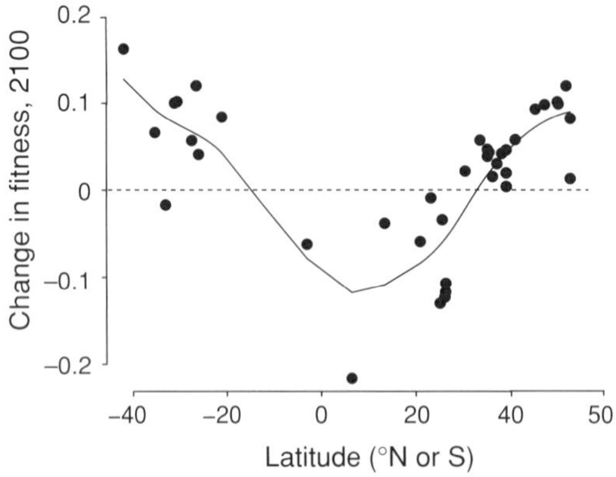

Figure 9.5 Predicted effect of twentieth century climate warming on change in fitness of 38 insect species (redrawn from Deutsch *et al.*, 2008). Warming is predicted to enhance fitness at mid-to-high latitudes, but reduce fitness in the tropics. The line is a smooth spline fit (df = 6).

warm-adapted) insects and will in fact benefit many higher-latitude (thus cold-adapted) insects. They noted that the impact of warming depends not only on the rate of temperature change, but also on the physiological sensitivity of organisms to that change. By combining climate and physiological models, they predicted that warming in the tropics (though small in magnitude relative to higher latitudes) will have very deleterious effects there, not only because tropical species are relatively sensitive to temperature change, but also because they are living in climates that are already very close to their physiologically optimal temperatures. Thus tropical insects exposed even to modest warming will very likely be subjected to stressful temperatures, at least in summer. In contrast, insects living at higher latitudes typically have optimal temperatures that are well above climate temperatures, such that warming is likely to enhance their fitness (all else being equal). Simulations by Deutsche *et al.* (2008) support this expectation (Fig. 9.5). (See Bradshaw and Holzapfel, Chapter 10, this volume for a fascinating discussion of why higher-latitude species also have a fitness advantage against invading species from lower latitudes, even during climate warming.)

I emphasize that the prediction that warming will enhance fitness of cold-adapted species (Fig. 9.5) ignores considerable biology! Warming-induced shifts in phenology could be highly detrimental, especially if this disrupts photoperiod cues (Bradshaw and Holzapfel, Chapter 10, this volume). Moreover, although insect activity seasons may lengthen because winter warming is proceeding more rapidly than is summer warming on some continents (IPPC, 2007), the warmer winter temperatures may actually be deleterious to insects. For one thing, the resultant longer growing seasons may increase activity-season mortality (Wilson and Cooke,

2001). Moreover, warmer winters may directly increase metabolic rates during inactivity and thus accelerate the depletion of stored resources, in turn resulting in reduced survival and fecundity (Irwin and Lee, 2003; Williams *et al.*, 2003; Zani, 2008). Thus the net impact of warming on high-latitude insects is challenging to predict.

If we assume that the Deutsch *et al.* (2008) model is largely correct, its implications for tropical insects and ecosystems are profound. The vast majority of insects are tropical, and thus the vast majority of insect biodiversity is at risk from warming. Moreover, the model predicts that many cold-adapted species will paradoxically benefit from warming (see also Bradshaw *et al.*, 2004). It is important to be explicit, however, that this simple model ignores the likelihood that warming will disrupt the synchrony of insects and host plants (Van Asch and Visser, 2007), photoperiodic timing (Bradshaw and Holzapfel, 2001; Bradshaw and Holzapfel, 2006; Ragland and Kingsolver, 2008, especially Bradshaw and Holzapfel, Chapter 10, this volume), and interspecific interactions (Davis *et al.*, 1998).

9.6 Summary

This chapter explores some physiological and ecological consequences of adaptation to low temperatures, such as those found in high-latitude areas. I first reviewed how the thermal sensitivity of insects evolves with respect to latitude and reiterate key conclusions:

First, CT_{max} and CT_{min} temperatures both decline with latitude, but the decline is much more dramatic for CT_{min} (Fig. 9.2, Addo-Bediako *et al.*, 2000; Zani *et al.*, 2005). This well-known pattern, which also holds for many ectotherms (Brett, 1970; Ghalambor *et al.*, 2006), probably reflects the more rapid drop in winter temperatures than in summer temperatures with latitude (Addo-Bediako *et al.*, 2000; Bradshaw and Holzapfel, 2006). In any case, an insect's tolerance range (that is, $CT_{max} - CT_{min}$) increases dramatically with latitude, no doubt reflecting adaptation to increased seasonality of temperature at high latitudes. (Addo-Bediako *et al.*, 2000; Bradshaw and Holzapfel, Chapter 10, this volume; Deutsch *et al.*, 2008).

Second, optimal temperatures for the intrinsic rate of population growth also drop significantly with latitude and thus with overall climate, but not very steeply (Fig. 9.3)! In fact, optimal temperatures of insects in our sample at least are surprisingly homogeneous: the interquartile range of T_o is only $\sim 5\,^{\circ}C$! This might reflect inadvertent sample bias in the insects studied here (see Introduction), but it might alternatively indicate that the evolution of optimal temperatures in ectotherms is evolutionarily very conservative (see Hertz *et al.*, 1983; Huey *et al.*, 2003). Specifically, insects may not be able to evolve an extremely low T_o in most environments simply because of fitness decrements caused by thermodynamic depression of rates of population growth (see below), or perhaps because operative temperatures during insect activity seasons don't change drastically with latitude.

Third, the range of temperatures over which insects perform at high levels (B_{80}) is surprisingly independent of latitude! A similar pattern has been described for lizards (van Berkum 1986), suggesting that this pattern might be general for terrestrial ectotherms. Because tolerance range increases with latitude (Fig. 9.2), but performance breadth does not, suggests that the primary adaptation of thermal fitness curves to cold and fluctuating environments (Fig. 9.1d) is seen in increased tolerance of extremely cold temperatures but with relatively minor shifts in T_o, performance breadth, or CT_{max}.

Fourth, I explored the debate as to whether insects evolving in low temperatures suffer a thermodynamic constraint, such that insects that have evolved low T_o would necessarily have lower maximal rates of population growth than insects that have evolved higher T_o: if so, then "colder is slower" or conversely, "warmer is better" (Bennett, 1987; Kingsolver and Huey, 2008). The debate here involves two opposing world views: in their extreme form, they can be called the "thermodynamic constraint hypothesis" versus the "perfect compensation hypothesis" (Frazier *et al.* 2006). The thermodynamic model (Savage *et al.*, 2004) predicts that log r_{max} should decline with "inverse" temperature ($1/KT$), which would support "warmer is better." A recent comparative study (Frazier *et al.*, 2006) with insects supports that prediction. Frazier *et al.* (2006) used a comparative, "strong inference" approach to evaluate whether the insect fitness data support warmer is better versus two competing hypotheses. The results were clear: only "warmer is better" is supported (Fig. 9.4). As a result, available data suggest that even though insects are able to live and even thrive in cold environments, their maximal rates of population growth appear slow relative to that of their more warm-adapted ancestors even when tested at their optimal body temperature. For insects evolving in the cold, colder seems slower.

Finally, I addressed the fate of cold-adapted insects in a globally warming world. Most insects are tropical and thus warm-adapted, but many insects live in the colder environments of high latitude and are relatively cold-adapted. Importantly, it is at high latitudes where rates of climate change are greatest (IPPC, 2007): consequently, cold-adapted insects might be expected to be especially vulnerable to climate warming. However, vulnerability to warming depends, not only on the magnitude of the climate shift, but also on an insect's sensitivity to temperature change (Tewksbury *et al.*, 2008; Deutsch *et al.*, 2008; Williams *et al.*, 2008). High-latitude insects might in fact benefit from warming (Fig. 9.5) because they are relatively less sensitive to climate change, because they retain relatively high T_o and CT_{max} (Figs. 9.2 and 9.3), and mainly because they are living in climates that are generally cool relative to physiologically optimal temperatures. The observed range shifts to higher latitudes and in lengthened growing seasons is consistent with this expectation that warming is benefiting many high-latitude insects (but see Bradshaw and Holzapfel, Chapter 10, this volume). Ironically, the "warm-adapted" tropical insects may be the most vulnerable to climate warming (Fig. 9.5, Tewksbury *et al.*, 2008; Fig. 9.5, Deutsch *et al.*, 2008). If so, the consequences could be disastrous, because the tropics are the epicenter of insect biodiversity. If this view is correct, cold-adapted insects may inherit a warming earth – though they may be physiologically excluded from the warmest areas.

Acknowledgements

I thank M. Angilletta, A. Bennett, W. Bradshaw, D. Berrigan, C. Deutsch, M. Frazier, C. Holzapfel, J. Kingsolver and J. Tewksbury for many constructive discussions on these issues. The research described here has been funded by several grants (most recently IOB-0416843) from the National Science Foundation.

References

Addo-Bediako, A., Chown, S. L. and Gaston, K. J. (2000). Thermal tolerance, climatic variability and latitude. *Proceedings of the Royal Society, Series B* **267**, 739–745.

Angilletta, M. J. (2009). *Thermal Adaptation: A Theoretical and Empirical Synthesis.* Oxford, U.K.: Oxford University Press.

Angilletta, M. J., Jr., Bennett, A. F., Guderley, H., Navas, C. A., Seebacher, F. and Wilson, R. S. (2006). Coadapation: a unifying principle in evolutionary thermal biology. *Physiological and Biochemical Zoology* **79**, 282–294.

Angilletta, M. J., Jr., Hill, T. and Robson, M. A. (2002). Is physiological performance optimized by thermoregulatory behavior?: a case study of the eastern fence lizard, *Sceloporus undulatus*. *Journal of Thermal Biology* **27**, 199–204.

Bakken, G. S. (1992). Measurement and application of operative and standard operative temperatures in ecology. *American Zoologist* **32**, 194–216.

Beck, S. D. (1980). *Insect Photoperiodism.* New York: Academic Press.

Bennett, A. F. (1987). Evolution of the control of body temperature: is warmer better? In *Comparative Physiology: Life in Water and on Land*, ed. P. Dejours, C. R. Taylor and E. R. Weibel. Padova, Italy: Liviana Press, pp. 421–431.

Bennett, A. F., Lenski, R. E. and Mittler, J. E. (1992). Evolutionary adaptation to temperature. I. Fitness responses of *Escherichia coli* to changes in its thermal environment. *Evolution* **46**, 16–30.

Bogert, C. M. (1949). Thermoregulation in reptiles, a factor in evolution. *Evolution* **3**, 195–211.

Bradshaw, W. E. and Holzapfel, C. M. (2001). Genetic shift in photoperiodic response correlated with global warming. *Proceedings of the National Academy of Sciences, USA* **98**, 14509–14511.

Bradshaw, W. E. and Holzapfel, C. M. (2006). Evolutionary response to rapid climate change. *Science* **312**, 1477–1478.

Bradshaw, W. E., Zani, P. A. and Holzapfel, C. M. (2004). Adaptation to temperate climates. *Evolution* **58**, 1748–1762.

Brett, J. R. (1970). Temperature, fishes. In *Marine Ecology* vol. 1, ed. O. Kinne. New York, NY: John Wiley & Sons, pp. 515–560.

Carey, J. R. (1993). *Applied Demography for Biologists.* Oxford: Oxford University Press.

Charlesworth, B. (1994). *Evolution in Age-structured Populations*, 2nd edn. Cambridge, UK: Cambridge University Press.

Chen, C.-P., Lee, R. E., Jr. and Denlinger, D. L. (1990). A comparison of the responses of tropical and temperate flies (Diptera: Sarcophagidae) to cold and heat stress. *Journal of Comparative Physiology B* **160**, 543–547.

Chown, S. L., Gaston, K. J. and Robinson, D. (2004). Macrophysiology: large-scale patterns in physiological traits and their ecological implications. *Functional Ecology* **18**, 159–167.

Chown, S. L., Jumbam, K. R., Sørensen, J. G. and Terblanche, J. S. (2008). Phenotypic variance, plasticity and heritability estimates of critical thermal limits depend on methodological context. *Functional Ecology* **22**, 1–8.

Chown, S. L. and Nicolson, S. W. (2004). *Insect Physiological Ecology. Mechanisms and Patterns.* Oxford: Oxford University Press.

Clarke, A. (2006). Temperature and the metabolic theory of ecology. *Functional Ecology* **20**, 405–412.

Cossins, A. R. and Bowler, K. (1987). *Temperature Biology of Animals.* New York, NY: Chapman & Hall.

Crozier, L. and Dwyer, G. (2006). Combining population-dynamic and ecophysiological models to predict climate-induced insect range shifts. *American Naturalist* **167**, 853–866.

Davis, A., Lawton, J., Shorrocks, B. and Jenkinson, L. (1998). Individualistic species responses invalidate simple physiological models of community dynamics under global environmental change. *Journal of Animal Ecology* **67**, 600–612.

Denlinger, D. L., Giebultowicz, J. M. and Saunders, D. S. (eds.) (2001). *Insect Timing: Circadian Rhythmicity to Seasonality*. Amsterdam: Elsevier.

Deutsch, C. A., Tewksbury, J. J., Huey, R. B., Sheldon, K. S., Ghalambor, C. K., Haak, D. C. and Martin, P. R. (2008). Impacts of climate warming on terrestrial ectotherms across latitude. *Proceedings of the National Academy of Sciences, USA.* **105**, 6668–6672.

Frazier, M. R., Huey, R. B. and Berrigan, D. (2006). Thermodynamics constrains the evolution of insect population growth rates: "warmer is better". *American Naturalist* **168**, 512–520.

Garland, T., Jr. (1994). Phylogenetic analyses of lizard endurance capacity in relation to body size and body temperature. In *Lizard Ecology: Historical and Experimental Perspectives*, ed. L. J. Vitt and E. R. Pianka. Princeton, NJ: Princeton University Press, pp. 237–259.

Ghalambor, C. K., Huey, R. B., Martin, P. R., Tewksbury, J. J. and Wang, G. (2006). Are mountain passes higher in the tropics? Janzen's hypothesis revisited. *Integrative and Comparative Biology* **46**, 5–17.

Gibert, P., Moreteau, B., Pla, E., Petavy, G., Karan, D. and David, J. R. (2001). Chill-coma tolerance, a major climatic adaptation among *Drosophila* species. *Evolution* **55**, 1063–1068.

Gilchrist, G. W. (1995). Specialists and generalists in changing environments. I. Fitness landscapes of thermal sensitivity. *American Naturalist* **146**, 252–270.

Gilchrist, G. W. (1996). Quantitative genetic analysis of the thermal sensitivity of locomotory performance curve of *Aphidius ervi*. *Evolution* **50**, 1560–1572.

Goto, S. G. and Kimura, M. T. (1988). Heat- and cold-shock responses and temperature adaptations in subtropical and temperate species of *Drosophila*. *Journal of Insect Physiology* **44**, 1233–1239.

Gould, S. J. and Lewontin, R. C. (1979). The spandrels of San Marcos and the Panglossian paradigm – a critique of the adaptationist program. *Proceedings of the Royal Society, Series B* **205**, 581–598.

Hamilton, W. J., III (1973). *Life's Color Code*. New York, NY: McGraw Hill.

Heinrich, B. (1981). *Insect Thermoregulation*. New York: John Wiley & Sons, Inc.

Helmuth, B., Kingsolver, J. G. and Carrington, E. (2005). Biophysics, physiological ecology and climate change: does mechanism matter? *Annual Review of Physiology* **67**, 177–201.

Hertz, P. E., Huey, R. B. and Nevo, E. (1983). Homage to Santa Anita: thermal sensitivity of sprint speed in agamid lizards. *Evolution* **37**, 1075–1084.

Hochachka, P. W. and Somero, G. N. (2002). *Biochemical Adaptation: Mechanism and Process in Physiological Evolution*. New York: Oxford University Press.

Hoffmann, A. A. and Blows, M. W. (1993). Evolutionary genetics and climate change: will animals adapt to global warming? In *Biotic Interactions and Global Change*, ed. P. M. Kareiva, J. G. Kingsolver and R. B. Huey. Sunderland., MA: Sinauer Assoc., pp. 165–178.

Hoffmann, A. A., Sørensen, J. G. and Loeschcke, V. (2003). Adaptation of *Drosophila* to temperature extremes: bringing together quantitative and molecular approaches. *Journal of Thermal Biology* **26**, 175–216.

Hoffmann, A. A. and Watson, M. (1993). Geographical variation in the acclimation responses of *Drosophila* to temperature extremes. *American Naturalist* **142**, S93-S113.

Huey, R. B. (1987). Phylogeny, history and the comparative method. In *New Directions in Ecological Physiology*, ed. M. E. Feder, A. F. Bennett, W. W. Burggren and R. B. Huey. Cambridge, UK: Cambridge University Press, pp. 76–98.

Huey, R. B. and Bennett, A. F. (1987). Phylogenetic studies of coadaptation: preferred temperatures versus optimal performance temperatures of lizards. *Evolution* **41**, 1098–1115.

Huey, R. B. and Berrigan, D. (2001). Temperature, demography and ectotherm fitness. *American Naturalist* **158**, 204–210.

Huey, R. B. and Hertz, P. E. (1984). Is a jack-of-all-temperatures a master of none? *Evolution* **38**, 41–50.

Huey, R. B., Hertz, P. E. and Sinervo, B. (2003). Behavioral drive versus behavioral inertia: a null model approach. *American Naturalist* **161**, 357–366.

Huey, R. B. and Kingsolver, J. G. (1989). Evolution of thermal sensitivity of ectotherm performance. *Trends in Ecology and Evolution* **4**, 131–135.

Huey, R. B. and Rosenzweig, F. (2009). Laboratory evolution meets Catch 22: balancing simplicity and realism. In *Experimental Evolution: Concepts, Methods and Applications*, ed. T. Garland, Jr. and M. R. Rose. Berkeley: University of California Press, pp. 671–701.

Huey, R. B. and Slatkin, M. (1976). Costs and benefits of lizard thermoregulation. *Quarterly Review of Biology* **51**, 363–384.

Huey, R. B. and Stevenson, R. D. (1979). Integrating thermal physiology and ecology of ectotherms: a discussion of approaches. *American Zoologist* **19**, 357–366.

IPPC (2007). *Climate Change 2007: The Physical Science Basis.* Cambridge: Cambridge University Press.

Irwin, J. T. and Lee, R. E., Jr. (2003). Cold winter microenvironments conserve energy and improve overwintering survival and potential fecundity of the goldenrod gall fly, *Eurosta solidaginis*. *Oikos* **100**, 71–78.

Izem, R. and Kingsolver, J. G. (2005). Variation in continuous reaction norms: quantifying directions of biological interest. *American Naturalist* **166**, 277–289.

Janzen, D. H. (1967). Why mountain passes are higher in the tropics. *American Naturalist* **101**, 233–249.

Kingsolver, J. G. and Gomulkiewicz, R. (2004). Environmental variation and selection on performance curves. *Integrative and Comparative Biology* **43**, 470–477.

Kingsolver, J. G. and Huey, R. B. (1998). Evolutionary analyses of morphological and physiological plasticity in thermally variable environments. *American Zoologist* **38**, 323–336.

Kingsolver, J. G. and Huey, R. B. (2008). Size, temperature, and fitness: three rules. *Evolutionary Ecology Research* **10**, 251–268.

Kingsolver, J. G. and Nagle, A. M. (2007). Rapid divergence of thermal sensitivity and diapause in field and laboratory populations of *Manduca sexta*. *Physiological and Biochemical Zoology* **80**, 473–479.

Kingsolver, J. G., Ragland, G. J. and Shlichta, J. G. (2004). Quantitative genetics of continuous reaction norms: thermal sensitivity of caterpillar growth rates. *Evolution* **58**, 1521–1529.

Klok, C. J. and Chown, S. L. (1997). Critical thermal limits, temperature tolerance and water balance of a sub-Antarctic caterpillar, *Pringleophaga marioni* Viette (Lepidoptera: Tineidae). *Journal of Insect Physiology* **43**, 685–694.

Lee, R. E. and Denlinger, D. L. (1991). *Insects at Low Temperature.* New York: Chapman & Hall.

Lee, R. E. and Denlinger, D. L. (2006). Entomology on the Antarctic Peninsula: the southernmost insect. *American Entomologist* **52**, 84–89.

Levins, R. (1968). *Evolution in Changing Environments.* Princeton, NJ: Princeton University Press.

MacArthur, R. H. (1984). *Geographical Ecology: Patterns in the Distribution of Species.* Princeton, NJ: Princeton University Press.

Palaima, A. and Spitze, K. (2004). Is a jack-of-all temperatures a master of none? An experimental test with *Daphnia pulicaria* (Crustacea: Cladocera). *Evolutionary Ecology Research* **6**, 215–225.

Parmesan, C. (2006). Ecological and evolutionary responses to recent climate change. *Annual Review of Ecology and Systematics* **37**, 637–669.

Pörtner, H. O. (2002). Climate variations and the physiological basis of temperature dependent biogeography: tradeoffs in muscle design and performance in polar ectotherms. *Journal of Experimental Biology* **205**, 2217–2254.

Pörtner, H. O., Peck, L. and Somero, G. (2007). Thermal limits and adaptation in marine Antarctic ectotherms: an integrative view. *Philosophical Transactions of the Royal Society B* **362**, 2233–2258.

Price, P. W., Fernandes, G. W., Lara, A. C. F. and Brawn, J. (1998). Global patterns in the local number of insect galling species. *Journal of Biogeography* **25**, 581–591.

Ragland, G. J. and Kingsolver, J. G. (2008). Evolution of thermotolerance in seasonal environments: the effects of annual temperature variation and life-history timing in *Wyeomyia smithii*. *Evolution* **62**, 1345–1357.

Root, T. L., Price, J. T., Hall, K. R., Schneider, S. H., Rosenzweig, C. and Pounds, J. L. (2003). Fingerprints of global warming on wild animals and plants. *Nature* **421**, 37–42.

Savage, V. M., Gillooly, J. F., Brown, J. H., West, G. B. and Charnov, E. L. (2004). Effects of body size and temperature on population growth. *American Naturalist* **163**, 429–441.

Tewksbury, J. J., Huey, R. B. and Deutsch, C. A. (2008). Putting the heat on tropical animals. *Science* **320**, 1296–1297.

van Asch, M. and Visser, M. E. (2007). Phenology of forest caterpillars and their host trees: the importance of synchrony. *Annual Review of Entomology* **52**, 37–55.

van Berkum, F. H. (1988). Latitudinal patterns of the thermal sensitivity of sprint speed in lizards. *American Naturalist* **132**, 327–343.

Williams, J. B., Shorthouse, J. D. and Lee, R. E., Jr. (2003). Deleterious effects of mild simulated overwintering temperatures on survival and potential fecundity of rose-galling *Diplolepis* wasps (Hymenoptera: Cynipidae). *Journal of Experimental Zoology* **298A**, 23–31.

Williams, S. E., Shoo, L. P., Isaac, J. L., Hoffmann, A. A. and Langham, G. (2008). Toward an integrated framework for assessing the vulnerability of species to climate change. *PLoS Biology* **6**, 2621–2626.

Wilson, B. S. and Cooke, D. E. (2001). Latitudinal variation in rates of overwinter mortality in the lizard *Uta stansburiana*. *Ecology* **85**, 3406–3417.

Zani, P. A. (2008). Climate-change trade-offs in the side-blotched lizard (*Uta stansburiana*): effects of growing-season length and mild temperatures on winter survival. *Physiological and Biochemical Zoology* **81**, 797–809.

Zani, P. A., Swanson, S. E. T., Corbin, D., Cohnstaedt, L. W., Agotsch, M. D., Bradshaw, W. E. and Holzapfel, C. M. (2005). Geographic variation in tolerance of transient thermal stress in the mosquito *Wyeomyia smithii*. *Ecology* **86**, 1206–1211.

10

Insects at not so low temperature: Climate change in the temperate zone and its biotic consequences

WILLIAM E. BRADSHAW AND CHRISTINA M. HOLZAPFEL

10.1 Introduction

Herein, we consider insects in the temperate zone where, in central and eastern continental land masses, favorable summers are interspersed with inamicable, often lethal, winters. At latitudes higher than 30°, fitness consists of the ability to exploit the warm season through growth, development and reproduction, the ability to avoid or mitigate the effects of winter cold through dormancy or migration, and the ability to make a timely transition between summer and winter lifestyles. Fitness is then defined by performance integrated through all four seasons, not just by a measure of performance in a single environment characteristic of a single time of year.

Below we discuss the geographical and seasonal patterns in light and temperature in the temperate zone, how climate change is affecting and will affect these patterns, and the actual and potential biotic responses by insects to climate warming. Physiological processes lie at the level of integration between the environment and the gene, and are important in regulating the acquisition, assimilation and allocation of resources, in regulating growth, development and reproduction, and in maintaining homeostasis in variable environments. Environmental change elicits a physiological response, either through non-genetic change within individual animals (homeostasis and phenotypic plasticity) or through genetic change in animal populations (evolution). Mechanism matters. Physiological processes enable animals to maximize survivorship and reproduction, and are primary determinants of fitness in environments that vary in time and space. In temperate regions, the major biotic patterns of response to climate change are clear: animals are

Low Temperature Biology of Insects, ed. David L. Denlinger and Richard E. Lee Jr. Published by Cambridge University Press.

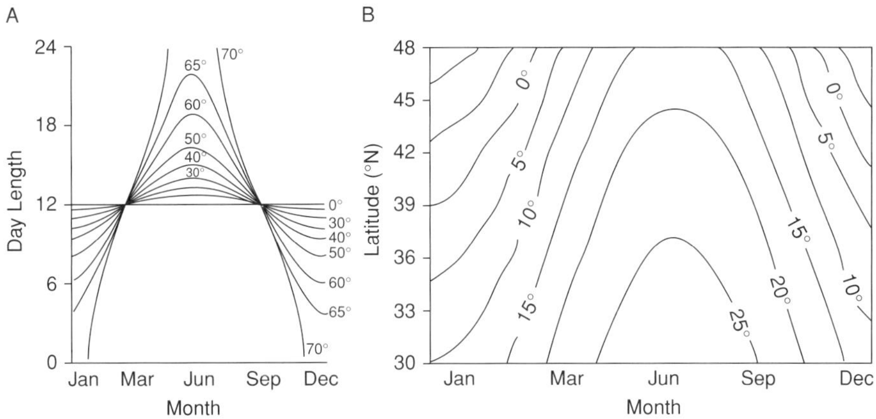

Figure 10.1 Geographic and seasonal variation in day length and temperature. (A) Seasonal patterns in day length (sunrise to sunset) at different latitudes (°N) in the northern hemisphere (Danilevskii, 1965). Day length at temperate and polar latitudes predicts future seasons more reliably than any other environmental cue. (B) Isotherms for mean monthly temperature in central and eastern North America (Bradshaw *et al.*, 2004). The latitudinal variation in climate is less a matter of summer warmth (June isotherms are far apart) than winter cold (January isotherms are close together) and northern populations experience a shorter growing season than southern populations. Hence, changes in season length, and the timing of spring and fall activities have a greater effect on animal populations than do the direct effects of temperature.

expanding their ranges poleward and are altering the timing of their seasonal development. Herein, we consider established effects of day length and temperature on development, reproduction and dormancy, and how these effects impede or facilitate response to rapid climate change at both the phenotypic and genetic (evolutionary) level.

10.2 Heat and light on a rotating earth

10.2.1 How does seasonality change with geography?

Because of the 23° tilt of the earth's axis of daily rotation relative to the plane of its annual rotation about the sun, northern and southern hemispheres experience opposite periods of summer and winter. Also because of this tilt, not only is winter day length shorter at higher than at lower latitudes (Fig. 10.1A), but also the angle of incident winter sunlight is more acute and imparts less heat per hour of daylight. Consequently, the latitudinal gradient in climate is primarily one of winter cold rather than summer heat (Fig. 10.1B). Concomitantly, cold-tolerance or lower lethal limits in insects are more closely associated with latitude than are heat-tolerance or upper lethal limits (Addo-Bediako *et al.*, 2000; Zani *et al.*, 2005).

As one proceeds polewards, spring arrives later and fall arrives earlier. Hence, the length of the favorable season declines regularly with increasing latitude. For example, in Fig. 10.1B, the 15 °C isotherm declines from $10\frac{1}{2}$ months at 30°N to 3 months at 48°N. When air masses encounter mountain ranges, they rise, expand and cool, so that the length of the favorable season also declines with increasing altitude (MacArthur, 1972; Critchfield, 1974). In eastern North America, the number of freeze-free days decreases by about 9.3 days with every degree of increasing latitude and 94 days with every 1000 m of increasing elevation (Bradshaw, 1976). Seasonal activities of temperate insects are therefore intimately related to the coming of spring and fall, and to the length of the growing season. As the length of the growing season and the coming of spring and fall vary with geography, so also do the optimal times to initiate growth, development, reproduction, dormancy, or migration.

10.2.2 How do insects time their seasonal activities?

A wide variety of animals use the length of day as an anticipatory cue (photoperiodism) to prepare in advance for the changing seasons. Photoperiodism is used to cue the seasonal timing of reproduction, migration, or dormancy in rotifers (Pourriot and Clément, 1975), annelids (Fong and Pearse, 1992; Last and Olive, 1999, 2004; Schierwater and Hauenschild, 1990), molluscs (Joosse, 1984; Ansart *et al.*, 2001; Hommay *et al.*, 2001), echinoderms (Halberg *et al.*, 1987; Pearse *et al.*, 1986), fish (Kemp, 1984; Bromage *et al.*, 2001), amphibians (Laurila, *et al.* 2001), lizards (Fox and Dessauer, 1957; Licht, 1973; Cuellar and Cuellar, 1977), birds (Dawson, 2002; Dawson *et al.*, 2001) and mammals (Goldman, 2001; Hofman, 2004), as well as arthropods, where there exists a vast literature on insects (Danilevskii, 1965; Tauber *et al.*, 1986; Danks, 1987; Saunders, 2002). From the insect citations, we make several generalizations about photoperiodism in temperate and more polar insect populations.

1. In general, day length provides the go/no-go cue for the direct timing of seasonal events or for the initiation of physiological, endocrinological and developmental cascades that once started are irrevocable or at least under natural conditions, usually not reversed before the completion of the seasonal event under selection. Day length may also modulate the rates of completion of these events through its interaction with food, temperature and moisture. When there is an interaction between photoperiod and temperature, high temperatures tend to reinforce long-day effects and low temperatures short-day effects.
2. Because of the tilt of the earth relative to its plane of rotation about the sun, day length varies with both time of year and latitude (Fig. 10.1A). At

the equator, the surface of the earth receives 12 hours of light per day (sunrise to sunset) all year long. As one proceeds north or south of the equator, the annual variation in day length becomes progressively more extreme, increasing from zero at the equator to 24 hours in the summer at latitudes greater than 67°. There are several consequences of these patterns for animals using photoperiod to time their seasonal activities. First, at tropical latitudes below about 15°, the annual change in day length is not sufficient to provide a reliable seasonal cue. Even so, a few insects are responsive to day lengths at latitudes as low as 9°, but probably not within 5° of the equator (Norris, 1965; Wolda and Denlinger, 1984; Denlinger, 1986). Second, in the temperate and polar zones above 30° latitude, wherein lies the greatest proportion of the earth's landmass, the annual change in day length provides a strong and highly reliable seasonal cue over evolutionary time. The day length at a given latitude is the same today as it was 20 000 years ago and will be 20 000 years from now. This reliability is important. Insects cannot wait for often highly variable temperatures to change and for winter to arrive before entering dormancy or preparing for migration; rather, they use day length as an anticipatory cue to prepare for future seasonal change. Insects tend to enter diapause at a specific stage of their life cycle. During the developmental stage that is responsive to day length, the physiological decision is made to enter diapause or delay entry into diapause for another generation. The optimal time to enter diapause is then just over one full generation before the actual onset of winter (Taylor, 1980), very often when the weather is warm and resources are still abundant. Hence, the adaptive significance of photoperiodism lies not only in the high reliability of day length as a cue, but also in the anticipatory nature of using day length to switch off development and reproduction when the immediate environmental conditions are still favorable for these activities.

3. In temperate environments with warm, moist summers and harsh winters, especially in mid- or eastern continental climates, "long-day" arthropods usually enter a hibernal diapause that is initiated by short or shortening days. Timing is of the essence. An early entry into diapause results in the consumption of resources while temperatures are warm; a late entry into diapause runs the risk of encountering lethal winter conditions (Leather *et al.*, 1993; Bradshaw *et al.*, 2004; Bergland *et al.*, 2005). Survivorship through winter, or possessing resources available for development and reproduction the following spring, are dependent on the timing of diapause the previous fall (Fig. 10.2). During the fall, insects are confronted with a go/no-go physiological decision, the consequences of which are

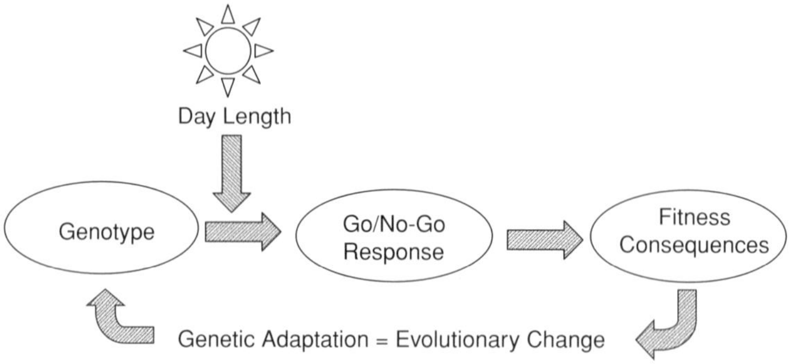

Figure 10.2 Selection on photoperiodic response. The photoperiodically mediated go/no-go physiological decision to diapause or to continue developing is based on the combination of an individual's genotype and the day length it sees. The resultant timing of diapause then has fitness consequences that impose selection on genotypes that respond to different day lengths. Since insects can enter diapause too early or too late in the fall, there is stabilizing selection on the optimal response to day length.

not realized until some time in the future. Each individual has its own, genetically determined, response to day length and, based on that response, makes the irrevocable physiological decision to diapause/continue development. Future environmental conditions (arrival of winter, duration of winter, resources available the next spring) then determine the fitness consequences of that decision. Since photoperiodism is the major physiological mechanism by which temperate insects time dormancy and migration in the fall (Danilevskii, 1965; Tauber *et al.*, 1986; Danks, 1987), fitness is dependent upon the proper response to day length and the target of selection in seasonal environments is the response to day length. The important point is that photoperiodism is an adaptation to length of the growing season, not to day length. As climates change over geography, so also does the optimal time to enter diapause and the optimal day length to use to switch from continuous development to diapause. Consequently, there is directional selection on the switching day length on a latitudinal scale, but stabilizing selection on a local scale (Hard *et al.*, 1993a).

4. Photoperiod is generally important for timing the switch from continuous development to diapause in the fall, but diapause is usually terminated spontaneously or in response to a prolonged exposure to cold temperatures (chilling). In some cases, chilling may combine with food, moisture, or day length to terminate diapause and, as we illustrate below (Section 10.5), modify rates of subsequent development.

10.2.3 Conclusion

The tilt of the earth relative to its plane of rotation about the sun generates a climate at higher latitudes where summer seasons are favorable for growth, development and reproduction, and alternate with winter seasons that are unfavorable for these activities. During winter, insects avoid or mitigate the exigencies of cold through migration and dormancy. Most insects are able to use the length of day (photoperiodism) to anticipate and prepare in advance for the arrival of winter. Photoperiodism provides a highly reliable environmental cue that is stable through evolutionary time and provides the go/no-go signal that initiates a physiological and endocrinological cascade that commits an individual to a developmental pathway that is generally irrevocable within the lifetime of the insect or at least within a seasonal context. At any given locality, there is an optimal time to enter diapause that is an adaptive compromise between avoiding winter cold and conserving resources for overwintering, and for development and reproduction the following spring. Hence, the proper, genetically determined response to day length is an important determinant of fitness at temperate and more polar latitudes. As the length of the growing season declines with increasing latitude or altitude, so also does the day length (critical photoperiod) that insects use to switch from active development to dormancy, so that, while there is stabilizing selection on the critical photoperiod on a local scale, there is directional selection on a geographic scale. Once in diapause, the termination of diapause is often dependent upon accumulated chilling that may interact with moisture, food, or day length. Along with ambient temperature, these same processes may affect the rates of post-diapause development in the spring. In the next section, we discuss the response of insects to climate change in the context of geographic patterns in day length and temperature.

10.3 Relative roles of day length and temperature in biotic response to rapid climate change

10.3.1 Climate change, day length and temperature

Climate change does not alter the rotation of the earth about its axis nor the rotation of the earth about the sun. Consequently, day length on any given day at any given latitude (Fig. 10.1A) does not change during periods of global warming. Climate warming at temperate and more polar latitudes does ameliorate the thermal environment, does alter the optimal timing of seasonal events, and therefore does alter selection on the day length insects use to time these events (Bradshaw *et al.*, 2004; Gomi *et al.*, 2007). The phenotypic "fingerprint" of biotic response to climate change in animals is seen in the northward expansion of species' ranges, the earlier migration and reproduction in the spring, and the later migration or entrance into hibernation in the fall (Hughes, 2000; Peñuelas and Filella, 2001; Walther *et al.*, 2002; Root *et al.*, 2003; Parmesan and Yohe, 2003; Warren, 2006; Parmesan, 2006, 2007). These changes impose selection on *response* to day length. The extension of the growing season later into the fall selects for later entry into diapause and, hence, a shorter, more southern, critical photoperiod. Indeed, evolutionary (genetic) responses to rapid climate change have resulted in shorter critical photoperiods (Bradshaw and Holzapfel, 2001a; Gomi *et al.*, 2007). This genetic response to recent rapid climate change can take place over as short a time span as five years and is greater at higher latitudes, where the climate is changing faster and selection is more intense than at lower latitudes (Bradshaw and Holzapfel, 2001a).

Recent rapid climate change has not occurred uniformly across latitudes and times of year. Climate warming has been proceeding faster with increasing latitude and in the winter than in the summer (IPCC, 2001, 2007). The net result is that winters are becoming milder, spring is arriving earlier, fall is arriving later, growing seasons are becoming longer, and the duration and severity of winter cold is abating. The northern thermal years are becoming more like southern thermal years (Fig. 10.1B). Consequently, climate warming in the temperate and more northern regions should be alleviating cold stress without imposing substantial heat stress (Bradshaw *et al.*, 2004; Bradshaw and Holzapfel, 2006). Indeed, over the next century, temperate and more northern insects are expected to achieve increasing, not decreasing fitness (*r*) due to the warmer temperatures alone and this effect increases with latitude (Deutsch *et al.*, 2008).

10.3.2 Photoperiodic versus thermal adaptation in a warming world

We used the pitcher-plant mosquito, *Wyeomyia smithii*, to determine the relative selective importance of temperature versus photoperiodism during rapid climate warming (Bradshaw *et al.*, 2004). *Wyeomyia smithii* lays its eggs and completes its entire pre-adult development only within the water-filled leaves of the purple pitcher plant, *Sarracenia purpurea*. The range of the mosquito follows that of its host plant from the Gulf of Mexico (30°N) to northern Canada from Labrador to northern Alberta (58°N). We simulated the natural, annual and daily progression of temperatures and day lengths of a more northern (Newfoundland, 50°N) and a more southern (New Jersey, 40°N) climate, using computer-driven, controlled-environment rooms. We then "transplanted" replicate northern populations to the more southern climate, representing a temperature change equivalent to 180–200 years of global warming at its present rate, and maintained controls in their native northern thermal year. We determined fitness as the year-long cohort replacement rate, integrating performance over all four seasons (Fig. 10.3A). All experiments were carried out in the leaves of intact pitcher plants using feeding schedules that mimicked prey capture by host plants in the field. The experiments pertinent to climate change involved three treatments:

- Northern populations in a *northern* thermal year programmed to enter and terminate diapause at the correct time for the northern thermal year (Fig. 10.3B).
- Northern populations in a *southern* thermal year programmed to enter and terminate diapause at the correct time for the southern thermal year (Fig. 10.3C).
- Northern populations in a southern thermal year allowed to express their northern, genetically determined response to day length in the foreign southern photic year (Fig. 10.3D).

The treatments in Figs. 10.3B versus 10.3C determine the comparative effects of temperature on fitness after genetically determined responses to day length have been factored out. As shown in Fig. 10.3E, northern populations achieved a greater fitness in the warmer, more southern thermal year (Fig. 10.3C) than in their native northern thermal year (Fig. 10.3B). These results are in accord with previous predictions that climate warming should alleviate, not exacerbate thermal stress on these northern populations (Bradshaw and Holzapfel, 2001a, 2006, 2008; Deutsch *et al.*, 2008).

The treatments in Figs. 10.3C versus 10.3D determine the comparative effects of day length on fitness. As shown in Fig. 10.3E, northern populations lost 88% of fitness when allowed to express their northern, genetically determined response to day length (Fig. 10.3D) as compared to their being programmed to enter and terminate diapause at the correct time of year (Fig. 10.3C). Hence, in a benign thermal environment, fitness was critically dependent upon possessing the correct, genetically determined response to day length. *Simply put, the imposition of 180–200 years of climate warming promoted increased fitness among northern populations, but when these same populations encountered the wrong day length (*Fig. 10.3D*), they lost 88% of fitness.*

The cause of the differential fitness between the two photoperiod treatments in the benign thermal year (Figs. 10.3C versus 10.3D) is apparent from the profiles of pupation during the same simulated southern thermal year (Fig. 10.4). The genetically determined critical photoperiods of these northern populations range from 15–16 hours of light per day (Bradshaw *et al.*, 2003). At the start of the experiment, there was insufficient day length to stimulate direct development until day lengths exceeded the critical photoperiod and some larvae did not develop until the following spring. Had we simply transplanted populations in nature from the

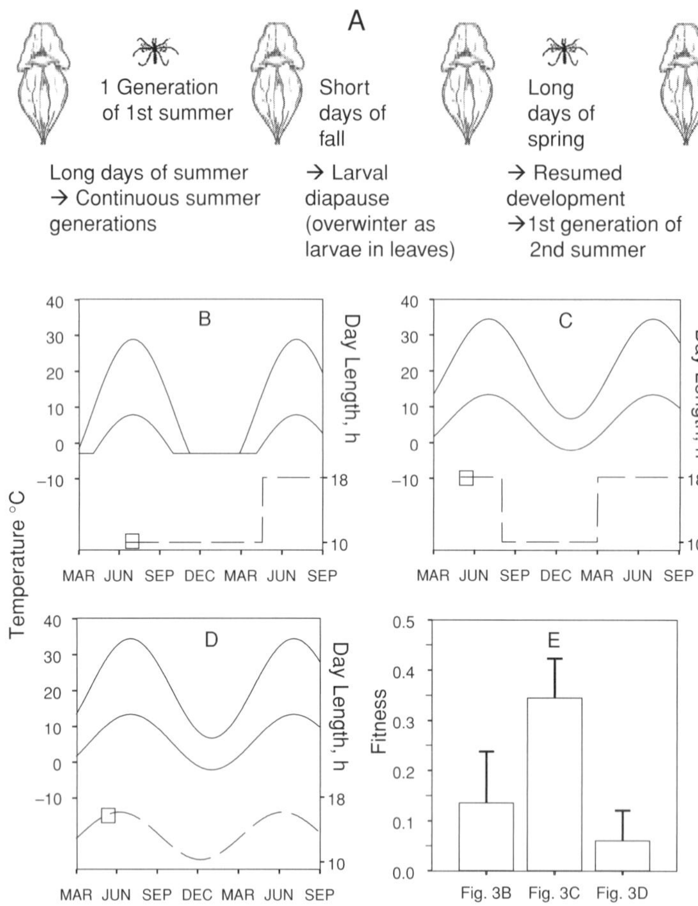

Figure 10.3 Determination of fitness integrated through all four seasons. (A) Determination of the year-long cohort replacement rate: Ry = (number of first instars hatching from the overwintering cohort as the first generation of the second summer) ÷ (number of first instars introduced into the pitcher plants leaves as the first generation of the first summer). (B–D) Simulated thermal years of 50 and 40°N showing maximum and minimum daily temperatures (upper and lower solid lines, respectively) and day length (dashed line). Experimental cohorts are started weekly during a one-month period indicated by the box □ and continue through the spring and early summer of the next year. An 18-h day length is a long day for all populations regardless of their latitudinally specific genetically programmed response to day length and, hence, stimulates development in all populations. Likewise, a 10-h day length is a short day and initiates and maintains diapause in all populations regardless of their latitudinally specific, genetically programmed response to day length. In this way, we are able to factor out genetic differences in response to day length and to induce and terminate diapause at the latitudinally appropriate time of year. (E) Fitness achieved by northern populations exposed to the three environments in Fig. 10.3B–D. A comparison between Fig. 10.3B versus 10.3C tests directly for thermal adaptation when response to day length is factored out; a comparison between Fig. 10.3C versus 10.3D tests directly for photoperiodic adaptation when temperature is held constant. Fitness is plotted as $\text{Log}(Ry + 1)$ + one standard error.

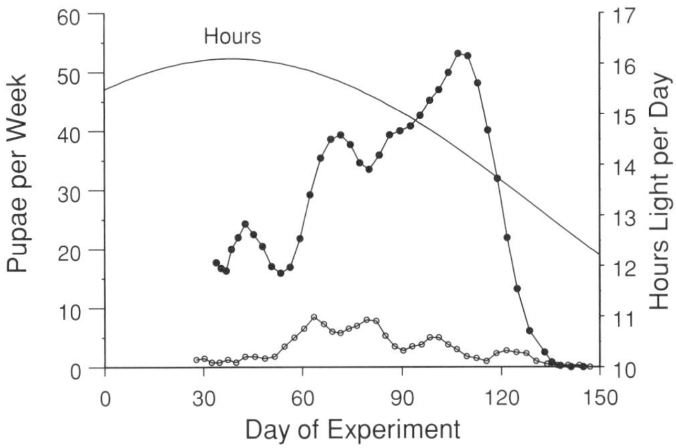

Figure 10.4 Seasonal trajectory of pupation by northern populations subjected to a benign southern thermal year and either (•) programmed to enter and terminate diapause at the correct time of year as in Fig. 10.3C or (○) allowed to express their northern, genetically determined response to day length in a foreign southern photic year as in Fig. 10.3D.

northern to the southern latitude (Fig. 10.3B versus 10.3D), to determine the potential effects of climate warming on these northern populations, and ignored day length, we would incorrectly have predicted that warmer temperatures associated with climate change would drive these populations to extinction. However, it was the inappropriate day length (Fig. 10.3C versus 10.3D), not the warmer temperature (Fig. 10.3B versus 10.3C) that resulted in the observed loss of fitness. The very important point here is that because the annual change in day length is fixed at each latitude (Fig. 10.1A), transplants from northern to southern localities in nature (e.g., Etterson and Shaw, 2001) cannot be used to evaluate the potential impact of climate warming, because the effects of the novel photic environment will always confound the effects of temperature and result in an overestimation of the effects of climate change. However, the effects of changing thermal or moisture environments can be made correctly by transplants across longitudes (e.g. Fenster and Galloway, 2000), between altitudes at the same latitude as in the historic experiments of Clausen *et al.* (1940) or across time at the same locality by comparing contemporary with ancestral plants resurrected from appropriately stored seed in their common habitat (e.g., Franks *et al.*, 2007).

10.3.3 Conclusion

In conclusion, when we compared directly the effects of temperature and photoperiod on the year-long cohort replacement rate that integrated fitness over all four seasons, possessing an inappropriate, genetically determined response to day length resulted in a drastic loss of fitness, while climate warming tended to improve fitness of northern populations in accord with "warmer is better" (Frazier *et al.*, 2006; Deutsch, *et al.*, 2008). We have therefore concluded that in temperate and more northern climates, the primary target of selection during rapid climate change will be on the timing of seasonal events rather than on heat-tolerance or thermal optima. Consequently, the primary genetic response of insect populations to rapid climate change will be photoperiodic rather than thermal adaptation; thermal adaptations would then be secondary adaptations that take place after populations are well established and have attained concordance with the local seasonal cycle. (Bradshaw *et al.*, 2000, 2004; Bradshaw and Holzapfel, 2001a, 2006, 2008). When considering the effects of climate change on fitness of populations *in situ* or of transplanted populations across latitudes, the effects of photoperiod need to be ruled out before any observed phenotypic or genetic shifts can be ascribed to the direct effects of temperature.

Mechanism does matter (Helmuth *et al.*, 2005; Biro *et al.*, 2007). Recent rapid climate warming at temperate latitudes has resulted in genetic shifts in photoperiodic response in insects, in migratory patterns or phenotypic plasticity in birds, in resource-tracking by squirrels and in the frequency of genetic markers in flies (Bradshaw and Holzapfel, 2008). Most of the reaction to recent rapid climate change has focused on the important issue of curtailing or reversing CO_2 emissions (IPCC, 2007). However, the climate will continue to warm for at least another century due to greenhouse gases and aerosols already present in the atmosphere (Warren, 2006; IPCC, 2007). It is therefore important to know how to mitigate the effects of future climate warming on biotic systems. In order to do so, we have to know the actual targets of selection in natural populations of insects and the genetic ability of those populations to respond to selection, particularly in large animals with longer generation times and smaller population sizes than insects.

10.4 Genetic architecture

10.4.1 Genetic variability and response to selection

From the above, it is clear that climate warming is imposing selection for altered timing of seasonal activities and, hence, altered response to day length. Response to selection is dependent primarily on the strength of selection, the genetic variation on which that selection can act and on the extent to which selection in one dimension of the fitness landscape affects other dimensions of that landscape.

More formally, response to selection (R) is the product of the heritability of a trait (h^2) and the strength of selection (S) applied to the trait: $R = h^2S$, where the heritability represents the genetic variation available for selection to act upon and can be thought of as a measure of the efficiency of response to selection. As discussed above, recent rapid climate change has resulted in longer growing seasons, with earlier springs and later falls, with concomitant selection on altered timing of seasonal activities. The magnitude of change in the length of the growing season then imposes a proportional selection pressure on the timing of seasonal activities and, consequently, response to day length. Since the rate of climate warming is increasing with latitude, so also is the rate of season lengthening and the strength of selection (S) on response to day length. In *Wyeomyia smithii*, the heritability (h^2) of critical photoperiod also increases with latitude; hence, in hindsight, it is no surprise that the genetic shift towards shorter, more southern critical photoperiods (R) has increased with latitude over the last 30 years (Bradshaw and Holzapfel, 2001ab). In fact, this genetic shift can be detected over a timescale as short as 5 years, illustrating the great genetic potential of insects to keep pace with rapid climate change and the fact that phenotypic plasticity is not sufficient to account for biotic response to climate change.

Response to selection can be modified due to genetic correlations between the trait under selection and other traits that have an opposite effect on fitness. Essentially, if one imposes selection on a single trait and obtains a genetic response in a trait not under selection, then the two traits are genetically correlated. Genetic correlations can impede the simultaneous maximization of two fitness traits (Stearns, 1976, 1992; Rose, 1991; Roff, 1992, 2002). However, genetic correlations can facilitate as well as impede response to selection. In *Wyeomyia smithii*, generation time and critical photoperiod are positively genetically correlated (Hard *et al.*, 1993b), i.e., animals with shorter generation times are genetically predisposed to enter dormancy later in the year, in accord with theoretical predictions (Istock *et al.*, 1976; Taylor, 1980). The genetic correlations are coordinated rather than antagonistic with respect to the timing of seasonal development. Individuals with genetically shorter generations can effectively develop later in the season and still be in diapause by the time winter arrives. In addition, critical photoperiod forms part of a “diapause syndrome”, where selection on shorter critical photoperiods results in fewer short days required to initiate diapause, a lower intensity of diapause and diapause in a later larval instar and vice versa for selection on longer critical photoperiods (Campbell and Bradshaw, 1992). In this case, the genetic correlations within populations are the same as the genetic differences among populations and represent a gradient from diapause-prone to diapause-averse genotypes, with the diapause-prone genotypes increasing with the severity and duration of winter. Response to selection on the timing of seasonal development is then reinforced rather than inhibited by the underlying genetic correlations.

10.4.2 Photoperiodism genes?

Genetic correlations immediately beg the question as to the genes underlying photoperiodic response and how they interact. Heretofore, identifying genes responsible for photoperiodism has been strongly biased by assuming a causal connection between the circadian clock and the photoperiodic timer (Bünning, 1936; Withrow, 1959; Menaker, 1971; Pittendrigh, 1981; Pittendrigh and Takamura, 1993; Takeda and Skopik, 1997; Vaz Nunes and Saunders, 1999; Tauber and Kyriacou, 2001). Terrestrial animals living in temperate parts of the world encounter and exhibit two major periodicities: Daily periodicities include everything from daily behavioral activities to metabolic function, cell division, hunger and sleep patterns. Primary control is by an internal, self-sustaining circadian clock that cycles or oscillates with a repeat period of about 24 hours. Circadian clocks orchestrate the daily transcription of hundreds of genes and are highly buffered against environmental variation and genetic background. The molecular genetic basis of circadian rhythmicity is well understood in insects, especially *Drosophila*. Seasonal periodicities include development, reproduction, migration and dormancy. Primary control is by response to day length. The molecular genetic basis of photoperiodism is not well understood in any animal, including *Drosophila*.

A causal relationship between the circadian clock and photoperiodism would mean that one or more genes are mediating both processes, i.e., their functional connection is due to pleiotropy. In this case, pleiotropy would be bi-directional: selection on one trait would generate a correlated response in the other trait, leading to potential tradeoffs between fitness-related traits (Rose, 1991; Roff, 1992; Stearns, 1992). The circadian clock orchestrates the daily temporal coordination of hundreds of genes in *Drosophila* (Claridge-Chang *et al.*, 2001; McDonald and Rosbash, 2001; Ceriani *et al.*, 2002; Duffield, 2003) and is insulated against environmental perturbation, especially temperature, it is re-set every day, and persists under constant conditions (Pittendrigh 1960, 1981). By contrast, photoperiodic time measurement enables organisms to anticipate and prepare in advance for future seasonal changes in their environment, and the correct, climate-specific photoperiodic response is essential for maintaining fitness in temperate seasonal environments (Bradshaw *et al.*, 2004). Photoperiodic response is genetically variable, evolutionarily flexible, is often affected by food, temperature or moisture, and usually signals a go/no-go switch that is usually irreversible and does not repeat under constant conditions. If circadian rhythmicity and the evolutionary modification of photoperiodic time measurement were causally connected through pleiotropy, then rapid evolution of photoperiodic time measurement by invading species (Hoy, 1978; Tauber *et al.*, 1986, pp. 238–245; Fochs *et al.*, 1994; Lounibos *et al.*, 2003) or in response to rapid climate change (Bradshaw and Holzapfel, 2001a;

Gomi *et al.*, 2007) would necessarily involve significant adjustments in the precise daily coordination of important metabolic events. The daily clock and the seasonal timer serve two, separate, adaptive functions and are affected by different suites of environmental inputs (Danks, 2005).

In *Drosophila littoralis*, critical photoperiod (photoperiodic response) and period of the eclosion rhythm (circadian response) are both correlated with latitude, but this correlation is due to independent evolution of both traits and the apparent genetic correlation between them is due to linkage and not pleiotropy (Lankinen, 1986a,b; Lankinen and Forsman, 2006). In North American *Wyeomyia smithii*, critical photoperiod is positively correlated with latitude and altitude, but is not correlated with either the period or amplitude of the circadian clock (Bradshaw *et al.*, 2006; Emerson *et al.* 2009a). Hence, in both *D. littoralis* and *W. smithii*, the photoperiodic timer has evolved independently of the circadian clock over the climatic gradients of their respective continents. It is difficult to argue a causal relationship between the circadian clock and the evolution of photoperiodic response when the two are not correlated over strong selection gradients through evolutionary time.

At the molecular level, all investigations of which we are aware, including some of our own, have treated circadian rhythm genes as candidate loci for photoperiod genes. The search for the molecular basis of photoperiodism through studies of specific circadian clock genes, especially in Diptera (Saunders, 1990; Saunders *et al.*, 1989; Mathias *et al.*, 2005, 2007; Goto *et al.*, 2006; Tauber *et al.*, 2007; Sandrelli *et al.*, 2007; Stehlik *et al.*, 2008) has revealed more about circadian rhythm genes themselves, but the genetic mechanisms underlying the evolution of photoperiodic response remain elusive (Bradshaw and Holzapfel, 2007, 2008; Emerson *et al.* 2009b). At the time of writing, there has not been a single photoperiod gene identified in a natural population of any animal. Future searches for the mechanistic basis of photoperiodism and its evolution should therefore focus on forward genetic approaches (working from phenotype to genotype) such as fine-scale mapping of genes on chromosomes or microarrays showing differential gene expression (Mathias *et al.*, 2007; Benfey and Mitchell-Olds, 2008; Stinchcombe and Hoekstra, 2008) that are unbiased by the assumption of a causal connection with the circadian clock (Bradshaw and Holzapfel, 2007; Tauber and Kyriacou, 2008). Towards this end, we have developed the first quantitative trait loci (QTL) map of photoperiodic response in any animal (Fig. 10.5). This map shows that there are some 6–9 regions of the *Wyeomyia smithii* genome involved in the evolution of response to day length. This map confirms prior quantitative genetic results that the evolution of critical photoperiod in *W. smithii* is a complex trait involving additive, dominance and epistatic effects of alleles within and between multiple loci, and a genetic correlation between critical photoperiod and stage of diapause (Hard *et al.*, 1993b; Campbell and Bradshaw, 1992; Lair *et al.*, 1997; Bradshaw and

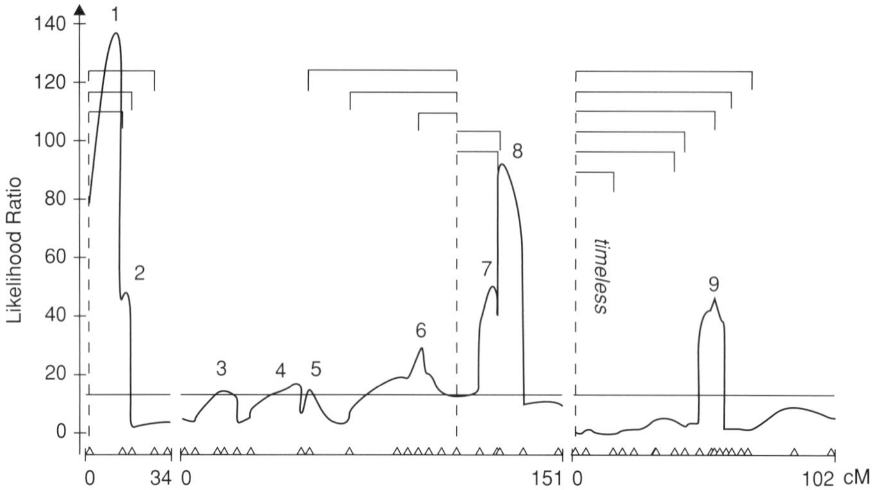

Figure 10.5 Quantitative trait loci (QTL) map for critical photoperiod in *Wyeomyia smithii* based on the F_2 critical photoperiods of a single-pair cross between a northern and a southern population (Mathias *et al.*, 2007). AFLP and gene markers are indicated by open triangles. Peaks with likelihood scores above the horizontal black line are indicative of a likely QTL. The two breaks in the horizontal black line indicate the three chromosomes. Digenic epistatic interactions between markers are shown as brackets subtended by a dashed line.

Holzapfel, 2001b; Bradshaw *et al.*, 2005). Concurrently, we were able to develop a QTL map for stage of diapause. Most intriguingly, QTL 8 (Fig. 10.5) coincides with a QTL for stage of diapause and is also centered over two genes involved in continuous development (ecdysteroid receptors). Hence, this region of the genome may be responsible for coordinating the external environment (day length), continuous development and diapause.

10.4.3 Conclusion

Insect populations harbor sufficient genetic variability in their response to day length to track selection on the timing of seasonal development imposed by recent rapid climate change. Genetic correlations involving critical photoperiod are coordinated to form a continuum between diapause-prone to diapause-averse genetic syndromes, so that response to selection on the timing of seasonal development and diapause is reinforced rather than inhibited by the underlying genetic correlations. The search for genes responsible for photoperiodism at the molecular level has provided tantalizing, but elusive connections between the daily circadian clock and evolution of the seasonal photoperiodic timer. Apart from the QTL map in *W. smithii*, all approaches at the molecular level have treated circadian genes as candidate loci for genes involved in photoperiodism. Forward genetic approaches starting with photoperiodic phenotypes themselves and working to the causative genes themselves are more likely to reveal the genes actually responsible for photoperiodism and its evolutionary response to selection along seasonal gradients or to rapid climate change. Finally, the coincidence in a single region of *W. smithii*'s genome of QTL affecting photoperiodism and diapause, as well as genes involved in continuous development indicate that this region may form a co-adapted gene complex responsible for coordinating the external environment, development and diapause.

10.5 Response to climate change when it's actually warming

10.5.1 Consequences of warming winters

The very term "global warming" invites the assumption that a warmer climate means heat when it's hot, i.e., during the summer, and that we can understand the main effects of climate change by determining the phenotypic and genetic limits, and evolvability of heat-tolerance and thermal optima. But, as we have pointed out, surface temperatures on the earth are increasing faster in the winter than in the summer (IPCC, 2001, 2007) and, at temperate and higher latitudes, climate warming is expected to increase, not decrease, insect fitness (Bradshaw *et al*., 2004; Bradshaw and Holzapfel, 2006; Deutsch *et al*., 2008). The question then remains as to the consequences of climate warming at the time of year when temperature is now increasing at its maximum rate, i.e., during the winter, when insects are in diapause.

A wide variety of temperate arthropods use day length to program the initiation of diapause but, once in diapause, most insects become refractory to day length and rely on other environmental factors, such as moisture, food and, most often, chilling, to terminate diapause (Andrewartha, 1952; Lees, 1955; Tauber *et al*., 1986; Leather *et al*., 1993; Koštál, 2006). The termination of diapause by chilling during the winter leaves insects in a quiescent state, where development is usually constrained by the direct effects of temperature. Once diapause is terminated, synchronization of vernal development is then achieved passively due to the cumulative effects of developing in a thermal environment of rising temperatures (Bradshaw, 1973; Danks, 1987). The reduction in intensity and ultimate termination of diapause by chilling is a temperature-dependent process. The effect of temperature on the termination of diapause is much like the effect of temperature on continuous development. There is an optimal temperature at which the intensity of diapause decreases at the fastest rate and lower and higher

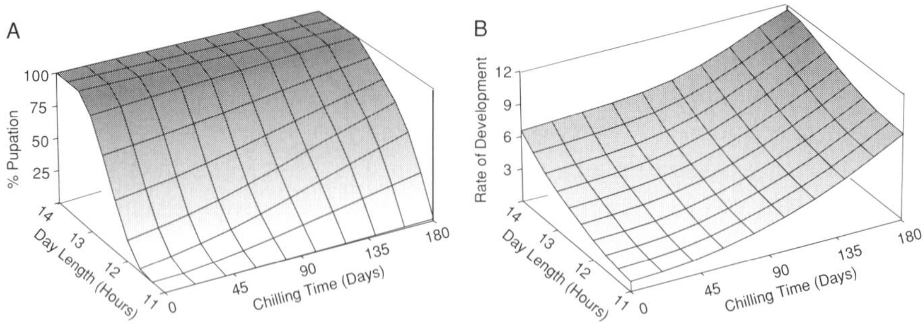

Figure 10.6 Effect of day length and chilling at 7 °C on subsequent termination of larval diapause and rate of development at 21 °C in the tree-hole mosquito, *Toxorhynchites rutilus* from 41°N (Bradshaw and Holzapfel, 1977). (A) Percent pupation within 50 days post-chilling at 21 °C ($R = 0.86$). (B) Rate of development (100/days to pupation) among larvae pupating within 50 days post-chilling at 21 °C ($R = 0.84$).

temperatures at which the intensity of diapause does not change. (Lees, 1955; Bradshaw and Holzapfel, 1977; Tauber *et al.*, 1986; Danks, 1987; Leather *et al.*, 1993).

While most insects become refractory to day length while in diapause, many do not. In those that do not, short days tend to maintain and long days tend to terminate diapause. However, chilling may still proceed, so that the termination of diapause and post-diapause development combine into a dynamic interaction between chilling time and day length (Fig. 10.6). Vernal development is then a function of the degree by which winter temperature departs from the chilling optimum, since departure from this optimum can determine the intensity of diapause, the date on which diapause terminates, and the response to day length. Assuming that winter temperatures are near the thermal optimum for chilling, the intensity and termination of diapause, and vernal development would normally proceed along the diagonals from lower left to upper right in Fig. 10.6, accelerating the termination of diapause (Fig. 10.6A) and potential rates of post-chilling development (Fig. 10.6B). As climate change has resulted in warmer winters and earlier springs, most insects have advanced the timing of their vernal development; however, there are a few butterflies whose vernal development has not changed or has been delayed, despite the warmer spring (Roy and Sparks, 2000; Forister and Shapiro, 2003). We propose that, in these cases, winter temperatures may have warmed enough to void or abate the effects of chilling and therefore have left those insects still in diapause at the end of winter and/or requiring longer days and more time to develop despite the warmer spring. This effect is already seen in trees where the warmer winters do not provide sufficient cold to “satisfy” the requirement for vernalization (Schwartz *et al.*, 2006).

10.5.2 Conclusion

Although much of the research related to climate warming has investigated the phenotypic range, heritability and evolvability of heat-tolerance and thermal optima during the growing season, the most rapid rate of climate warming is occurring during the dormant, winter season. Warmer winter temperatures that retard the chilling process and, hence, the termination of diapause may provide the reason why, despite earlier, warmer springs, some species have not advanced or have even delayed vernal development. Understanding the impact of recent rapid climate change on insect populations is going to involve understanding the effects of warmer winter temperatures on the maintenance and termination of diapause, and on post-diapause development and reproduction in the spring.

10.6 Unexplored implications of climate warming and gene flow

10.6.1 Range expansion and the depletion of heterozygosity

The two major overt patterns of biotic response to climate change in the temperate zone have been the altered timing of seasonal events and the poleward expansion of species' ranges (10.3, above). In the latter case, individuals are expanding their ranges into a vacant niche, i.e., unoccupied by members of their own species. In the northern hemisphere, this migratory pattern is not new, but has been taking place since recession of the Laurentide Ice Sheet over the last 8000–20 000 years. As immigrants from a more southern area found new, more northern populations, they take with them only a fraction of the genetic variation of the ancestral source population. Isolation and drift then leave the new population with lower heterozygosity than the source population. The new population grows and ultimately produces emigrants who found even newer populations, but bring with them an even smaller fraction of the alleles from the original source population. A northward succession of founder events then creates a latitudinal gradient of decreasing heterozygosity in plants (Schwaegerle and Schaal, 1979; Cwynar and MacDonald, 1987), earth-bound animals (Highton and Webster, 1976; Bellemin *et al.*, 1978; Green *et al.*, 1996) and insects (Stone and Sunnuck, 1993; Armbruster *et al.*, 1998). The question remains as to why heterozygosity remains depleted in the new populations and is not replenished from more southern populations as the latter continue to grow and flourish. We propose that the answer may relate: (a) to the distinction between the original founder event, which took place into a vacant niche, previously unoccupied by conspecifics, and subsequent immigration of individuals, which takes place into a niche already occupied by conspecifics and (b) to the difference in temperature-dependent fitness between the

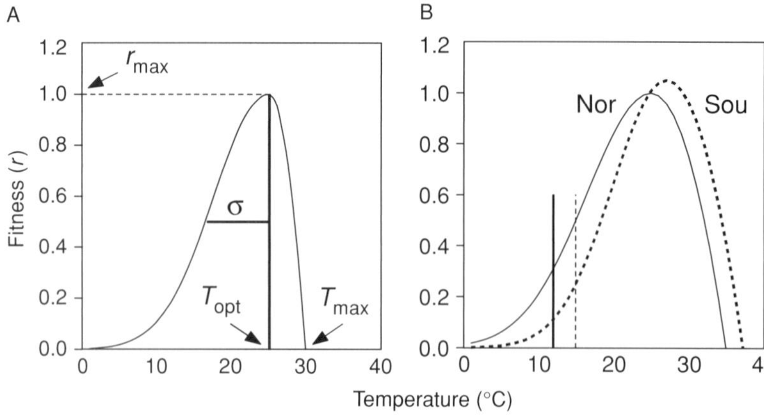

Figure 10.7 Effect of temperature on ectotherm fitness. (A) The performance curves are derived from empirically estimated optimal temperatures (T_{opt}) at which maximum fitness occurs (r_{max}), the upper thermal threshold (T_{max}) and performance breadth (σ, the standard deviation of the rising logistic function) (Deutsch *et al.*, 2008). (B) Simulated comparison of temperature-dependent fitness in a northern, cold-adapted population (solid line) and a slightly more southern, warm-adapted population (dashed line). The vertical lines indicate temperature before (solid) and after (dashed) a period of climate warming. (see text for equations; North: $T_{opt} = 25$; $r_{max} = 1.0$; $T_{max} = 35$; $\sigma = 5$; South: $T_{opt} = 27$; $r_{max} = 1.1$; $T_{max} = 37$; $\sigma = 6$).

southern, more warm-adapted immigrants and the northern, more cold-adapted residents.

10.6.2 Fitness at a single locality

Based on both theoretical and empirical observations (Frazier *et al.*, 2006), fitness increases exponentially with temperature up to maximum fitness (r_{max}) at the thermal optimum (T_{opt}) (Fig. 10.7A) and declines precipitously to zero at higher temperatures (T_{max}). Fitness for any temperature T (W_T) can then be described (Deutsch *et al.*, 2008) from a Gaussian function for the rising portion of the curve, from a quadratic function for the falling portion of the curve, from the width of the curve (performance breadth) in units of phenotypic standard deviation (σ), and from r_{max}, fitness at T_{opt}:

$$W_T = r_{max}\left[\exp\left\{-\left(\frac{T - T_{opt}}{2\sigma}\right)^2\right\}\right] \text{ for } T \leq T_{opt}$$

$$W_T = r_{max}\left\{1 - \left(\frac{T - T_{opt}}{T_{opt} - T_{max}}\right)^2\right\} \text{ for } T > T_{opt}$$

Frazier *et al.* (2006) make several important empirical generalizations. First, cold-adapted insects reach their upper limits of tolerance (T_{max}) at lower temperatures than do warm-adapted populations. Second, cold-adapted insects achieve lower fitness (r_{max}) at their thermal optimum (T_{opt}) than warm-adapted insects do at their thermal optimum. Third, higher-latitude populations experience a wider range of daily and annual temperatures than do low-latitude populations. Consequently, northern, more cold-adapted populations tend to have a wider range of thermal tolerance (performance breadth, σ).

10.6.3 Relative fitness of immigrants and residents

Figure 10.7B compares the fitness functions between two conspecific populations, a more northern, cold-adapted population (solid line) and a more southern, warm-adapted population (dashed line), in close enough proximity so that there could be gene flow between them. In accord with Frazier *et al.*'s (2006) conclusions, the northern population: (a) has a lower thermal maximum, (b) a lower thermal optimum at which it achieves lower fitness, and (c) a wider range of thermal tolerance. Note that, as long as the warming temperatures are below the optimal temperature (T_{opt}) for the northern resident population, southern immigrants will always be at a fitness disadvantage. During periods of rapid climate warming (ambient temperature shifts from the solid to the dashed vertical line), gene flow from southern, more warm-adapted populations into northern, more cold-adapted populations, will continue to be inhibited so long as the increase in summer temperature does not exceed T_{opt} of the more northern population. To put this last conclusion into perspective, the increase in temperature over the next 100 years is not expected to exceed T_{opt} of temperate and more northern insects in general (Deutsch *et al.*, 2008). If summer warming as a consequence of rapid climate change imposes selection on the thermal optimum or on heat-tolerance, evolutionary response to that selection at temperate and more northern latitudes is going to take place primarily by changes in gene frequencies in extant populations rather than by gene flow from more southern populations.

These conclusions are eminently amenable to experimental testing. Frazier *et al.* (2006) list a large number of insects with a wide range of thermal performance curves relating fitness to temperature. If two populations in the same species can be identified that satisfy the conditions in Fig. 10.7B, it is then straightforward to determine whether an immigrant from a more southern population introduced into a more northern population is able to invade that population successfully at temperatures below T_{opt} for the more northern population. The "success" could be evaluated from molecular markers that are now feasible for non-model organisms.

10.6.4 Conclusion

Range expansion accompanying recent rapid climate change has often involved migrant populations entering into a vacant niche, unoccupied by conspecifics. However, the situation is very different for gene flow within the range of existing conspecific populations in temperate and more northern latitudes. If northern populations are more cold-adapted and have a wider range of thermal-tolerance, immigrants from a more southern, warm-adapted population will be at a consistent fitness disadvantage compared to the more northern population, even if the southern population has a warmer thermal optimum and higher tolerance of heat. This lower fitness of immigrants may explain why the decrease in genetic variability with increasing latitude in post-glacial populations persists, despite the opportunity for immigration from more southern, more genetically variable source populations.

10.7 When you aren't always photoperiodic – the special case of *Drosophila*

10.7.1 Photoperiodism in Drosophila

"Although *D. melanogaster* with its unrivaled genetic background has provided a foundation for uncovering the molecular basis of the circadian mechanism . . . it is probably less useful as a model for photoperiodism" (Saunders, 2002). *Drosophila melanogaster* likely originated in central Africa and has a very recent history in the temperate zone. In Europe, where it has invaded during the last ~10 000–15 000 years, after the last glaciation (Lachaise *et al.* 1988, Lachaise and Silvain, 2004; Baudry *et al.*, 2004), *D. melanogaster* is photoperiodic. However, European populations exhibit a linear, graded response to day length over a range of 8–18 hours of light per day (Tauber *et al.*, 2007), rather than a steep, sigmoid response that is characteristic of most insect populations (Danilevskii, 1965; Danks 1987; Saunders, 2002). In these European populations, photoperiod is acting more like a continuous modulator of development, rather than a physiological go/no-go trigger of diapause. In North America, where *D. melanogaster* likely originated from Africa within the last 300–400 years (Lachaise *et al.* 1988; David and Capy, 1988), only the Canton-S strain of *D. melanogaster* is unambiguously photoperiodic and only over a very restricted range of temperatures; all other strains tested are dubiously photoperiodic or clearly not photoperiodic (Saunders and Gilbert, 1990; Tatar *et al.* 2001). Even in Canton-S, the narrow range of temperature over which they are photoperiodic suggests that photoperiod would have little influence on the timing of diapause in natural populations. At a constant low temperature, North American *D. melanogaster* from Florida to Maine show a clear linear cline in incidence of diapause (Schmidt *et al.*, 2005). It would therefore appear that the primary determinant of diapause in *D. melanogaster* is low temperature, consistent with the control of diapause by temperature in tropical and subtropical insects in general (Denlinger, 1986).

Despite a lack of a strong photoperiodically induced diapause, nonphotoperiodic *Drosophila* exhibit seasonal and geographic variation in allozymes and chromosomal inversions that have shifted in their frequency over the last 40 years in a manner consistent with a genetic response to recent climate change.

Dobzhansky (1948) observed that chromosomal inversions in Californian populations of *Drosophila pseudoobscura* changed with seasonality such that the frequency of certain inversions increased during the summer and decreased during the winter. At the same time, the frequency of the winter inversions increased with latitude and altitude. Hence, the inversions are under seasonal selection.

Seasonal cycling and a latitudinal cline in inversion frequencies also occur in European *D. subobscura*. With increased climate warming, the frequency of the northern inversion declined over 16 years in Spanish populations (Rodríguez-Trelles and Rodríguez, 1998). During the 1970s, *D. subobscura* has also independently invaded western North and South America and on both continents established a latitudinal cline in inversion frequencies, reflecting the ancestral cline in Europe (Ayala *et al.*, 1989; Prevosti *et al.*, 1988). After 24 years between samplings, climates had warmed significantly in 22 of 26 populations and, concomitantly, in 21 of these populations, there has been a shift towards inversions characteristic of low latitudes on all three continents (Balanyá *et al.*, 2006). In central and eastern North America, there has also been an increasing frequency of southern inversions in *D. robusta* that parallels increasing local minimum air temperatures (Levitan, 2003; Levitan and Etges, 2005). Finally, over a 20-year period in eastern Australia, there has been a northern shift in both inversion and allozyme frequencies in populations of *D. melanogaster* (Umina *et al.*, 2005). Hence, at the local, continental, and global scales, there has been a shift towards equatorial or low-elevation inversions or allozyme frequencies that parallels recent climate warming across four species of *Drosophila* and four continents.

The increase in the "summer" or "equatorial" inversions in *Drosophila* means that in populations as a whole, there has been a genetic shift towards genotypes associated with warmer climate. However, it is not known whether this genetic shift has been due to hotter summer temperatures per se, or due to a longer warm and a shorter cool season, during which the alternative selective forces can act. In the first case, we would expect there to be greater thermal tolerance or a higher thermal optimum for the summer or winter inversions in contemporary than ancestral populations. In the second case, we would expect there to be no change in thermal tolerance or optima associated with specific inversions and would conclude that the shift in inversion frequency has been due to the length of the growing season rather than higher temperatures per se (Rodríguez-Trelles and Rodríguez, 2007). In addition, there is a latitudinal cline in genetic tendency to diapause among populations of *D. melanogaster* (Schmidt *et al.*, 2005; Tauber *et al.*,

2007), and a Dutch population of *D. melanogaster* has a longer critical photoperiod than Italian populations (Tauber *et al.*, 2007). The changes in inversion frequencies may therefore also be related to selection for reduced incidence of diapause or a shorter critical photoperiod. In short, we know that equatorial inversion and allozyme frequencies are positively correlated with recent, rapid climate change, but not whether this pattern results from selection on thermal optima, thermal tolerance or season length.

10.7.2 Conclusion

Different populations in different species of *Drosophila* that are not known to be photoperiodic undergo seasonal cycling of allozyme and inversion frequencies. The "summer" allozymes or inversion types also decrease with increasing latitude, indicating a high genetic sensitivity of *Drosophila* populations to seasonal selection. During recent, rapid climate change, there has been a pervasive tendency for "summer" or "equatorial" allozyme or inversion frequencies to increase in extant populations, indicating a persistent genetic change in these populations. These shifting frequencies could result from selection on thermal optima, heat-tolerance, season length or incidence of diapause. Thermal rather than photoperiodic control of diapause appears to be the norm for *D. melanogaster, D. subobscura* and *D. robusta* under natural conditions. These species are the most likely candidates for demonstrating an evolutionary (genetic) response in thermal tolerance or thermal optima to rapid climate change.

10.8 Summary

Global warming is proceeding fastest during the winter rather than in the summer and the rate of climate warming is fastest at more polar latitudes, where the gradient in winter cold is steepest, thereby alleviating cold stress without imposing appreciable summer heat stress and improving fitness of temperate and polar ectotherms. Earlier springs and later falls have resulted in longer growing seasons, and the major biotic responses to recent rapid climate change at temperate and polar latitudes have altered timing of seasonal activities and a poleward extension of species' ranges. The target of selection is then on traits related to the seasonal activities of animals, not thermal adaptations. There are now examples of genetic shifts in the seasonal timing of development and reproduction in plants, insects, birds and mammals in response to recent rapid climate warming in the temperate zone; but, to our knowledge, there are no genetically based examples of any animal increasing its thermal optimum or tolerance of higher summer temperatures. A wide variety of animals, from rotifers to rodents, use day length (photoperiodism) to time their seasonal activities and a genetic change in critical photoperiod during recent rapid climate change has been demonstrated in at least two insects.

In accord with these considerations, when northern populations of *Wyeomyia smithii* are transplanted to a warmer more southern thermal year, fitness increases; however, when the northern populations are exposed to the more southern day lengths as well as the more southern thermal year, fitness declines by 88%. These results mean first that exposure to the incorrect day length, not the warmer temperatures, resulted in the dramatic loss of fitness. Second, to assess the potential impact of climate change on populations, transplantation from northern to southern localities in nature will always confound the adverse effects of genetically programmed responses to the foreign day length with the effects of a warmer climate, thereby overestimating the effects of temperature. Third, the effects of photoperiod need to be ruled out before any observed phenotypic or genetic changes associated with climate change can be ascribed to the direct effects of warmer temperature. Mitigating the effects of continued climate change will depend crucially on knowing the actual targets of selection in natural insect populations and the ability of those populations to respond to selection. Insects generally consist of large populations with short generation times and high heritabilities for response to day length. The genetic correlations related to photoperiodism can be coordinated rather than antagonistic, so that response to selection on the timing of seasonal development is reinforced rather than inhibited by the underlying genetic architecture.

The search for genes involved in photoperiodism at the molecular level has largely used circadian clock genes as candidate loci with little definitive results. Forward genetic approaches, such as fine-scale quantitative trait loci mapping and/or cDNA microarrays to assess expressed genes, are more likely to reveal genes involved in photoperiodism and its adaptive evolution over climatic gradients and in response to rapid climate change. The sole extant QTL map for photoperiodism in any animal indicates 6–9 QTL and a region of the genome containing genes affecting critical photoperiod, stage of diapause and receptors for hormones involved in continuous development suggests spatial coordination of genes involved in interpreting the external environment and making the go/no-go switch between continuous development and diapause.

Temperate and more polar climates are warming primarily during the winter and not the summer, but little consideration has been given to the potential effects of climate warming on the progression of diapause during the winter months. We show that the intensity of diapause, rates of response to day length and the completion of post-diapause development are all affected by the duration of winter cold (chilling). Recent, warmer winters may then not provide sufficient cold to reduce the intensity of diapause and may thereby explain why, despite earlier, warmer springs, the timing of vernal development has not changed or may even be retarded for some insects during recent climate change.

Although species are expanding their ranges poleward, this expansion primarily involves immigration into vacant niches. However, in those cases when the northern regions are currently occupied by conspecifics that are already cold-adapted, immigrants from a southern population will consistently be at a fitness disadvantage, based on strictly thermal considerations. This lower fitness of conspecific immigrants may explain the persistence of low genetic variability in post-glacial populations, despite the opportunity for southern, genetically variable source populations to contribute to these impoverished gene pools, despite thousands of years of opportunity to do so.

Finally, we consider the various *Drosophila* that are not photoperiodic, weakly photoperiodic, or photoperiodic over only a narrow range of temperatures. With the very short generation times and large population sizes, *Drosophila* populations change genetically over the time span of the four seasons or over a few years of climate change and are therefore highly responsive to seasonal selection. Whether this responsiveness results from selection on thermal optimum, thermal tolerance, season length or incidence of diapause is yet to be determined.

10.9 Some predictions and implications

1. As more southern species invade more northern latitudes, biodiversity at northern latitudes should increase. This prediction is consistent with the observation that biodiversity increases towards the tropics, so that, as more northern latitudes become warmer, species richness should increase.
2. Since biotic interactions tend to be more important than physical limiting factors in benign environments, and since predators tend to be larger, have longer generations and smaller population sizes than their prey, we can expect that competitive interactions among insect herbivores and that the impact of these herbivores on plants will intensify with increased climate warming. These relationships will alter the timing and intensity of biotic interactions, leading to altered community structure and ecological "surprises," communities that lack contemporary analogs (Williams and Jackson, 2007).
3. Along with other insects, vectors of disease will expand their ranges northwards and, as the favorable season increases, there will be more time for pathogens to complete the extrinsic incubation cycle in their insect hosts. Consequently, there will be an increased diversity of vectors, with

greater competence to transmit pathogens among wildlife, livestock and humans.

4. The genes controlling photoperiodism and, hence, seasonal timing of many insects will consist of a network of interacting genes that coordinate the interpretation of the environment with continuous development and diapause.
5. Identification of photoperiod gene networks and their interaction with hormonal control of development and diapause may provide a means for controlling the expected increase in the abundance of agricultural pests and disease vectors, much as the use of juvenile hormone analogs have provided "third generation" pesticides for the control of mosquitoes.

Acknowledgements

We are grateful to Kevin Emerson and Ray Huey for stimulating discussion while preparing this manuscript, to Kevin Emerson for reading previous versions of this manuscript and to the John Simon Guggenheim Memorial Foundation, the Fulbright Commission, the National Geographic Society and the National Science Foundation for support of our research in the field and lab, most recently through NSF grants DEB-0412573 and IOB-0445710.

References

Addo-Bediako, A., Chown, S. L. and Gaston, K. J. (2000). Thermal tolerance, climatic variability and latitude. *Proceedings of the Royal Society, Series B* **267**, 739–745.

Andrewartha, H. G. (1952). Diapause in relation to the ecology of insects. *Biological Reviews* **27**, 50–107.

Ansart, A., Vernon, P. and Daguzan, J. (2001). Photoperiod is the main cue that triggers supercooling ability in the land snail, *Helix aspersa* (Gastropoda: Helicidae). *Cryobiology* **42**, 266–273.

Armbruster, P. A., Bradshaw, W. E. and Holzapfel, C. M. (1998). Effects of postglacial range expansion on allozyme and quantitative genetic variation in the pitcher-plant mosquito, *Wyeomyia smithii*. *Evolution* **52**, 1697–1704.

Ayala, F. J., Serra, L. L. and Prevosti, A. (1989). A grand experiment in evolution: the *Drosophila subobscura* colonization of the Americas. *Genome* **31**, 246–255.

Balanyá, J., Oller, J. M., Huey, R. B., Gilchrist, G. W. and Serra, L. (2006). Global genetic change tracks global climate warming in *Drosophila subobscura*. *Science* **313**, 1773–1775.

Baudry, E., Viginier, B. and Veuille, M. (2004). Non-African populations of *Drosophila melanogaster* have a unique origin. *Molecular Biology and Evolution* **21**, 1482–1491.

Bellemin, J., Adest, G. and Gorman, G. C. (1978). Genetic uniformity in northern populations of *Thamnophis sirtalis* (Serpentes: Colubridae). *Copeia* **1978**, 150–151.

Benfey, P. N. and Mitchell-Olds, T. (2008). From genotype to phenotype: systems biology meets natural variation. *Science* **320**, 495–497.

Bergland, A. O., Agotsch, M., Mathias, D., Bradshaw, W. E. and Holzapfel, C. M. (2005). Factors influencing the seasonal life history of the pitcher-plant mosquito, *Wyeomyia smithii*. *Ecological Entomology* **30**, 129–137.

Biro, P. A., Post, J. R. and Booth, D. J. (2007). Mechanisms for climate-induced mortality of fish populations in whole-lake experiments. *Proceedings of the National Academy of Sciences, USA* **104**, 9715–9719.

Bradshaw, W. E. (1973). Homeostasis and polymorphism in vernal development of *Chaoborus americanus*. *Ecology* **54**, 1247–1259.

Bradshaw, W. E. (1976). Geography of photoperiodic response in a diapausing mosquito. *Nature* **262**, 384–386.

Bradshaw, W. E., Fujiyama, S. and Holzapfel, C. M. (2000). Adaptation to the thermal climate of North America by the pitcher-plant mosquito, *Wyeomyia smithii*. *Ecology* **81**, 1262–1272.

Bradshaw, W. E., Haggerty, B. P. and Holzapfel, C. M. (2005). Epistasis underlying a fitness trait within a natural population of the pitcher-plant mosquito, *Wyeomyia smithii*. *Genetics* **169**, 485–488.

Bradshaw, W. E. and Holzapfel, C. M. (1977). Interaction between photoperiod, temperature, and chilling in dormant larvae of the tree-hole mosquito, *Toxorhynchites rutilus* Coq. *Biological Bulletin* **152**, 147–158.

Bradshaw, W. E. and Holzapfel, C. M. (2001a). Genetic shift in photoperiodic response correlated with global warming. *Proceedings of the National Academy of Sciences, USA* **98**, 14509–14511.

Bradshaw, W. E. and Holzapfel, C. M. (2001b). Phenotypic evolution and the genetic architecture underlying photoperiodic time measurement. *Journal of Insect Physiology* **47**, 809–820.

Bradshaw, W. E. and Holzapfel, C. M. (2006). Evolutionary response to rapid climate change. *Science* **312**, 1477–1478.

Bradshaw, W. E. and Holzapfel, C. M. (2007). Tantalizing *timeless*. *Science* **316**, 1851–1852.

Bradshaw, W. E. and Holzapfel, C. M. (2008). Genetic response to rapid climate change: it's seasonal timing that matters. *Molecular Ecology* **17**, 157–166.

Bradshaw, W. E., Holzapfel, C. M. and Mathias, D. (2006). Circadian rhythmicity and photoperiodism in the pitcher-plant mosquito: can the seasonal timer evolve independently of the circadian clock? *The American Naturalist* **167**, 601–605.

Bradshaw, W. E., Quebodeaux, M. C. and Holzapfel, C. M. (2003). Circadian rhythmicity and photoperiodism in the pitcher-plant mosquito: adaptive response to the photic environment or correlated response to the seasonal environment? *The American Naturalist* **161**, 735–748.

Bradshaw, W. E., Zani, P. A. and Holzapfel, C. M. (2004). Adaptation to temperate climates. *Evolution* **58**, 1748–1762.

Bromage, N., Porter, M. and Randall, C. (2001). The environmental regulation of maturation in farmed finfish with special reference to the role of photoperiod and melatonin. *Aquaculture* **197**, 63–98.

Bünning, E. (1936). Die endogene Tagesrhythmik als Grundlage der photoperiodischen Reaktion. *Beriche der Deutschen botanischen Gesellschaft* **54**, 590–607.

Campbell, M. D. and Bradshaw, W. E. (1992). Genetic coordination of diapause in the pitcherplant mosquito, *Wyeomyia smithii* (Diptera: Culicidae). *Annals of the Entomological Society of America* **85**, 445–451.

Ceriani, M. F., Hogenesch, J. B., Yanovsky, M., Panda, S., Straume, M. and Kay, S. A. (2002). Genome-wide expression analysis in *Drosophila* reveals genes controlling circadian behavior. *The Journal of Neuroscience* **22**, 9305–9319.

Claridge-Chang, A., Wijnen, H., Nacef, F., Boothroyd, C., Rajewsky, N. and Young, M. W. (2001). Circadian regulation of gene expression systems in the *Drosophila* head. *Neuron* **37**, 657–671.

Clausen, J., Keck, D. D., and Hiesey, W. M. (1940). *Experimental Studies on the Nature of Species. 1. Effect of Varied Environments on Western North American Plants.* Washington, DC: Carnegie Institute of Washington.

Critchfield, H. J. (1974). *General Climatology*. Englewood Cliffs, NJ: Prentice-Hall.

Cuellar, H. S. and Cuellar, O. (1977). Evidence for endogenous rhythmicity in the reproductive cycle of the parthenogenetic lizard *Cnemidophorus uniparens* (Reptilia: Teiidae). *Copeia* **1977**, 554–557.

Cwynar, L. C. and MacDonald, G. M. (1987). Geographical variation in lodgepole pine in relation to population history. *The American Naturalist* **129**, 463–469.

Danilevskii, A. S. (1965). *Photoperiodism and Seasonal Development in Insects.* Edinburgh: Oliver and Boyd.

Danks, H. V. (1987). *Insect Dormancy: an Ecological Perspective.* Ottawa: Biological Survey of Canada (Terrestrial Arthropods).

Danks, H. V. (2005). How similar are daily and seasonal biological clocks? *Journal of Insect Physiology* **51**, 609–619.

David, J. R. and Capy, P. (1988). Genetic variation of *Drosophila melanogaster* natural popualtions. *Trends in Genetics* **4**, 106–111.

Dawson, A. (2002). Photoperiodic control of the annual cycle in birds and comparison with mammals. *Ardea* **90**, 355–367.

Dawson, A., King, V. M., Bentley, G. E. and Ball, G. F. (2001). Photoperiodic control of seasonality in birds. *Journal of Biological Rhythms* **16**, 365–380.

Denlinger, D. L. (1986). Dormancy in tropical insects. *Annual Review of Entomology* **31**, 239–264.

Deutsch, C. A., Tewksbury, J. L., Huey, R. B., Sheldon, K. S., Ghalambor, C. K., Haak, D. C. and Martin, P. R. (2008). Impacts of climate warming on terrestrial ectotherms across latitude. *Proceedings of the National Academy of Sciences, USA* **105**, 6668–6672.

Dobzhansky, T. (1948). Genetics of natural populations. XVI. Altitudinal and seasonal changes produced by natural selection in certain populations of *Drosophila pseudoobscura* and *Drosophila persimilis*. *Genetics* **33**, 158–176.

Duffield, G. E. (2003). DNA microarray analyses of circadian timing: the genomic basis of biological time. *Journal of Neuroendocrinology* **15**, 991–1002.

Etterson, J. R. and Shaw, R. G. (2001). Constraint to adaptive evolution in response to global warming. *Science* **294**, 151–154.

Emerson, K. J., Dake, S. J., Bradshaw, W. E. and Holzapfel, C. M. (2009a). Evolution of photoperiodic time measurement is independent of the circadian clock in the pitcher-plant mosquito, *Wyeomyia smithii*. *Journal of Comparative Physiology A* **195**, 385–391.

Emerson, K. J., Bradshaw, W. E. and Holzapfel, C. M. (2009b). Complications of complexity: integrating environmental, genetic and hormonal control of insect diapause. *Trends in Genetics* **25**, 217–225.

Fenster, C. B. and Galloway, L. F. (2000). Population differentiation in an annual legume: genetic architecture. *Evolution* **54**, 1157–1172.

Fochs, D. A., Linda, S. B., Craig Jr., G. B., Hawley, W. A. and Pumpuni, C. B. (1994). *Aedes albopictus* (Diptera: Culicidae): a statistical model of the role of temperature, photoperiod, and geography in the induction of egg diapause. *Journal of Medical Entomology* **31**, 278–286.

Fong, P. P. and Pearse, J. S. (1992). Evidence for a programmed circannual life cycle modulated by increasing daylengths in *Neanthes limnicola* (Polychaeta: Nereidae) from central California. *Biological Bulletin* **182**, 289–297.

Forister, M. L. and Shapiro, A. M. (2003). Climatic trends and advancing spring flight of butterflies in lowland California. *Global Change Biology* **9**, 1130–1135.

Fox, W. and Dessauer, H. C. (1957). Photoperiodic stimulation of appetite and growth in the male lizard, *Anolis carolinensis*. *Journal of Experimental Zoology* **134**, 557–575.

Franks, S. J., Sim, S. and Weis, A. E. (2007). Rapid evolution of flowering time by an annual plant in response to climate fluctuation. *Proceedings of the National Academy of Sciences, USA* **104**, 1278–1282.

Frazier, M. R., Huey, R. B. and Berrigan, D. (2006). Thermodynamics constrains the evolution of insect population growth rates: "Warmer is better". *The American Naturalist* **168**, 512–520.

Goldman, B. D. (2001). Mammalian photoperiodic system: formal properties and neuroendocrine mechanisms of photoperiodic time measurement. *Journal of Biological Rhythms* **16**, 283–301.

Gomi, T., Nagasaka, M., Fukuda, T. and Hagihara, H. (2007). Shifting of the life cycle and life-history traits of the fall webworm in relation to climate change. *Entomologia Experimentalis et Applicata* **125**, 179–184.

Goto, S. G. and Denlinger, D. L. (2002). Short-day and long-day expression patterns of genes involved in the flesh fly clock mechanism: *period*, *timeless*, *cycle* and cryptochrome. *Journal of Insect Physiology* **48**, 803–816.

Goto, S. G., Han B. and Denlinger, D. L. (2006). A nondiapausing variant of the flesh fly, *Sarcophaga bullata*, that shows arrhythmic adult eclosion and elevated expression of two circadian clock genes, *period* and *timeless*. *Journal of Insect Physiology* **52**, 1213–1218.

Green, D. M., Sharbel, T. F., Kearsley, J. and Kaiser, H. (1996). Postglacial range fluctuation, genetic subdivision and speciation in the western North American spotted frog complex, *Rana pretiosa*. *Evolution* **50**, 374–390.

Halberg, F., Shankaraiah, K., Giese, A. C. and Halberg, F. (1987). The chronobiology of marine invertebrates: Methods of analysis. In *Reproduction of Marine*

Invertebrates, ed. A. C. Giese, J. S. Pearse and V. B. Pearse. Palo Alto, CA: Blackwell. pp. 331–384.

Hard, J. J., Bradshaw, W. E. and Holzapfel, C. M. (1993a). The genetic basis of photoperiodism and evolutionary divergence among populations of the pitcher-plant mosquito, Wyeomyia smithii. *The American Naturalist* **142**, 457–473.

Hard, J. J., Bradshaw, W. E. and Holzapfel, C. M. (1993b). Genetic coordination of demography and phenology in the pitcher-plant mosquito, *Wyeomyia smithii*. *Journal of Evolutionary Biology* **6**, 707–723.

Helmuth, B., Kingsolver, J. G. and Carrington, E. (2005). Biophysics, physiological ecology, and climate change: Does mechanism matter? *Annual Review of Physiology* **67**, 177–201.

Highton, R. and Webster, T. P. (1976). Geographic protein variation and divergence in populations of the salamander *Plethodon cinereus*. *Evolution* **30**, 33–45.

Hofman, M. A. (2004). The brain's calendar: neural mechanisms of seasonal timing. *Biological Reviews* **79**, 61–77.

Hommay, G., Kienlen, J. C., Gertz, C. and Hill, A. (2001). Growth and reproduction of the slug *Limax ventianus* Férussac in experimental conditions. *Journal of Molluscan Studies* **67**, 191–207.

Hoy, M. A. (1978). Variability in diapause attributes of insects and mites: some evolutionary and practical implications. In *Evolution of Insect Migration and Diapause*, ed. H. Dingle. New York, NY.: Springer-Verlag. pp. 101–126.

Hughes, L. (2000). Biological consequences of global warming: is the signal already apparent? *Trends in Ecology and Evolution* **15**, 56–61.

IPCC (2001). *Climate Change 2001: The Scientific Basis. Contribution of Working Group I to the Third Assessment Report of the Intergovernmental Panel on Climate Change*. Cambridge, UK: Cambridge University Press.

IPCC (2007). *Climate Change 2007: The Physical Basis. Contribution of Working Group I to the Fourth Assessment of the Intergovernmental Panel on Climate Change*. Geneva, Switzerland: IPCC Secretariat.

Istock, C. A., Zisfein, J. and Vavra, K. J. (1976). Ecology and evolution of the pitcher-plant mosquito. 2. The substructure of fitness. *Evolution* **30**, 535–547.

Joosse, J. (1984). Photoperiodicity, rhythmicity and endocrinology of reproduction in the snail *Lymnaea stagnalis*. In *Photoperiodic Regulation of Insect and Molluscan Hormones*, ed. R. Porter and G. M. Collins. London: Pitman. pp. 204–220.

Kemp, A. (1984). Spawning of the Australian lungfish, *Neoceratodus fosteri* (Krefft) in the Brisbane River and Enoggera Reservoir, Queensland. *Memoirs of the Queensland Museum* **21**, 391–399.

Koštál, V. (2006). Eco-physiological phases of insect diapause. *Journal of Insect Physiology* **52**, 113–127.

Lachaise, D. and Silvain, J.-F. (2004). How two Afrotrpoical endemics made two cosmopolitan human commensals: the *Drosophila melanogaster-D. simulans* palaeogeographic riddle. *Genetica* **120**, 17–39.

Lachaise, G., Cariou, M. L. D. J. R., Lemeunier, F., Tsacas, L. and Ashburner, M. (1988). Historical biogeography of the *Drosophila melanogaster* species subgroup. *Evolutionary Biology* **22**, 159–225.

Lair, K. P., Bradshaw, W. E. and Holzapfel, C. M. (1997). Evolutionary divergence of the genetic architecture underlying photoperiodism in the pitcher-plant mosquito, *Wyeomyia smithii*. *Genetics* **147**, 1873–1883.

Lankinen, P. (1986a). Genetic correlation between circadian eclosion rhythm and photoperiodic diapause in *Drosophila littoralis*. *Journal of Biological Rhythms* **1**, 101–118.

Lankinen, P. (1986b). Geographical variation in circadian eclosion rhythm and photoperiodic adult diapause in *Drosophila littoralis*. *Journal of Comparative Physiology A* **159**, 123–142.

Lankinen, P. and Forsman, P. (2006). Independence of genetic geographical variation between photoperiodic diapause, circadian eclosion rhythm, and Thr-Gly repeat region of the *period* gene in *Drosophila littoralis*. *Journal of Biological Rhythms* **21**, 3–12.

Last, K. S. and Olive, P. J. W. (1999). Photoperiodic control of growth and segment proliferation by *Nereis (Neanthes) virens* in relation to state of maturity and season. *Marine Biology* **134**, 191–199.

Last, K. S. and Olive, P. J. W. (2004). Interaction between photoperiod and an endogenous seasonal factor influencing the diel locomotor activity of the benthic polychaete *Nereis virens* Sars. *Biological Bulletin* **206**, 103–112.

Laurila, A. Pakkasmaa, S. M. J. and Merilä, J. (2001). Influence of seasonal time constraints on growth and development of common frog tadpoles: a photoperiod experiment. *Oikos* **95**, 451–460.

Leather, S. R., Walters, K. F. A. and Bale, J. S. (1993). *The Ecology of Insect Overwintering*. Cambridge, UK: Cambridge University Press.

Lees, A. D. (1955). *Physiology of Diapause in Arthropods*. Cambridge, UK: Cambridge at the University Press.

Levitan, M. (2003). Climatic factors and increased frequencies of 'southern' chromosome forms in natural populations of *Drosophila robusta*. *Evolutionary Ecology Research* **5**, 597–604.

Levitan, M. and Etges, W. J. (2005). Climate change and recent genetic flux in populations of *Drosophila robusta*. *BMC Evolutionary Biology* **5**, 4.

Licht, P. (1973). Influence of temperature and photoperiod on the annual ovarian cycle in the lizard *Anolis carolinensis*. *Copeia* **1973**, 465–472.

Lounibos, L. P., Escher, R. L. and Lorenço-De-Oliveira, R. (2003). Asymmetric evolution of photoperiodic diapause in temperate and tropical invasive populations of *Aedes albopictus* (Diptera: Culicidae). *Annals of the Entomological Society of America* **96**, 512–518.

MacArthur, R. H. 1972. *Geographical Ecology*. New York, NY: Harper & Row.

Mathias, D., Jacky, L., Bradshaw, W. E. and Holzapfel, C. M. (2005). Geographic and developmental variation in expression of the circadian rhythm gene, *timeless*, in the pitcher-plant mosquito, *Wyeomyia smithii*. *Journal of Insect Physiology* **51**, 661–667.

Mathias, D., Jacky, L., Bradshaw, W. E. and Holzapfel, C. M. (2007). Quantitative trait loci associated with photoperiodic response and stage of diapause in the pitcher-plant mosquito, *Wyeomyia smithii*. *Genetics* **176**, 391–402.

McDonald, M. J. and Rosbash, M. (2001). Microarray analysis and organization of circadian gene expression in *Drosophila*. *Cell* **107**, 567–578.

Menaker, M. (1971). *Biochronometry*. Washington, DC: National Academy of Sciences.

Norris, M. J. (1965). The influence of constant and changing photoperiods on imaginal diapause in the red locus (*Nomadacris septemfasciata* Serv.). *Journal of Insect Physiology* **11**, 1105–1119.

Parmesan, C. (2006). Ecological and evolutionary responses to recent climate change. *Annual Review of Ecology, Evolution and Systematics* **37**, 637–669.

Parmesan, C. (2007). Influences of species, latitudes and methodologies in estimates of phenological response to global warming. *Global Change Biology* **13**, 1860–1872.

Parmesan, C. and Yohe, G. (2003). A globally coherent fingerprint of climate change impacts across natural systems. *Nature* **421**, 37–42.

Pearse, J. S., Eernisse, D. J., Pearse, V. B. and Beauchamp, K. A. (1986). Photoperiodic regulation of gametogenesis in sea stars, with evidence for an annual calendar independent of fixed daylength. *American Zoologist* **26**, 417–431.

Peñuelas, J. and Filella, I. (2001). Response to a warming world. *Science* **294**, 793–795.

Pittendrigh, C. S. (1960). Circadian rhythms and the circadian organization of living systems. *Cold Spring Harbor Symposia in Quantitative Biology* **25**, 159–184.

Pittendrigh, C. S. (1981). Circadian organization and the photoperiodic phenomena. In *Biological Clocks in Seasonal Reproductive Cycles*, ed. B. K. Follett and D. E. Follett. Bristol, UK: John Wright. pp. 1–35.

Pittendrigh, C. S. and Takamura, T. (1993). Homage to Sinzo Masaki: Circadian components in the photoperiodic responses of *Drosophila auraria*. In *Seasonal Adaptation and Diapause in Insects (in Japanese)*, ed. M. Takeda and S. Tanaka. Tokyo: Bun-ichi Sôgô Shuppan. pp. 288–305.

Pourriot, R. and Clément, P. (1975). Influence de la durée de l'éclairement quotidien sur le taux de femelles mictiques chez *Notommata copeus* Ehr. (Rotifère). *Oecologia (Berlin)* **22**, 67–77.

Prevosti, A., Ribo, G., Serra, L., Aguade, M., Balanyá, J., Monclus, M. and Mestres, F. (1988). Colonization of America by *Drosophila subobscura*: Experiment in natural populations that supports the adaptive role of chromosomal-inversion polymorphism. *Proceedings of the National Academy of Sciences, USA* **85**, 5597–5600.

Rodríguez-Trelles, F. and Rodríguez, Á. (2007). Comment on "Global genetic change tracks global climate warming in *Drosophila subobscura*". *Science* **315**, 1497a.

Rodríguez-Trelles, F. and Rodríguez, M. A. (1998). Rapid micro-evolution and loss of chromosomal diversity in *Drosophila* in response to global warming. *Evolutionary Ecology* **12**, 829–838.

Roff, D. (1992). *The Evolution of Life Histories*. New York: Chapman & Hall.

Roff, D. (2002). *Life History Evolution*. Sunderland, MA: Sinauer Associates.

Root, T. L., Price, J. T., Hall, K. R., Schneider, S. H., Rosenzweig, C. and Pounds, J. A. (2003). Fingerprints of global warming on wild animals and plants. *Nature* **421**, 57–60.

Rose, M. R. (1991). *Evolutionary Biology of Aging*. New York: Oxford University Press.

Roy, D. B. and Sparks, T. H. (2000). Phenology of British butterflies and climate change. *Global Change Biology* **6**, 407–416.

Sandrelli, F., Tauber, E., Pegoraro, M., Mazzotta, G., Cisotto, P., Landskrom, J., Stanewsky, R., Piccin, A., Rosato, E., Zordan, M., Costa, R. and Kyriacou, C. P. (2007). A molecular basis for natural selection at the *timeless* locus in *Drosophila melanogaster*. *Science* **316**, 1898–1900.

Saunders, D. S. (1990). The circadian basis of ovarian diapause regulation in Drosophila melanogaster: is the *period* gene causally involved in photoperiodic time measurement? *Journal of Biological Rhythms* **5**, 315–331.

Saunders, D. S. (2002). *Insect Clocks*. Amsterdam: Elsevier.

Saunders, D. S. and Gilbert, L. I. (1990). Regulation of ovarian diapause in *Drosophila melanogaster* by photoperiod and moderately low temperature. *Journal of Insect Physiology* **36**, 195–200.

Saunders, D. S., Henrich, V. C. and Gilbert, L. I. (1989). Induction of diapause in *Drosophila melanogaster*: photoperiodic regulation and the impact of arrhythmic clock mutants on time measurement. *Proceedings of the National Academy of Sciences, USA* **86**, 3748–3752.

Schierwater, B. and Hauenschild, C. (1990). A photoperiod determined life-cycle in an oligochate worm. *Biological Bulletin* **178**, 111–117.

Schmidt, P. S., Matzkin, L. M., Ippolito, M. and Eanes, W. F. (2005). Geographic variation in diapause incidence, life history traits and climatic adaptation in *Drosophila melanogaster*. *Evolution* **59**, 1721–1732.

Schwaegerle, K. E. and Schaal, B. A. (1979). Genetic variability and founder effect in the pitcher plant *Sarracenia purpurea L. Evolution* **33**, 1210–1218.

Schwartz, M. D., Ahas, R. and Aasa, A. (2006). Onset of spring starting earlier across the Northern Hemisphere. *Global Change Biology* **12**, 343–351.

Stearns, S. C. (1976). Life history tactics: A review of the ideas. *Quarterly Review of Biology* **51**, 3–47.

Stearns, S. C. (1992). *The Evolution of Life Histories*. Oxford, UK: Oxford University Press.

Stehlík, J., Závodská, S. K., Šauman, I. and Koštál, V. (2008). Photoperiodic induction of diapause requires regulated transcription of *timeless* in the larval brain of *Chymomyza costata*. *Journal of Biological Rhythms* **23**, 129–139.

Stinchcombe, J. R. and Hoekstra, H. E. (2008). Combining population genomics and quantitative genetics: finding the genes underlying ecologically important traits. *Heredity* **100**, 158–170.

Stone, G. N. and Sunnuck, P. (1993). Genetic consequences of an invasion through a patchy environment – the cynipid gallwasp *Andrecus quercuscalicis*. *Molecular Ecology* **2**, 251–268.

Takeda, M. and Skopik, S. D. (1997). Photoperiodic time measurement and related physiological mechanisms in insects and mites. *Annual Review of Entomology* **42**, 323–349.

Tatar, M., Chien, S. A. and Priest, N. K. (2001). Negligible senescence during reproductive dormancy in *Drosophila melanogaster*. *The American Naturalist* **158**, 248–258.

Tauber, E. and Kyriacou, C. P. (2001). Insect photoperiodism and circadian clocks: models and mechanisms. *Journal of Biological Rhythms* **16**, 381–390.

Tauber, E. and Kyriacou, C. P. (2008). Genomic approaches for studying biological clocks. *Functional Ecology* **22**, 19–29.

Tauber, E., Zordan, M., Sandrelli, F., Pegoraro, M., Osterwalder, N., Breda, C., Daga, A., Selmin, A., Monger, K., Benna, C., Rosato, E., Kyriacou, C. P. and Costa, R. (2007). Natural selection favors a newly derived *timeless* allele in *Drosophila melanogaster*. *Science* **316**, 1895–1898.

Tauber, M. J., Tauber, C. A. and Masaki, S. (1986). *Seasonal Adaptations of Insects.* New York, NY: Oxford University Press.

Taylor, F. (1980). Optimal switching to diapause in relation to the onset of winter. *Theoretical Population Biology* **18**, 125–133.

Umina, P. A., Weeks, A. R., Kearney, M. R., McKechnie, S. W. and Hoffmann, A. A. (2005). A rapid shift in a classic clinal pattern in *Drosophila* reflecting climate change. *Science* **308**, 691–693.

Vaz Nunes, M. and Saunders, D. (1999). Photoperiodic time measurement in insects: a review of clock models. *Journal of Biological Rhythms* **14**, 84–104.

Walther, G.-R., Post, E., Convey, P., Menzel, A., Parmesan, C., Beebee, T. J. C., Fromentin, J.-M., Hoegh-Guldberg, O. and Bairlein, F. (2002). Ecological response to recent climate change. *Nature* **416**, 389–395.

Warren, R. (2006). Impacts of global climate change at different annual mean global temperature increases. In *Avoiding Dangerous Climate Change*, ed. H. J. Schellnhuber, W. Cramer, N. Nakicenovic, T. Wigley and G. Yohe. Cambridge, UK: Cambridge University Press. pp. 93–131.

Williams, J. W. and Jackson, S. T. (2007). Novel climates, no-analog communities, and ecological surprises. *Frontiers in Ecology and the Environment* **5**, 475–482.

Withrow, R. B. (1959). *Photoperiodism and Related Phenomena in Plants and Animals.* Washington, DC: American Association for the Advancement of Science.

Wolda, H. and Denlinger, D. L. (1984). Diapause in a large aggregation of a tropical beetle. *Ecological Entomology* **9**, 217–230.

Zani, P., Swanson, S. E. T., Corbin, D., Cohnstaedt, L. W., Agotsh, M. D., Bradshaw, W. E. and Holzapfel, C. M. (2005). Geographic variation in tolerance of transient thermal stress in the mosquito *Wyeomyia smithii*. *Ecology* **86**, 1206–1211.

11

Genetic variability and evolution of cold-tolerance

JOHANNES OVERGAARD, JESPER G. SØRENSEN AND VOLKER LOESCHCKE

11.1 Introduction

Temperature has a large impact on the distribution and abundance of insect species. This can be seen in the varying degrees of behavioral, physiological, or biochemical adaptations to endure exposure to seasonal and acute thermal fluctuations. Variance in thermal tolerance limits between species and/or individuals is also highly influenced by acclimation/acclimatization effects and the underlying genetic variability for these traits (Ayrinhac *et al.*, 2004; Hoffmann *et al.*, 2005a). Thus, cold-tolerance is constrained by the underlying genetic background, including genetic correlations among traits, but also by various factors, such as the thermal history of the individual (acclimation), life stage, age, light cycle and diapause. The contributions of genetic and environmental factors to a trait such as cold-tolerance are difficult to separate if the appropriate design and methods are not used, and this is highlighted by the ongoing discussion of the relative importance of genetic variation versus phenotypic plasticity for thermal adaptation (Gienapp *et al.*, 2008). Ayrinhac *et al.* (2004) and Hoffmann *et al.* (2005a) both found genetic variation in cold-tolerance of *Drosophila melanogaster*, but they also found that the contribution to cold-tolerance of acclimation was greater than the contribution of genetic variation between both species and populations within species. Differences in cold-tolerance between and within insect species therefore may relate more to factors in the surrounding environment than to differences in the genetic background. However, the ability to mount a response (i.e. show phenotypic plasticity) and the extent of this response also has a genetic component

Low Temperature Biology of Insects, ed. David L. Denlinger and Richard E. Lee Jr. Published by Cambridge University Press.

and thus the degree of phenotypic plasticity also will be under selection in natural populations (Gabriel, 2005).

Insect cold-tolerance is diverse, both in terms of the absolute thermal limits endured and in terms of the underlying strategies that are used to optimize cold-tolerance. In general, cold-tolerant insects are broadly classified as being either freeze-tolerant or freeze-avoiding (Zachariassen, 1985; Lee, 1991; Bale, 1993; Sinclair *et al.*, 2003a). However, the distinction between freeze-tolerant and freeze-avoiding is probably overly simplistic, since cold-tolerance strategies can be split into several additional subcategories (see Nedved, 2000). However, many insects die at temperatures far above the temperature of crystallization of their body fluids and such insects should generally be defined as chill-susceptible. It is also important to recognize that insect development occurs over different life stages that are often ecologically and physiologically very different. Thus, the thermal environment experienced by adults and juvenile stages may be very different and it is possible that some life stages are under stronger selection for cold-tolerance. Clearly, this complicates the distinction as to which life stage/resistance traits show adaptive variation. For example, cold-shock resistance and the ability to rapidly acclimate differ markedly during ontogeny in *D. melanogaster* and *S. crassipalpis* (Lee and Denlinger, 1985; Jensen *et al.*, 2007). Still, most studies of cold resistance and adaptation investigate a single life stage, often the adult stage, even though this life stage might rely more on behavioral avoidance of low temperatures. Kimura and Beppu (1993), for example, showed that altitudinal distributions of *D. curvicaps* and *D. immigrans* were limited during winter periods, suggesting that adults escape cold by local migration to lower altitudes and warmer conditions. Similarly, Jenkins and Hoffmann (2001) suggest that the latitudinal distribution of *D. serrata* changes seasonally, due to thermal constraints on this species at low temperatures.

Although different insect species or life stages may have different sensitivity to cold, they all have thermal limits for performance and survival that are ultimately determined by gene–environment interactions. In the case of cold-tolerance, the most frequently investigated traits are "resistance" traits, such as survival, knock-down (also called critical thermal limit), recovery time and supercooling point. However, fitness may be constrained by other traits such as growth rate, activity pattern, or reproductive capacity at low temperature. Indeed it is possible that the thermal-tolerance limits of these latter "capacitance" traits are more important for cold-adaptation. Clearly, to understand adaptation and responses to cold it is important to identify the genetic components that are most important for driving the evolution of cold-tolerance, and this may be linked to traits other than just thermal resistance of adults. In this chapter, we discuss the role of genetic variability in insect cold-tolerance, but a major aim of this chapter is also to discuss

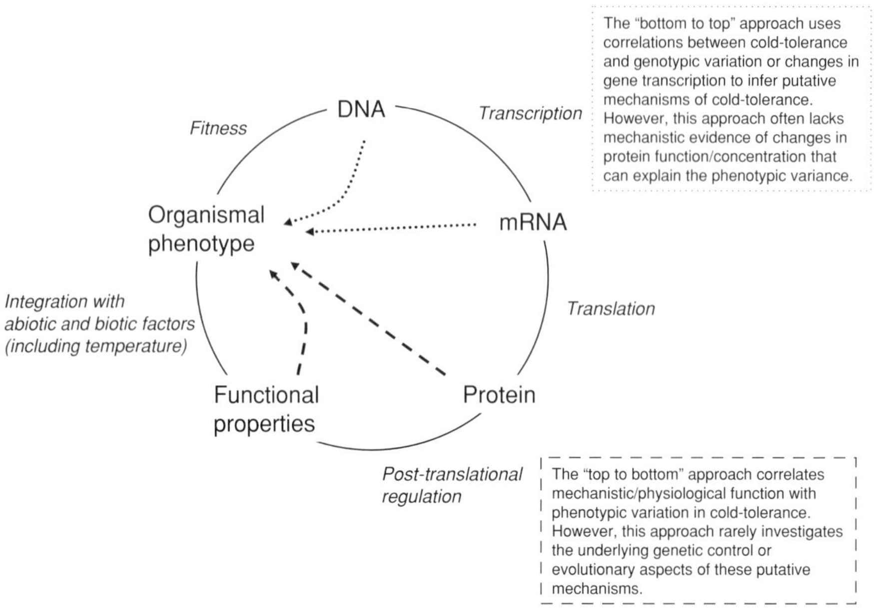

Figure 11.1 Genetic variation or variation in transcription levels of specific genes may putatively influence a phenotype such as cold-tolerance. To understand the importance of such relationships one should ideally also understand how the gene-product influences physiological function, and ultimately how the resulting variance at these levels influences variation in fitness. Studies of genetic variation therefore demand an integrative approach where genetic variation is studied at several levels of organization including transcriptional, biochemical, physiological and ecological levels.

the experimental approaches used to address these questions, as well as potential avenues for future research that could increase our knowledge and understanding of the genetic components of cold-adaptation and cold-resistance.

11.2 Studying genetic variability and cold-tolerance

When studying the genetic variability or evolution of cold-tolerance, one should ideally demonstrate a connection between genetic differences or differences in gene transcription to differences in phenotype and function (see Fig. 11.1). This should be done through measurable differences in physiological parameters and/or whole-animal phenotypic differences that affect fitness. As discussed below, many studies have found variation in cold-tolerance among populations or species, but the physiological mechanisms behind this variation are, in most cases, still illusive. Similarly, there are numerous studies that describe important mechanisms of cold-tolerance, but the underlying genetic control or genetic variability of these mechanisms is rarely investigated (see Fig. 11.1). Clearly, to improve our understanding of cold-tolerance we need studies that better connect the variability at the genetic level with variability at the phenotypic level. Although this goal is often not possible within a single study, we believe this approach should be highlighted as the ultimate goal for increasing our understanding of the evolution of cold-tolerance.

This chapter is generally biased towards studies of cold-tolerance of drosophilid species. The rationale for this is clearly not due to any extraordinary cold-tolerance of *Drosophila* species, since most drosophilids are in fact quite sensitive to cold (Chen *et al.*, 1991). However, due to historical and practical reasons, drosophilids, and in particular *Drosophila melanogaster*, have often been the chosen study organism for evolutionary and/or genetic studies on cold-adaptation (Gibert *et al.*, 2001; Hoffmann *et al.*, 2003a). This choice mainly relates to three important factors. Firstly, drosophilid species are generally easy to maintain in laboratory cultures, and, combined with short generation times, this makes them well suited for laboratory experiments, including selection studies. Secondly, drosophilid species have a wide distribution range in the tropic and temperate parts of the world, and several species have well-described clinal distributions along latitudinal and altitudinal gradients that are well suited for studies of evolutionary responses to temperature (Gibert *et al.*, 2001; Hoffmann *et al.*, 2002; Sørensen *et al.*, 2005; Balanya *et al.*, 2006; Hoffmann and Weeks, 2007). Finally, *D. melanogaster* was the first insect species to have its entire genome mapped (Adams *et al.*, 2000). Genomes have now been sequenced for 12 *Drosophila* species (Clark *et al.*, 2007), and this knowledge base is clearly of importance when investigating evolutionary responses and the relation between variation in cold-tolerance and variation at the RNA/DNA level. However, the development of powerful new markers such as single nucleotide polymorphisms (SNP) and the growing applicability of cDNA libraries for non-model animals will allow for easier study of genetic variability in non-model animals in the near future. Furthermore, gene chips developed for one species can also be used successfully for closely related species (e.g. Laayouni *et al.*, 2007).

The down side of focusing primarily on drosophilid species in studies of cold-tolerance is that some of the classical responses associated with cold-tolerance in insects are generally not very well developed in drosophilids. Antifreeze proteins are, for example, not expressed naturally in drosophilids, although artificial expression of insect antifreeze proteins may improve cold-tolerance (Nicodemus *et al.*, 2006). Similarly, it seems that mobilization of putative cryoprotectants only occurs to a moderate degree in drosophilids. Even though these moderate increases have been correlated with improved cold-resistance, it is unlikely that such changes will alter either the melting or the supercooling point (Kimura, 1982; Overgaard *et al.*, 2007). Although freeze-tolerance has been reported for the drosophilid, *Chymomyza costata* (Koštál *et al.*, 2003), drosophilids are generally sensitive to freezing, and certainly *D. melanogaster*, the species most commonly used in studies of thermal adaptation, is not freeze-tolerant.

In *D. melanogaster*, general transcriptomic changes have been investigated using commercially available gene chips after cold-hardening (Qin *et al.*, 2005) and after selection for increased cold-resistance (Sørensen *et al.*, 2007). Qin *et al.* (2005) found changes in transcript abundance for 36 genes after cold-hardening. These genes included the heat-shock genes *hsp23*, *hsp26* and *hsp83*, as well as *frost* and a number of membrane-associated proteins. Sørensen *et al.* (2007) found no significant differences in gene regulation between control flies and flies that had been exposed to 10 generations of cold selection. In the study by Sørensen *et al.* (2007), there was ample variation in gene expression patterns when selecting for other stress-resistance traits, which may indicate that variation in cold-tolerance in *D. melanogaster* is not so much a matter of variation in gene expression, but rather is due to post-transcriptional regulation and/or allele/allozyme differences. In addition to studies screening the entire genome of an insect species, a number of other studies have reported altered transcription levels of a wide range of genes in different insect species. These include antifreeze proteins, stress proteins, membrane desaturases, transcription factors and metabolic genes of importance for ATP production rates (Goto *et al.*, 1998; Goto, 2001; Duman 2001; Rineheart *et al.*, 2006; Sinclair *et al.*, 2007a; Kayukawa *et al.*, 2007; McMullen and Storey, 2008). Although none of these studies investigated genetic variability per se, they may help to identify putative biochemical and physiological pathways that are of importance to cold-tolerance, and future experiments will be able to determine if genetic variability in these genes is linked to variance of cold-tolerance.

In species where crosses can be performed easily in the lab, one can also search for quantitative trait loci (QTL), and thereby identify chromosomal regions that have an impact on cold-tolerance. Such regions, however, usually contain many genes, e.g. up to 100 genes within 1 cM in *Drosophila*. Thus, going from a QTL study to the precise gene(s) underlying a particular QTL can be a huge task. Morgan and Mackay (2006) and Norry *et al.* (2007) have identified independently and in different genetic backgrounds QTL for chill-coma recovery in *D. melanogaster*. These QTL were found to be located on chromosome 2, with polymorphic markers used approximately 10 cM apart, in a region free of inversions, but containing several candidate genes. This chill-coma QTL co-localized with an antagonistic QTL for heat knockdown resistance (Norry *et al.*, 2007).

11.3 Interspecific variation in insect cold-tolerance

Comparative studies have yielded information about cold-adaptation strategies and the repeated evolution of these strategies across the arthropod phylum. For example, it has been shown using phylogenetic correction that freeze-avoidance is basal within arthropods, and that freeze-tolerance has evolved repeatedly within different taxa (Sinclair *et al.*, 2003b). There seems to be a general difference in the level of freeze-tolerance between species from the northern and southern hemispheres (Sinclair *et al.*, 2003b; Sinclair and Chown, 2005). Overall, these authors found a significantly higher frequency of freeze-tolerance among investigated species in the southern hemisphere (77%, $n = 27$) than species from the northern hemisphere (29%, $n = 258$). Northern hemisphere freeze-tolerant species generally tolerate temperatures much below the temperature at which they freeze, while southern species on average only tolerate moderate freezing. This difference might very well relate to general differences in the climate of the two hemispheres and can therefore be considered adaptive. The northern hemisphere cold climate is dominated by continental conditions, with extended periods of low subzero temperatures, and freeze-tolerance is not always present, but acquired prior to winter. Many southern cold environments are milder, with unpredictable cold snaps, and freeze-tolerance of southern species does not vary seasonally to the same extent as in the north. Thus, environmental predictability is probably important for the evolution of insect cold-adaptation (Sinclair *et al.*, 2003b; Sinclair and Chown, 2005). This hypothesis was confirmed in studies of cold-tolerance in five different mite species from sub-Antarctic Marion Island. Here the two species from marine (predictable) habitats showed larger acclimation responses than the three species from terrestrial (unpredictable) habitats (Deere *et al.*, 2006).

Much of the information about insects from different geographical origins comes from studies of *Drosophila* species. Species comparisons were reviewed and discussed extensively by Hoffmann *et al.* (2003a, see Table 6). Here it was concluded that *Drosophila* comparisons generally conformed to expectations based upon adaptation to cold conditions. This is also the general picture that emerges from the extensive studies comparing differential thermal tolerance of East Asian *Drosophila* species from either temperate or tropical origin (Kimura, 1988; Goto and Kimura, 1998; Goto *et al.*, 1999; Goto *et al.*, 2000; Kimura, 2004). Although there are exceptions to the general pattern, these data suggest that cold-tolerance is an important trait for maintaining the distribution pattern of *Drosophila* species. The interspecific differences in drosophilid cold-tolerance correlated with differences in, for example, Hsp expression pattern and membrane phospholipid composition, which suggests that these mechanisms may be of importance for cold-adaptation in drosophilids (Goto and Kimura, 1998; Ohtsu *et al.*, 1998; 1999).

Many early comparative studies were based on only a few species (represented by few/single lines) and with no control for phylogenetic effects (Felsenstein, 1995), thus making evolutionary interpretations about cold-tolerance and responses to cold uncertain. The issue of phylogenetic inertia is an emerging problem in all fields of comparative and evolutionary physiology and arises when a seemingly adaptive trait occurs repeatedly due to a single evolutionary event prior to speciation, and not as an independent evolutionary adaptive process (Garland *et al.*, 2005). When phylogenetic information is incorporated, it can give very strong support to evolutionary hypotheses. Gibert *et al.* (2001) investigated chill-coma recovery time in *Drosophila* species of temperate and tropical origin. Temperate species generally showed greater cold-tolerance by showing much faster recovery after exposure to 0 °C. More interesting, by including phylogenetic information, they showed that chill-coma resistance has evolved repeatedly in temperate species (Gibert *et al.*, 2001). Chill-coma temperature (or critical thermal limit), the temperature where flies lose their righting response after progressive cooling, also showed evidence of adaptive differences among species from different latitudes (Gibert and Huey, 2001). This pattern was also evident after the inclusion of phylogenetic information, suggesting that these adaptations have evolved independently several times.

11.4 Variation in cold-tolerance within species

As discussed above, there is considerable variation in cold-tolerance between species, but variation may also be observed within species. The source of this variation may be difficult to identify, unless appropriate experimental methods are used. Intraspecific variation is often manifested in variable cold-tolerance among local populations collected along a climatic gradient, and this type of geographic variation in insect cold-tolerance generally correlates with altitude and latitude in the expected direction (Gaston and Chown, 1999; Chown, 2001; Ayrinhac *et al.*, 2004; Hoffmann *et al.*, 2005a). However, in some of these studies, the data are generated from field collected (F0) generations, which does not allow for an evaluation of whether differences are due to genetic differentiation or due to local acclimation (see Fig. 11.2). Indeed, several studies have suggested that the role of phenotypic plasticity is much larger than that of the genetic differences along these gradients for drosophilids (Ayrinhac *et al.*, 2004; Hoffmann *et al.*, 2005a). As a result, latitudinal clines may be “steeper” when phenotypic traits are measured on wild-caught animals relative to experiments where the animals have developed under common laboratory conditions (James *et al.*, 1997). Nevertheless, populations collected along a climatic gradient (latitudinal or altitudinal) can be a powerful tool to demonstrate genetic variability in cold-tolerance, if the populations are investigated in a common-garden design or using reciprocal transplantation (Fig. 11.2) (Hoffmann and Weeks, 2007). To our knowledge, there have been no studies that evaluate the power of reciprocal transplantation versus common-garden experiments with regard to cold-tolerance of insects. However, in

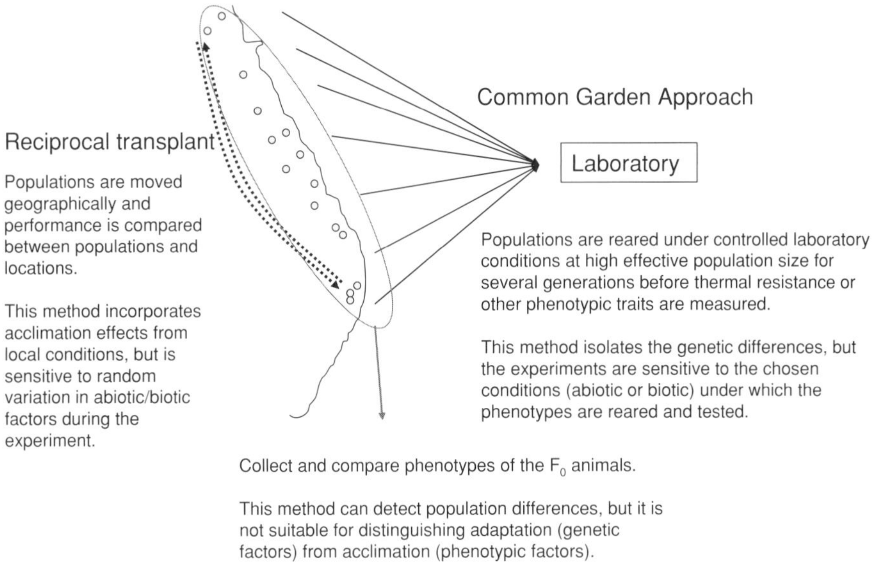

Figure 11.2 Geographical variation in cold-tolerance is due to local acclimatization and/or genetic variation between subpopulations. To distinguish the role of genetic factors from local acclimation appropriate methods, such as common garden, or reciprocal transplant studies, should be used.

a study of plant–pathogen interactions, Laine (2007) showed that these two experimental approaches may render different results. It was, therefore, concluded that common-garden studies should aim at including "natural" variance in factors such as temperature and light, to avoid confounding factors from "standard" laboratory conditions. A practical concern regarding reciprocal transplant studies in animals is, however, that it is difficult to control/limit migration, and studies of reciprocal transplantations might therefore better be approached in a restricted form, as, for example, in field release studies (see Hoffmann and Loeschcke, 2006; Loeschcke and Hoffmann, 2007; Kristensen *et al.*, 2008a) or by using field cage experiments (Hoffmann *et al.*, 2003b; Kristensen *et al.*, 2008b).

Another way of assessing genetic variation for cold-tolerance can be based on simple pedigree information, as in parent–offspring comparisons, and this approach is also applicable to non-model organisms. Such data can be used to estimate field heritabilities by correlating responses between field-collected individuals with their laboratory-reared offspring (Jenkins and Hoffmann, 1994). Similarly, creating isofemale lines provides a way of investigating the level of variation among and within lines and can be used to get an estimate of broad- or narrow-sense heritabilities and variance components (Hoffmann and Parsons, 1988). Using isofemale lines, Hoffmann *et al.* (2005b) tested for a trade-off between starvation and chill-coma recovery, by scoring correlated responses in these traits among lines selected for increased starvation resistance. The result was sex-specific, since females showed a genetically robust trade-off, while males did not.

Several studies have demonstrated clinal variation in phenotypic traits related to thermal adaptation of drosophilid species and, in particular, *D. melanogaster*. Cold-tolerance traits varied clinally for *D. melanogaster* and *D. serrata* in east Australian populations and a similar latitudinal correlation with chill-coma recovery exists across a group of populations collected from North and South America, Europe and Africa (Stanley and Parsons, 1981; Davidson, 1990; Hallas *et al.*, 2002; Hoffmann *et al.*, 2002; David *et al.*, 2004; Ayrinhac *et al.*, 2004). In the case of the east Australian cline of *D. melanogaster*, these data have recently been reviewed in great detail by Hoffmann and Weeks (2007). Thermal resistance traits co-occur with other phenotypic traits, such as size traits. Thus, according to Bergmann's rule, low temperature has repeatedly been associated with larger size in distributions of ectothermic animals, including insects in general and *D. melanogaster* in particular (Capy *et al.*, 1993; James *et al.*, 1997; Hallas *et al.*, 2002). The ratio of wing to thorax size also varies with latitude/temperature (Azevedo *et al.*, 1998), although recent studies do not seem to indicate that this ratio affects flight performance differently at high or low temperatures (Hoffmann *et al.*, 2007). Finally, there is also ample genetic variation in ovary number, egg size and egg-retention capacity in drosophilids collected along thermal gradients, and this indicates that the reproductive strategy/capacity also shows adaptive variation due to thermal variation (David and Bocquet, 1975; Bouletreaumerle *et al.*, 1992; Azevedo *et al.*, 1996).

Clinal variation in both cold-tolerance and reproductive stategy/capacity may be tightly connected to the occurrence of reproductive diapause. This state is elicited by environmental cues, particularly temperature and photoperiod, and its underlying genetic and neuroendocrine control is fairly well described (Denlinger, 2002; Allen, 2007). Clinal variation in *D. melanogaster* genotypes coding for the occurrence of reproductive diapause exists, with genotypes that undergo reproductive diapause more frequent at high latitudes (Schmidt *et al.*, 2005). This clinal variation is tightly connected with reproductive capacity, as non-diapause genotypes have faster development from egg to adult and also have higher fecundity in the early life stages (Schmidt and Conde, 2006). In contrast, the diapause genotypes have higher stress-tolerance, including cold-tolerance, than non-diapause genotypes, and this increases fitness under conditions where flies are exposed to stressful environments. In accordance with this, experiments in population cages demonstrated that diapause genotypes became dominant under stressful conditions, whereas the non-diapause genotypes were most successful under benign conditions (Schmidt and Conde, 2006). Given the relationship between the diapause genotypes and both cold-tolerance and reproductive strategy/capacity, it is apparent that clinal variation in diapause genotypes will have direct impact on clinal variation in both cold-tolerance traits and traits related to reproductive strategy.

The clinal variation found in cold-tolerance and other phenotypic traits co-exists with clinal variation in a number of karyotypes and frequencies of specific genes. Hoffmann and Weeks (2007) recently reviewed the numerous genetic patterns of clinal variations, which include several chromosomal inversions, allozymes, and specific genes, such as *hsp23*, *hsp26*, *hsp70* and *hsr-omega* (McColl and McKechnie, 1999; Bettencourt *et al.*, 2002; Frydenberg *et al.*, 2003; see Table 1 in Hoffmann and Weeks, 2007 for further details). Similar genetic clines also exist for other geographic distributions of *D. melanogaster*, as illustrated by the North American clines in *Adh* allele frequencies (Simmons *et al.*, 1989) and in genes coding for other metabolic enzymes (Sezgin *et al.*, 2004). Despite several attempts to link cold-tolerance traits directly to these genetic clines, it has only been shown that chill-coma recovery is associated with the chromosomal inversion *In(3L)Payne* (Weeks *et al.*, 2002). No other clinal variations in genotypes are directly linked to cold-resistance traits (see Weeks *et al.*, 2002; Rako *et al.*, 2006, 2007; Hoffmann and Weeks, 2007). Thus, some genotypes may simply be "hitchhiking" with genes of importance, and genetic clines may also arise as a result of historical events, genetic drift, gene flow, or spatial isolation (Endler, 1977). Clearly, it is important that potential candidate genes for cold-tolerance are linked both through correlations of phenotypic and genotypic variation, but more importantly through mechanistic/physiological investigations of these putative "cold" gene products (i.e. proteins/enzymes/transcription factors etc.). Nevertheless, indirect evidence suggests an association with thermal adaptation since genetic clines have shifted over the last 20 years, probably in response to changing global climates (Umina *et al.*, 2005). This is, for example, the case for *Adh* and the chromosomal inversion *In(3R)Payne*.

Even though there are at present only a few clear connections between cold-tolerance traits and clinal variations in candidate genes, this experimental approach represents a future avenue to identify such correlations. It could be especially interesting if this approach were applied to insect species where cold-survival is more likely to directly determine population distribution patterns. In the case of *D. melanogaster*, it is plausible that the relevant traits to examine would be ones important for maintenance of growth and reproduction at low temperature, rather than traits associated to cold-"resistance," such as, for example, survival or chill-coma recovery (see Cossins and Bowler, 1987). It is, for example, likely that traits associated with reproductive strategy, size, or feeding strategy are more important for local adaptation than actual resistance in terms of cold-survival of individuals.

Correlations between phenotypic traits and genetic variation often lack conclusive experimental evidence for mechanistic links. There are, however, some exceptions to this rule, and an example of this is the well-described geographic variation in the phosphoglucose isomerase (PGI) gene. Experimental work on beetles and, in particular, butterflies has demonstrated biochemical differences between PGI genotypes that result in differential flight performance, which can be directly linked to fitness. Although this example does not address a specific cold-resistance trait/mechanism, it illustrates how an integrative approach can demonstrate clear connections between thermal conditions, genetic variability, biochemical and physiological performance and ultimately fitness (Watt, 1977; 1992; Wheat *et al.*, 2006.

PGI (sometimes also called GPI) catalyses the second step of glycolysis by conversion of glucose-6-phosphate into fructose-6-phosphate. The kinetic properties of this enzymatic reaction are probably of particular importance during insect flight, where glycolysis is vital for sufficient ATP production. Allele frequency variation at the PGI locus is found in both butterflies and the montane beetle, *Chrysomela aeneicollis*, and this variation seems to correlate with local thermal conditions (Watt, 1983; Hughes and Zalucki, 1993; Dahlhoff and Rank, 2000; Hanski and Saccheri, 2006). The different PGI allozymes have different thermal stability and kinetic properties, and, in the case of *Colias* butterflies, heterozygosity is associated with superior kinetics of the PGI enzyme (Watt *et al.*, 1983). This superiority is related to structural interactions of the two monomers in the dimeric PGI enzyme (Wheat *et al.*, 2006). However, there is also evidence for functional differences in both kinetics and thermal stability among different homozygotes (Watt *et al.*, 1983) and these differences are directly related to differential performance in flight behavior, mobility and consequently in fecundity of butterflies experiencing fluctuating environmental conditions (Watt *et al.*, 1983; Watt, 1992; Wheat *et al.*, 2006).

Although the number of examples that establish clear connections between cold-resistance/performance and genetic variation in a single gene is limited, the development of new technologies that can be used to identify thousands of single nucleotide polymorphisms (SNP) will enable much easier identification of genetic variation of importance for cold-resistance. It is therefore likely that emerging molecular techniques will enhance our possibilities to identify both genetic variation of importance to cold-tolerance, but also to identify yet unknown candidate genes of importance for cold-tolerance.

11.5 Artificial selection for cold-tolerance and correlated responses

Selection experiments for increased or decreased performance at low-temperature conditions can be used to estimate the evolutionary potential in the trait or the amount of standing genetic variation available in the population in question. Artificial-selection experiments require relatively large effective population sizes, and replicated selection lines are needed to establish that results are due to selection. High effective population sizes also reduce the effects of genetic drift relative to the effects of the selection treatment. Selection intensity is also important, as too weak selection may yield changes too small to be detected, while too strong selection will create linkage disequilibrium and "hitch-hiking" effects not directly related to selection. As a consequence, most artificial-selection experiments with insects are restricted to *Drosophila*.

To our knowledge, only a few studies have used a selection approach to investigate the role of genetic variability for cold-tolerance. Tucic (1979) selected for cold-shock survival independently in several life stages (egg, early larvae, late larvae, pupae, or adult) and found correlated responses to selection for cold-resistance among developmental stages, although cold-resistance showed some degree of specificity to the developmental stage selected. Most other selection studies were made on adult flies. Watson and Hoffmann (1996) found a rapid response to selection for cold-resistance in *D. melanogaster* lines. However, after more generations of selection, the response in terms of increased cold-resistance reached a plateau or was even negative. After rearing for a generation without selection, increased cold-resistance was re-established suggesting parental/carry-over effects from the cold-selection treatment. Cold exposures might also affect fertility (David *et al.*, 1983) and induce maternal effects, influencing the outcome of selection (see also Anderson *et al.*, 2005). Thus, a relaxed generation between selection events might ensure that the population recovers and is able to express its genetic potential, so that it can be selected upon.

Different cold-resistance traits may respond differently to selection. This was the case in several studies selecting for either chilling-resistance (4 or 0 °C), cold-shock survival, combinations of chilling followed by cold-shock treatments and chill-coma recovery in *D. melanogaster* and *D. simulans* (Chen and Walker, 1993, 1994; Watson and Hoffmann, 1996; Anderson *et al.*, 2005). Selection responses were found to have levels of realized heritability from 30%, for the directly selected flies, to 6–12% for the lines that were pre-exposed to chilling (Watson and Hoffmann, 1996). An even higher realized heritability (33–46%) was found for chill-coma recovery. Substantial genetic variation for cold-tolerance thus seems to exist within populations.

Chen and Walker (1993, 1994) found that both cold-shock selection and selection for chill-resistance caused increased survival of cold-shock. However, only lines selected for chilling-resistance showed increased resistance to chilling, making the correlated responses asymmetric among cold-resistance traits. It seems, therefore, that resistance to different cold-resistance traits/different temperature exposures is made up of at least partly independent genetic mechanisms. In contrast Anderson *et al.* (2005) showed improved survival following cold-shock in flies that had been selected for chill-coma recovery, which could indicate a common mechanism underlying these two traits. Correlated responses to selection may also be found in relation to other traits. In a highly replicated study with high population sizes, Bubliy and Loeschcke (2005) investigated responses to selection for six different resistance traits, including cold-resistance (acclimation at 11 °C followed by survival test at 0 °C). Cold-selected lines, as expected, had superior cold-shock resistance, but increased cold-resistance was also found in lines selected for desiccation- and heat-resistance. Similarly, lines selected for increased cold-resistance showed a correlated, but asymmetric, response exhibiting increased starvation resistance (i.e. no increased cold-tolerance in starvation-selected flies) (Bubliy and Loeschcke, 2005). Sinclair *et al.* (2007b) investigated the potential correlated responses of selection for desiccation-tolerance on cold-tolerance traits, including the rapid cold-hardening response, cold-shock survival and chill-coma recovery. Only chill-coma recovery responded slightly to desiccation selection. Similarly, there seems to be a correlation between selection for longevity and cold-stress survival, although this result depends on genotype-by-sex interactions in both traits (Norry and Loeschcke, 2002).

11.6 Conclusions and perspectives

Molecular tools to study genetic variation are rapidly changing, with the technology available for model organisms becoming available to non-model organisms. The new technology of 454 sequencing allows the screening for thousands of single nucleotide polymorphisms (SNP), markers that represent neutral genetic variation, as well as variation under selection (Vera *et al.*, 2008). Also, DNA sequencing of thousands of species and the construction of cDNA gene chips is becoming a real and affordable possibility. Thus, in the near future, the study of genetic variation for cold-tolerance can more easily be extended to non-model species, particularly those that may display more pronounced adaptations for cold-tolerance, and these new possibilities will boost our general understanding of genetic variation and evolution of cold-tolerance in insects.

Genetic investigations of cold-tolerance, as done on *Drosophila*, are largely lacking for other insect groups, and it is still unknown how much of the information from *Drosophila* can be extrapolated to other insects. The experiments done so far have, however, invariably demonstrated substantial genetic variation for cold-resistance within populations. Although this genetic variation should enable expansion of species or population borders (Jenkins and Hoffmann, 1999), some seasonal changes in distribution patterns are best explained assuming that cold-stress is limiting the distribution towards colder climate conditions (Jenkins and Hoffmann, 2001). Some of this controversy may arise because cold-resistance is typically determined under laboratory conditions, which are certainly not always representative for the complex thermal constraints that animals experience in the natural environment. A number of recent studies have shown that both artificial selection and laboratory acclimation may affect field performance differently than would be expected from laboratory experiments (Kristensen *et al.*, 2007; 2008a). Clearly there are still many unresolved questions regarding the relationship between thermal performance under natural and artificial conditions.

The ultimate goal of research on genetic variation for cold-tolerance is to relate variation at the DNA level to variation in phenotype and function, i.e. to connect the corners of the "magic triangle", phenotype, genotype and fitness. As a genotype can express different phenotypes in different environments, the study of genetic variation for cold-tolerance and its evolution is complex, and pinpoints the importance of separating plastic effects from genetic effects. Detailed information on the role of genetic variation for adaptation to cold climate conditions is still scarce, and only a few examples demonstrate clear links between genetic variation in specific genes and cold-tolerance. However, the list of candidate genes is growing, making future studies more likely to demonstrate such connections, as well as their dependency on environment, sex and genetic background. As cold-tolerance is a quantitative trait, affected by a number of loci, another important step will be to unravel its genetic architecture and inheritance.

Acknowledgements

We are grateful to Torsten N. Kristensen and Stuart Barker for helpful comments on the text and the Danish Natural Sciences Research Council for financial support through center, frame and post-doctoral grants.

References

Adams, M. D., Celniker, S. E., Holt, R. A., Evans, C. A., Gocayne, J. D., Amanatides, P. G., Scherer, S. E., Li, P. W., Hoskins, R. A., Galle, R. F., George, R. A., Lewis, S. E. and Richards, S. (2000). The genome sequence of *Drosophila melanogaster*. *Science* **287**, 2185–2195.

Allen, M. J. (2007). What makes a fly enter diapause? *Fly* **1**, 307–310.

Anderson, A. R., Hoffmann, A. A. and McKechnie, S. W. (2005). Response to selection for rapid chill-coma recovery in *Drosophila melanogaster*: physiology and life-history traits. *Genetical Research* **85**, 15–22.

Ayrinhac, A., Debat, V., Gibert, P., Kister, A. G., Legout, H., Moreteau, B., Vergilino, R. and David, J. R. (2004). Cold adaptation in geographical populations of *Drosophila melanogaster*: phenotypic plasticity is more important than genetic variability. *Functional Ecology* **18**, 700–706.

Azevedo, R. B.R., French, V. and Partridge, L. (1996). Thermal evolution of egg size in *Drosophila melanogaster*. *Evolution* **50**, 2338–2345.

Azevedo, R. B.R., James, A. C., McCabe, J. and Partridge L. (1998). Latitudinal variation of wing:thorax size ratio and wing-aspect ratio in *Drosophila melanogaster*. *Evolution* **52**, 1353–1362.

Balanya, J., Oller, J. M., Huey, R. B., Gilchrist, G. W. and Serra, L. (2006). Global genetic change tracks global climate warming in *Drosophila subobscura*. *Science* **313**, 1773–1775.

Bale, J. S. (1993). Classes of insect cold hardiness. *Functional Ecology* **7**, 751–753.

Bettencourt, B. R., Kim, I., Hoffmann, A. A. and Feder, M. E. (2002). Response to natural and laboratory selection at the *Drosophila* hsp70 genes. *Evolution* **56**, 1796–1801.

Bouletreaumerle, J., Fouillet, P. and Terrier, O. (1992). Clinal and seasonal variations in initial retention capacity of virgin *Drosophila melanogaster* females as a strategy for fitness. *Evolutionary Ecology* **6**, 223–242.

Bubliy, O. A. and Loeschcke, V. (2005). Correlated responses to selection for stress resistance and longevity in a laboratory population of *Drosophila melanogaster*. *Journal of Evolutionary Biology* **18**, 789–803.

Capy, P., Pla, E. and David, J. R. (1993). Phenotypic and genetic variability of morphometrical traits in natural populations of *Drosophila melanogaster* and *D. simulans*. 1. Geographic variations. *Genetics Selection Evolution* **25**, 517–536.

Chen, C. P. and Walker, V. K. (1993). Increase in cold-shock tolerance by selection of cold resistant lines in *Drosophila melanogaster*. *Ecological Entomology* **18**, 184–190.

Chen, C. P. and Walker, V. K. (1994). Cold-shock and chilling tolerance in *Drosophila*. *Journal of Insect Physiology* **40**, 661–669.

Chen, C. P., Lee, R. E. and Denlinger, D. L. (1991). Cold shock and heat shock – a comparison of the protection generated by brief pretreatment at less severe temperatures. *Physiological Entomology* **16**, 19–26.

Chown, S. L. (2001). Physiological variation in insects: hierarchical levels and implications. *Journal of Insect Physiology* **47**, 649–660.

Clark, A. G., Eisen, M. B., Smith, D. R., Bergman, C. M., Oliver, B., Markow, T. A. *et al.* (2007). Evolution of genes and genomes on the *Drosophila* phylogeny. *Nature* **450**, 203–218.

Cossins, A. R. and Bowler, K. (1987). *Temperature Biology of Animals*. New York: Chapman and Hall.

Dahlhoff, E. P. and Rank, N. E. (2000). Functional and physiological consequences of genetic variation at phosphoglucose isomerase: Heat shock protein expression is related to enzyme genotype in a montane beetle. *Proceedings of the National Academy of Sciences, USA* **97**, 10056–10061.

David, J. R., Allemand, R., Capy, P., Chakir, M., Gibert, P., Pétavy, G. and Moreteau, B. (2004). Comparative life histories and ecophysiology of *Drosophila melanogaster* and *D. simulans*. *Genetica* **120**, 151–163.

David, J. R., Allemand, R., Van Herrewege, J. and Cohet, Y. (1983). Ecophysiology: abiotic factors. In *The Genetics and Biology of* Drosophila, ed. Ashburner, M., Carson, H. L. and Thompson, J. N. London: Academic Press, pp. 105–170.

David, J. R. and Bocquet, C. (1975). Evolution in a cosmopolitan species – Genetic latitudinal clines in *Drosophila melanogaster* wild populations. *Experientia* **31**, 164–166.

Davidson, J. K. (1990). Non-parallel geographic pattern for tolerance to cold and desiccation in *Drosophila melanogaster* and *D. simulans*. *Australian Journal of Zoology* **38**, 155–161.

Deere, J. A., Sinclair, B. J., Marshall, D. J. and Chown, S. L. (2006). Phenotypic plasticity of thermal tolerances in five oribatid mite species from sub-Antarctic Marion Island. *Journal of Insect Physiology* **52**, 693–700.

Denlinger, D. L. (2002). Regulation of diapause. *Annual Review of Entomology* **47**, 93–122.

Duman, J. G. (2001). Antifreeze and ice nucleator proteins in terrestrial arthropods. *Annual Review of Physiology* **63**, 327–357.

Endler, J. A. (1977). *Geographic Variation, Speciation and Clines*. Princeton, NJ: Princeton University Press.

Felsenstein, J. (1995). Phylogenies and the comparative method. *American Naturalist* **125**, 1–15.

Frydenberg, J., Hoffmann, A. A. and Loeschcke, V. (2003). DNA sequence variation and latitudinal associations in hsp23, hsp26 and hsp27 from natural populations of *Drosophila melanogaster*. *Molecular Ecology* **12**, 2025–2032.

Gabriel, W. (2005). How stress selects for reversible phenotypic plasticity. *Journal of Evolutionary Biology* **18**, 873–883.

Garland, T., Bennett, A. F. and Rezende, E. L. (2005). Phylogenetic approaches in comparative physiology. *Journal of Experimental Biology* **208**, 3015–3035.

Gaston K. J. and Chown S. L. (1999). Elevation and climatic tolerance: a test using dung beetles. *Oikos* **86**, 584–590.

Gibert, P. and Huey, R. B. (2001). Chill-coma temperature in *Drosophila*: Effects of developmental temperature, latitude, and phylogeny. *Physiological and Biochemical Zoology* **74**, 429–434.

Gibert, P., Moreteau, B., Petavy, G., Karan, D. and David, J. R. (2001). Chill-coma tolerance, a major climatic adaptation among *Drosophila* species. *Evolution* **55**, 1063–1068.

Gienapp, P., Teplitsky, C., Alho, J. S., Mills, J. A. and Merila, J. (2008) Climate change and evolution: disentangling environmental and genetic responses. *Molecular Ecology* **17**, 167–178.

Goto, S. G. (2001). A novel gene that is up-regulated during recovery from cold shock in *Drosophila melanogaster*. *Gene* **270**, 259–264.

Goto, S. G., Yoshida, K. M. and Kimura, M. T. (1998). Accumulation of Hsp70 mRNA under environmental stresses in diapausing and nondiapausing adults of *Drosophila triauraria*. *Journal of Insect Physiology* **44**, 1009–1015.

Goto, S. G. and Kimura, M. T. (1998). Heat- and cold-shock responses and temperature adaptations in subtropical and temperate species of *Drosophila*. *Journal of Insect Physiology* **44**, 1233–1239.

Goto, S. G., Yoshida, T., Beppu, K. and Kimura, M. T. (1999). Evolution of overwintering strategies in Eurasian species of the *Drosophila obscura* species group. *Biological Journal of the Linnean Society* **68**, 429–441.

Goto, S. G., Kitamura, H. W. and Kimura, M. T. (2000). Phylogenetic relationships and climatic adaptations in the *Drosophila takahashii* and *montium* species subgroups. *Molecular Phylogenetics and Evolution* **15**, 147–156.

Hallas, R., Schiffer, M. and Hoffmann, A. A. (2002). Clinal variation in *Drosophila serrata* for stress resistance and body size. *Genetical Research* **79**, 141–148.

Hanski, I. and Saccheri, I. (2006). Molecular-level variation affects population growth in a butterfly metapopulation. *PloS Biology* **4**, 719–726.

Hoffmann, A. A. and Parsons, P. A. (1988). The analysis of quantitative variation in natural populations with isofemale strains. *Genetics Selection Evolution* **20**, 87–98.

Hoffmann, A. A., Anderson, A. and Hallas, R. (2002). Opposing clines for high and low temperature resistance in *Drosophila melanogaster*. *Ecology Letters* **5**, 614–618.

Hoffmann, A. A., Sørensen, J. G. and Loeschcke, V. (2003a). Adaptation of *Drosophila* to temperature extremes: bringing together quantitative and molecular approaches. *Journal of Thermal Biology* **28**, 175–216.

Hoffmann, A. A., Scott, M., Partridge, L. and Hallas, R. (2003b). Overwintering in *Drosophila melanogaster*: outdoor field cage experiments on clinal and laboratory selected populations help to elucidate traits under selection. *Journal of Evolutionary Biology* **16**, 614–623.

Hoffmann, A. A., Hallas, R., Anderson, A. R. and Telonis-Scott, M. (2005a). Evidence for a robust sex-specific trade-off between cold resistance and starvation resistance in *Drosophila melanogaster*. *Journal of Evolutionary Biology* **18**, 804–810.

Hoffmann, A. A., Shirriffs, J. and Scott, M. (2005b). Relative importance of plastic vs. genetic factors in adaptive differentiation: geographical variation for stress resistance in *Drosophila melanogaster* from eastern Australia. *Functional Ecology* **19**, 222–227.

Hoffmann, A. A. and Loeschcke, V. (2006). Are fitness effects of density mediated by body size? Evidence from *Drosophila* field releases. *Evolutionary Ecology Research* **8**, 813–828.

Hoffmann A. A. and Weeks, A. R. (2007). Climatic selection on genes and traits after a 100 year-old invasion: a critical look at the temperate-tropical clines in *Drosophila melanogaster* from eastern Australia. *Genetica* **129**, 133–147.

Hoffmann, A. A., Ratna, E., Sgro, C. M., Barton, M., Blacket, M., Hallas, R., De Garis, S. and Weeks, A. R. (2007). Antagonistic selection between adult thorax and wing size in field released *Drosophila melanogaster* independent of thermal conditions. *Journal of Evolutionary Biology* **20**, 2219–2227.

Hughes, J. M. and Zalucki, M. P. (1993). The relationship between the pgi locus and the ability to fly at low temperature in the monarch butterfly *Danaus plexippus*. *Biochemical Genetics* **31**, 521–532.

James, A. C., Azevedo, R. B. and Partridge, L. (1997). Genetic and environmental responses to temperature of *Drosophila melanogaster* from a latitudinal cline. *Genetics* **146**, 881–90

Jenkins, N. L. and Hoffmann, A. A. (1994). Genetic and maternal variation for heat resistance in *Drosophila* from the field. *Genetics* **137**, 783–789.

Jenkins, N. L. and Hoffmann, A. A. (1999). Limits to the southern border of *Drosophila serrata*: Cold resistance, heritable variation and trade-offs. *Evolution* **53**, 1823–1834.

Jenkins, N. L. and Hoffmann, A. A. (2001). Distribution of *Drosophila serrata* Malloch (Diptera: Drosophilidae) in Australia with particular reference to the southern border. *Australian Journal of Entomology* **40**, 41–48.

Jensen, D., Overgaard, J. and Sørensen, J. G. (2007). The influence of developmental stage on cold shock resistance and ability to cold-harden in *Drosophila melanogaster*. *Journal of Insect Physiology* **53**, 179–186.

Kayukawa, T. Chen, B. Hoshizaki, S. and Ishikawa, Y. (2007). Upregulation of a desaturase is associated with the enhancement of cold hardiness in the onion maggot, *Delia antiqua*. *Insect Biochemistry and Molecular Biology* **37**, 1160–1167.

Kimura, M. T. (1982). Inheritance of cold hardiness and sugar contents in 2 closely related species, *Drosophila takahashii* and *Drosophila lutescens*. *Japanese Journal of Genetics* **57**, 575–580.

Kimura, M. T. (1988). Adaptations to temperate climates and evolution of overwintering strategies in the *Drosophila melanogaster* species group. *Evolution* **42**, 1288–1297.

Kimura, M. T. and Beppu, K. (1993). Climatic adaptations in the *Drosophila immigrans* species group – seasonal migration and thermal tolerance. *Ecological Entomology* **18**, 141–149.

Kimura, M. T. (2004). Cold and heat tolerance of drosophilid flies with reference to their latitudinal distributions. *Oecologia* **140**, 442–449.

Koštál, V., Berko, P. and Šimek, P. (2003). Remodelling of membrane phospholipids during transition to diapause and cold-acclimation in the larvae of *Chymomyza costata* (Drosophilidae). *Comparative Biochemistry and Physiology B* **135**, 407–419.

Kristensen, T. N., Loeschcke, V. and Hoffmann, A. A. (2007). Can artificially selected phenotypes influence a component of field fitness? Thermal selection and fly performance under thermal extremes. *Proceedings of the Royal Society, Series B* **274**, 771–778.

Kristensen, T. N., Hoffmann, A. A., Overgaard, J., Sørensen, J. G., Hallas, R. and Loeschcke, V. (2008a). Costs and benefits of cold acclimation in field-released *Drosophila*. *Proceedings of the National Academy of Sciences, USA* **105**, 216–221.

Kristensen, T. N., Barker, J. S.F., Pedersen, K. S. and Loeschcke, V. (2008b). Extreme temperatures increase the deleterious consequences of inbreeding under laboratory and semi-natural conditions. *Proceeding of the Royal Society B* **275**, 2055–2061.

Laayouni, H., Garcia-Franco, F., Chavez-Sandoval, B. E., Trotta, V., Beltran, S., Corominas, M. and Santos, M. (2007). Thermal evolution of gene expression profiles in *Drosophila subobscura*. *BMC Evolutionary Biology* **7**, 42.

Laine, A. L. (2007). Detecting local adaptation in a natural plant-pathogen metapopulation: a laboratory vs. field transplant approach. *Journal of Evolutionary Biology* **20**, 1665–1673.

Lee, R. E. and Denlinger, D. L. (1985). Cold tolerance in diapausing and non-diapausing stages of the flesh fly, *Sarcophaga crassipalpis*. *Physiological Entomology* **10**, 309–315.

Lee, R. E. (1991). Principles of insect low temperature tolerance. In *Insects at Low Temperature*, ed. Lee, R. E. and Denlinger, D. L. New York: Chapman and Hall, pp. 17–47.

Loeschcke, V. and Hoffmann, A. A. (2007). Heat hardening benefits and costs on field fitness of *Drosophila* depend on environmental temperature. *American Naturalist* **169**, 175–183.

McColl, G. and McKechnie, S. W. (1999). The *Drosophila* heat shock hsr-omega gene: An allele frequency cline detected by quantitative PCR. *Molecular Biology and Evolution* **16**, 1568–1574.

McMullen, D. C. and Storey, K. B. (2008). Mitochondria of cold hardy insects: Responses to cold and hypoxia assessed at enzymatic, mRNA and DNA levels. *Insect Biochemistry and Molecular Biology* **38**, 367–373.

Morgan, T. J. and Mackay, T. F.C. (2006). Quantitative trait loci for thermotolerance phenotypes in *Drosophila melanogaster*. *Heredity* **96**, 232–242.

Nedved, O. (2000). Snow white and the seven dwarfs: A multivariate approach to classification of cold tolerance. *CryoLetters* **21**, 339–348.

Nicodemus, J., O'Tousa, J. E. and Duman, J. G. (2006). Expression of a beetle, *Dendroides canadensis*, antifreeze protein in *Drosophila melanogaster*. *Journal of Insect Physiology* **52**, 888–896.

Norry, F. M. and Loeschcke, V. (2002). Longevity and resistance to cold stress in cold-stress selected lines and their controls in *Drosophila melanogaster*. *Journal of Evolutionary Biology* **15**, 775–783.

Norry, F. M., Gomez, F. H. and Loeschcke, V. (2007). Knock down resistance to heat stress and slow recovery from chill coma are genetically associated in a quantitative trait locus region of chromosome 2 in *Drosophila melanogaster*. *Molecular Ecology* **16**, 3274–3284.

Ohtsu, T., Kimura, M. T. and Katagiri, C. (1998). How *Drosophila* species acquire cold tolerance – Qualitative changes of phospholipids. *European Journal of Biochemistry* **252**, 608–611.

Ohtsu, T., Katagiri, C. and Kimura, M. T. (1999). Biochemical aspects of climatic adaptations in *Drosophila curviceps*, *D. immigrans*, and *D. albomicans* (Diptera: Drosophilidae). *Environmental Entomology* **28**, 968–972.

Overgaard, J., Malmendal, A., Sørensen, J. G., Bundy, J. G., Loeschcke, V., Nielsen, N. C. and Holmstrup, M. (2007). Metabolomic profiling of rapid cold hardening and cold shock in *Drosophila melanogaster*. *Journal of Insect Physiology* **53**, 1218–1232.

Qin, W., Neal, S. J., Robertson, R. M., Westwood, J. T. and Walker, V. K. (2005). Cold hardening and transcriptional change in *Drosophila melanogaster*. *Insect Molecular Biology* **14**, 607–613.

Rako, L., Anderson, A. R., Sgro, C. M., Stocker, A. J. and Hoffmann, A. A. (2006). The association between inversion In(3R)Payne and clinally varying traits in *Drosophila melanogaster*. *Genetica* **128**, 373–384.

Rako, L., Blacket, M. J., McKechnie, S. W. and Hoffmann, A. A. (2007). Candidate genes and thermal phenotypes: identifying ecologically important genetic variation for thermotolerance in the Australian *Drosophila melanogaster* cline. *Molecular Ecology* **16**, 2948–2957.

Rinehart, J. P., Hayward, S. A. L., Elnitsky, M. A., Sandro, L. H., Lee, R. E. and Denlinger, D. L. (2006). Continuous up-regulation of heat shock proteins in larvae, but not adults, of a polar insect. *Proceedings of the National Academy of Sciences, USA* **103**, 14223–14227.

Schmidt, P. S., Matzkin, L., Ippolito, M. and Eanes, W. F. (2005). Geographic variation in diapause incidence, life-history traits, and climatic adaptation in *Drosophila melanogaster*. *Evolution* **59**, 1721–1732.

Schmidt, P. S. and Conde, D. R. (2006). Environmental heterogeneity and the maintenance of genetic variation for reproductive diapause in *Drosophila melanogaster*. *Evolution* **60**, 1602–1611.

Schmidt, P. S. and Paaby, A. B. (2008). Reproductive diapause and life-history clines in North American populations of *Drosophila melanogaster*. *Evolution* **62**, 1204–1215.

Sezgin, E., Duvernell, D. D., Matzkin, L. M., Duan, Y. H., Zhu, C. T., Verrelli, B. C. and Eanes, W. F. (2004). Single-locus latitudinal clines and their relationship to temperate adaptation in metabolic genes and derived alleles in *Drosophila melanogaster*. *Genetics* **168**, 923–931.

Simmons, G. M., Kreitman, M. E., Quattlebaum, W. F. and Miyashita, N. (1989). Molecular analysis of the alleles of alcohol dehydrogenase along a cline in *Drosophila melanogaster*. 1. Maine, North Carolina, and Florida. *Evolution* **43**, 393–409.

Sinclair, B. J., Vernon, P., Klok, C. J. and Chown, S. L. (2003a). Insects at low temperatures: an ecological perspective. *Trends in Ecology & Evolution* **18**, 257–262.

Sinclair, B. J., Addo-Bediako, A. and Chown, S. L. (2003b). Climatic variability and the evolution of insect freeze tolerance. *Biological Reviews* **78**, 181–195.

Sinclair, B. J. and Chown, S. L. (2005). Climatic variability and hemispheric differences in insect cold tolerance: support from southern Africa. *Functional Ecology* **19**, 214–221.

Sinclair, B. J., Gibbs, A. G. and Roberts, S. P. (2007a). Gene transcription during exposure to, and recovery from, cold and desiccation stress in *Drosophila melanogaster*. *Insect Molecular Biology* **16**, 435–443.

Sinclair, B. J., Nelson, S., Nilson, T. L., Roberts, S. P. and Gibbs, A. G. (2007b). The effect of selection for desiccation resistance on cold tolerance of *Drosophila melanogaster*. *Physiological Entomology* **32**, 322–327.

Sørensen, J. G., Nielsen, M. M. and Loeschcke, V. (2007). Gene expression profile analysis of *Drosophila melanogaster* selected for resistance to environmental stressors. *Journal of Evolutionary Biology* **20**, 1624–1636.

Sørensen, J. G., Norry, F. M., Scannapieco, A. C. and Loeschcke, V. (2005). Altitudinal variation for stress resistance traits and thermal adaptation in adult *Drosophila buzzatii* from the New World. *Journal of Evolutionary Biology* **18**, 829–837.

Stanley, S. M. and Parsons, P. A. (1981). The response of the cosmopolitan species, *Drosophila melanogaster*, to ecological gradients. *Proceedings of the Ecological Society of Australia* **11**, 121–130.

Tucic, N. (1979). Genetic capacity for adaptation to cold resistance at different developmental stages of *Drosophila melanogaster*. *Evolution* **33**, 350–358.

Umina, P. A., Weeks, A. R., Kearney, M. R., McKechnie, S. W. and Hoffmann, A. A. (2005). A rapid shift in a classic clinal pattern in *Drosophila* reflecting climate change. *Science* **308**, 691–693.

Vera, J. C., Wheat, C. W., Fescemyer, H. W., Frilander M. J., Crawford, D. L., Hanski, I. and Narden, J. H. (2008). Rapid transcriptome characterization for a nonmodel organism using 454 pyrosequencing. *Molecular Ecology* **17**, 1636–1647.

Watson, M. J. O. and Hoffmann, A. A. (1996). Acclimation, cross-generation effects, and the response to selection for increased cold resistance in *Drosophila*. *Evolution* **50**, 1182–1192.

Watt, W. B. (1977). Adaptation at specific loci. 1. Natural selection on phosphoglucose isomerase of *Colias* butterflies – Biochemical and population aspects. *Genetics* **87**, 177–194.

Watt, W. B. (1983). Adaptation at specific loci. 2. Demographic and biochemical elements in the maintenance of the *Colias* polymorphism. *Genetics* **103**, 691–724.

Watt, W. B., Cassin, R. C. and Swan, M. S. (1983). Adaptation at specific loci. 3. Field behaviour and survivorship differences among *Colias* genotypes are predictable from in vitro biochemistry. *Genetics* **103**, 725–739.

Watt, W. B. (1992). Eggs, enzymes and evolution – natural genetic variants change insect fecundity. *Proceedings of the National Academy of Sciences, USA* **89**, 10608–10612.

Weeks, A. R., McKechnie, S. W. and Hoffmann, A. A. (2002). Dissecting adaptive clinal variation: markers, inversions and size/stress associations in *Drosophila melanogaster* from a central field population. *Ecology Letters* **6**, 756–763.

Wheat, C. W., Watt, W. B., Pollock, D. D. and Schulte, P. M. (2006). From DNA to fitness differences: Sequences and structures of adaptive variants of *Colias* phosphoglucose isomerase (PGI). *Molecular Biology and Evolution* **23**, 499–512.

Zachariassen, K. E. (1985). Physiology of cold tolerance in insects. *Physiological Reviews* **65**, 799–832.

12

Life-history adaptations to polar and alpine environments

PETER CONVEY

12.1 Introduction: extremes in the terrestrial environment

This chapter is concerned with the life-history features of terrestrial invertebrates inhabiting the cold regions of the world. It predominantly focuses on the Antarctic continent and the Arctic elements of the large northern continents, also drawing parallels with the alpine regions of the world's major mountain ranges. To human perception, these polar and montane regions of the planet are clearly challenging regions in which to live. They face environmental stresses that operate on a range of timescales, for example from chronic exposure to low temperature, high winds, freezing, or desiccation, to extreme or short-term acute events. At northern or southern latitudes beyond the polar circles, the sun remains permanently below the horizon for a period of days to months, depending on latitude, each winter, inevitably imposing considerable seasonality on organisms and ecosystems.

Focusing simply on temperature, in the absence of solar-energy input, terrestrial habitats of both regions face comparable extremely low air temperatures during winter. But the two regions are far from identical, with the Antarctic also enduring much lower typical summer temperatures than those of the Arctic (Convey, 1996a; Danks, 1999); hence lack of available energy provides a major constraint on biological activity here. However, the biological impacts of temperature are not well described simply by standard meteorological measures of mean air temperature, and scales and patterns of physical and temporal variation are also important. This is immediately clear in montane and alpine environments, where diurnal temperature ranges of tens of degrees centigrade are

Low Temperature Biology of Insects, ed. David L. Denlinger and Richard E. Lee Jr. Published by Cambridge University Press.

typical (Sømme, 1989). The importance of large variability in the thermal environment experienced in polar terrestrial habitats is also increasingly recognized (Peck *et al.*, 2006), and it can be of comparable or even greater magnitude to that seen at lower latitudes. Furthermore, at microhabitat scale, absorption of solar energy can result in short-term temperature maxima of 30–40 °C, sometimes more, at both High Arctic and Antarctic continental locations (Smith, 1988; Hodkinson, 2005a) – considerably greater than the average soil surface temperature under tropical rainforest canopies. Key variables of biological importance include upper and lower temperature extremes, temperature ranges over short and long timescales, rates of change and predictability (for general discussion see Gaines and Denny, 1993; Sinclair, 2001; Vasseur and Yodzis, 2004; Convey *et al.*, 2006a).

Polar and alpine habitats and their biota face other challenges from the physical environment. Central amongst these is absence of liquid water in many habitats. This is either because it remains biologically unavailable in the form of ice, even in the summer months, or because the low relative humidity of the air that is typical of both the polar latitudes and high altitudes results in rapid evaporation or sublimation of liquid water or snow, respectively. Indeed, much of the total area of the entire Antarctic continent receives so little precipitation on an annual basis that it can be classified as a frigid desert (Sømme, 1995), including by far the largest ice-free areas of the continent in the Dry Valleys of Victoria Land. Understanding patterns of water availability is now seen as being as, if not more, important than temperature in the biology of polar and alpine terrestrial biota (Kennedy, 1993; Sømme, 1995; Block, 1996; Hodkinson *et al.*, 1998, 1999).

The third major physical variable that can act as an environmental selection pressure is solar radiation. As indicated above, at high latitudes, this is a key driver of annual patterns of temperature while, at smaller physical and temporal scales, absorption of solar radiation by rock, soil and vegetation substrata leads to microclimatic conditions that can be considerably different from the general macroclimate. Here, two elements may be significant to terrestrial invertebrates. Solar radiation levels at ground level are obviously influenced by variables, including cloud, snow cover and albedo. While irradiance itself is clearly of primary importance to autotrophic primary producers, the latter provide the nutrients and, often, the habitat required by decomposers, consumers and higher trophic levels. Also, parts of the solar-radiation spectrum, in particular high-energy shorter-wavelength ultraviolet (UV) radiation, are potentially biologically damaging, both to autotrophs and heterotrophs (Caldwell *et al.*, 1998). In a polar context, attention has been focused in recent decades on changes in exposure to UV-B radiation as a consequence of anthropogenic ozone depletion (Farman *et al.*, 1985; Rozema, 1999). However, exposure to these wavelengths also increases naturally with increasing altitude (Hodkinson, 2005b).

The various environmental conditions experienced in polar and alpine regions, and in particular the Antarctic, are at one extreme of the range seen across the planet (Peck *et al.*, 2006; Chown and Convey, 2007). This leads to an evolutionary prediction that the strong selection pressures generated by these environments may lead to clearer adaptations than elsewhere. Any such adaptations are also particularly pertinent in identifying potential species and higher-level (ecosystem) responses to current trends of global environmental change, as the polar regions are both facing some of the fastest rates of change currently seen, and are proposed to provide particularly sensitive biological indicators of the consequences of these changes (e.g. Callaghan and Jonasson, 1995; Freckman and Virginia, 1997; Walther *et al.*, 2002; Convey *et al.*, 2003; Convey, 2006) – for instance, a small increase in temperature in an environment where organisms are near to their physiological thresholds for activity will have a greater consequence than the same magnitude of change in a non-threshold environment (Convey, 2001a).

12.2 General life-history strategy models

The search for general patterns has long been a feature of ecological research (Lawton, 1999), including more specifically soil ecology (Hodkinson and Wookey, 1999). The concept of 'life-history strategies' has proved fruitful in this context. Building on the ideas of MacArthur and Wilson (1967), Southwood (1977, 1988), Greenslade (1983) and Heal and Ineson (1984) are amongst those who have proposed evolutionary models linking environmental factors and the evolution of adaptive life-history strategies. These almost exactly paralleled developments in the botanical literature (see Grime, 1988). These general models recognize three major drivers of life-history evolution, relating to the potential for population growth, competitive ability and survival of adversity. Ricklefs and Wikelski (2002) suggest a different and simple rationalization, noting that most life-history variation lies along a "slow–fast continuum", with slow development, low reproductive rate and long lifespan at one end, and the opposite characteristics at the other. Hodkinson and Wookey (1999), noting that many classification systems are "organism-centered", highlighted the need for, and proposed, a "process-centered" or functional classification scheme applicable to tundra ecosystems. Their scheme is centered on four functional axes, relating to resource-use flexibility, ecophysiological flexibility, dispersability and population responsiveness.

Attempts to place the life-history strategies of polar terrestrial invertebrates within general models generally results in the identification of "adversity" or "stress" selected strategies (Convey, 1996b; Peck *et al.*, 2006; see also Panikov (1995) for application of analogous concepts to tundra soil microbial communities). These are characterized in invertebrates by considerable physiological/biochemical investment in stress-tolerance tactics, reduced or no competitive or dispersal ability, low reproductive investment and extended life cycles (Cannon and Block, 1988; Block, 1990; Sømme, 1995; Convey, 1996b, 1997, 2000). A common feature in the strategies of many polar invertebrates is considerable flexibility in various aspects of life history and physiology (Convey, 1996b; Hodkinson and Wookey, 1999; Hodkinson, 2005a). This feature, while not providing an example of a "neat" single adaptation, appears to be particularly advantageous in environments that are both towards the environmental limits permitting biological activity, and also experience typically rapid and unpredictable variation, at least at a spatial and temporal scale relevant to the biota.

An important foundation of these general life-history models is that they have an evolutionary basis; in other words, it is assumed that natural variation in the life-history features identified is underlain at least in part by a heritable component of genetic variation, without which the evolutionary process of natural selection cannot operate. It should be recognized at the outset that no studies exist of the heritability of traits argued to be components of adaptive life-history strategies in polar invertebrates. However, in some areas there is strong circumstantial evidence including, for instance, the typically considerable ecophysiological investment permitting cold- and desiccation-tolerance that indicates enhancement of ancestral capabilities. In other areas an evolutionary explanation may be inappropriate, even for features that appear consistent with adversity or stress selection. It is often assumed that life-cycle extension is a response to extreme polar environments. However, a much simpler explanation is that low levels of energy available when invertebrates can be active, combined with extended winter inactivity, are a direct and sufficient constraint to development (Norton, 1994).

12.3 Life-cycle features

12.3.1 Life-cycle length

The highly seasonal terrestrial environments of the Antarctic continent (usually described as including the maritime Antarctic and continental Antarctic; Smith, 1984; Convey, 2001b) and also the High Arctic typically have short summers with conditions suitable for invertebrate growth and development, and very long winters of inactivity, likely including enforced resource depletion. However, even the most northern ice-free ecosystems in the Arctic (80–84°N, including Ellesmere Island, northern Greenland, Svalbard, Franz Josef Land, Novaya and Severnaya Zemlya) experience positive summer monthly mean air temperatures several degrees above 0 °C. This is considerably greater than mean summer air temperatures throughout the continental Antarctic (typically well below 0 °C,

(a)

Figure 12.1(a) Ice-free ground at high latitudes in the Antarctic is generally restricted to small "islands", typically nunataks, surrounded by thick glaciers and ice sheets; even where larger exposures exist, such as here in Davis Valley, Pensacola Mts (*c.* 83°S), soil and rock surfaces are visually barren and even on close examination biodiversity is very low, limited to very small lichens, soil microbiota and microinvertebrates (Photos: P. Convey).

and only reaching 0–1 °C at certain locations along the continental coastline at *c.* 65–70°S) or even at much lower latitudes in the maritime Antarctic (55–72°S), where mean temperatures are typically 1–3 °C for 2–4 months during the short summer. Summer temperatures at lower Arctic latitudes are warmer again. Thus, there is much more limited comparability between the environmental conditions and constraints faced by the two polar regions than is commonly thought (see Fogg *et al.*, 2008, Chapter 1, for an overview of climate and energy balance of the two polar regions). As a result, both the plant and animal elements of terrestrial ecosystems show much greater diversity, complexity and biomass at comparable latitudes in the Arctic than in the Antarctic (Fig. 12.1).

Thus, in much of the Antarctic and parts of the Arctic, severe energy limitation means that only a proportion of the development required to complete the life cycle is possible within a single summer season, and extended life cycles are the obvious consequence. Any extension of development times (i.e. adult-adult) beyond the annual cycle leads automatically to a requirement for overwintering ability in more than one developmental stage. This is an unusual though not unknown feature in insects generally. If the overwintering stage is considered across all insects,

(b)

Figure 12.1(b) In contrast, considerably greater community development and biomass is present at the highest Arctic latitudes, here illustrated near Ny Ålesund, Svalbard (*c.* 79°N).

it is clear that there are examples (including Arctic species) of overwintering in any instar. For instance, different European Lepidoptera overwinter as the egg, larva (different instars), pupa, or adult. Although there are obvious and well-known exceptions (e.g. tropical cicadas, wood-feeding beetles), the vast majority of temperate and tropical insects complete at least one and sometimes two or more full generations in a year, meaning that a constraint of overwintering (or aestivating) in a single specific instar does not preclude their presence in a particular location.

Overwintering in a specific instar also automatically results in a degree of developmental synchronization at population level. With progression toward higher-latitude temperate regions, the potential for life-history flexibility is illustrated by, for instance, northern hemisphere Lepidoptera and Odonata species that have annual life cycles towards the south of their range, and biennial or longer life cycles towards the north. This is achieved by possessing two or occasionally more life stages with overwintering ability. However, in both polar and alpine regions considerably longer and typically "free running" life cycles are the norm (see below), often lasting at least several years and requiring overwintering in several if not all life stages, and sometimes more than once between moults (Kevan and Kukal, 1993; Sømme, 1989; Danks, 1992, 2004; Convey, 1996b; Bale *et al.*, 1997; Hodkinson and Wookey, 1999) (Fig. 12.2). This is exceptional in the context of general arthropod biology, and appears to indicate the loss of a developmental feature that is "fixed" in the life histories of most species, and to be an illustration

(a)

Figure 12.2 Life-cycle extension is a common feature of polar terrestrial invertebrate life histories: (a) SEM photograph of an adult of the common maritime Antarctic mite, *Alaskozetes antarcticus*, whose life cycle is thought to extend to at least 7 years (photo: BAS). (b) Mature larva of the High Arctic lymantriid moth, *Gynaephora groenlandica*, preparing its cocoon before pupation; the life cycle of this species may extend to more than 10 years under certain combinations of environmental and biological (parasitoid) stresses (photo: R.E. Lee). (c) The sub-Antarctic midge *Eretmoptera murphyi*, native and endemic to South Georgia, is an obligate parthenogenetic species, a characteristic that has facilitated its successful establishment on the more extreme maritime Antarctic Signy Island following inadvertent human introduction; it is likely to have been introduced as larvae amongst soil and moss transferred in plant transplant experiments in the 1960s (photo: BAS). (d) The wingless chironomid midge, *Belgica antarctica* (endemic to the Antarctic Peninsula and South Shetland Islands; photo: R.E. Lee) and (e) larvae of the sub-Antarctic perimylopid beetle, *Hydromedion sparsutum*, are examples of species showing flexibility in the expression of cold-tolerance strategy use (photo: BAS). (f) Introduced (non-native) invertebrates can lead to considerable impacts on ecosystem structure and function. For instance, predatory carabid beetles (here illustrated by *Trechisibus antarcticus*) have been introduced to sub-Antarctic South Georgia and Kerguelen, where the native terrestrial ecosystems do not contain this functional group of predators. Impacts have included local extinction of native (including endemic) prey invertebrates, and life-history alterations in prey species (photo: BAS).

Figure 12.2 *(cont.)*

(e)

(f)

Figure 12.2 *(cont.)*

of the increased flexibility in polar life-history strategies highlighted above. This can also be understood as a selective or biodiversity filter (Bergstrom *et al*., 2006), which limits successful establishment in polar terrestrial environments largely to species with life cycles that do not contain specific overwintering instars.

12.3.2 Diapause and quiescence

In species of high northern latitudes, and also of alpine environments, true diapause is a common feature of life cycles (Bale *et al*., 1997; Danks 2004), and serves an important function in synchronizing life cycles. This is not the case in southern polar regions (Convey, 1996b), and this difference may have several plausible explanations. In the more extreme regions of the maritime and continental Antarctic, summer conditions are typically considerably colder and nearer to physiological and activity thresholds than is the case even in the High Arctic. Even on a daily basis during summer there may be a risk of challenge from freezing conditions at microhabitat level, which means that it is difficult to identify a reliable cue for diapause initiation (with associated costs) without there being a high risk of this being triggered inappropriately. The potential costs of this are compounded in some cases by the extremely short period during which environmental conditions may permit any activity in terrestrial invertebrates, which may be as little as days to only a few weeks in the entire season (e.g. Sømme, 1986; Moorhead *et al.*, 2002). Alternatively, the very limited environmental variation seen on an annual basis in the sub-Antarctic islands may remove the need for any form of structured overwintering strategy, and there are few examples demonstrated of true diapause in the insects of this region (see Arnold and Convey, 1998). Finally, the general lack of reports of diapause in more extreme Antarctic regions may be a consequence of there being few representatives of "higher" invertebrate species within their fauna, whose higher taxonomic groups provide most reported examples of this feature globally.

12.3.3 Synchronization

With or without the presence of defined overwintering stages or true diapause, environmental seasonality can lead to synchronization of important life-cycle events, such as molting or reproduction. In the highly seasonal habitats of the maritime Antarctic, where invertebrates typically have multi-year life cycles and repeat overwintering, seasonality leads to early summer synchronization of molting by most individuals of some oribatid mite species, as the extended pre-molt preparation phase that is a characteristic of this mite group means that they almost inevitably enter this state in the latter part of summer and are then unable to molt until the following spring (Convey, 1994). Recent research further indicates that molt preparation (in this case in Collembola) is also advantageously, but possibly incidentally, linked with development of appropriate levels of cold-tolerance, thus assisting in winter survival (Worland & Convey, 2008). Molt synchronization is also seen, this time at a later point in the season, in some High Arctic oribatids (Søvik and Leinaas, 2003; Søvik *et al.*, 2003). Synchronized emergence of adult insects is frequently seen in Arctic species, as is a peak of reproduction early in the summer season (Danks, 2004). However, life cycles of the majority of invertebrates in these regions are also described as free running, with considerable overlap of generations, and populations with mixed-age structures (Convey, 1996b; Bale *et al.*, 1997; Hodkinson and Wookey, 1999; Søvik *et al.*, 2003; Hodkinson, 2005a).

The intense seasonality combined with low growth potential during short summers is not the only set of environmental features that may underlie free-running life cycles in invertebrates of the polar regions. Terrestrial ecosystems of the sub-Antarctic islands generally display much depressed levels of seasonality in comparison with other polar regions, either north or south (Convey, 1996a; Danks, 1999). Within this region, different islands experience different levels of seasonal variation depending on whether they are located north (less seasonality) or south (more seasonality) of the oceanic Antarctic Polar Frontal Zone. This permits comparison of life-history features of species occurring on islands on both sides of this Zone. Chown and Klok (2003) and Chown *et al.* (2006) report examples of a weevil (*Palirhoeus eatoni*) and a moth (*Embryonopsis halticella*), that occur on both Marion and Heard Islands, with overlapping generations typical of both species on the former, but not on the latter. Here, the very limited seasonal variation at microhabitat level on the northern Marion Island provides no pressure leading to synchronization, while the relatively greater amount of energy available within this region generally, combined with the greater seasonality on the southern Heard Island, permits synchronized annual life cycles in these higher insect species. Nevertheless, even in the sub-Antarctic, long life cycles are typical, with concomitant low reproductive rates, and shallow metabolic rate–temperature relationships (indicating limited capacity for more rapid development rates) (Crafford *et al.*, 1986; Davies, 1987; Chown and Scholtz, 1989; Barendse and Chown, 2000; Addo-Bediako *et al.*, 2002).

Analogous examples of univoltine life cycles within regions where life cycle extension is more typical are also well known from studies in Arctic and alpine regions (Sømme, 1989; Danks, 1992; Hodkinson, 2005a). There are very few examples of univoltine or even semivoltine invertebrates in the more extreme maritime and continental Antarctic regions, although some do exist including the mesostigmatid mite, *Gamasellus racovitzai*, and the endemic chironomid midge, *Belgica antarctica*, both thought to have biennial life cycles (Sugg *et al.*, 1983; Lister, 1984; Convey and Block, 1996).

At species level, although relatively few autecological studies of polar terrestrial invertebrates have been completed on year-round timescales (Hogg *et al.* 2006), there is evidence for seasonal variation in abundance, ranging from species showing little or no variation over the annual cycle, to examples of others peaking in virtually any season (West, 1982; Bellido and Cancela da Fonseca, 1988; Barendse and Chown, 2001). However, these examples refer to the sub-Antarctic islands, and it is not clear whether these patterns indicate flexibility within the life cycle, or are an element of a programmed strategy (Convey, 1996a,b; Danks, 1999; Barendse and Chown, 2001). While year-round studies of maritime Antarctic and High Arctic arthropods have been undertaken, the practicalities of winter sampling of inactive invertebrates in these environments means that such studies cannot provide abundance estimates over the annual cycle (e.g. Convey, 1994; Søvik *et al.*, 2003).

12.3.4 Reproduction

As a broad generalization, and consistent with the features of adversity-selected life-history strategies, reproductive output of polar and alpine invertebrates is lower than in the same or related species from less stressful environments (Danks, 1981; Convey, 1996b; Hodkinson, 2005b), although in reality this statement is based on remarkably few examples studied in detail.

The ability to reproduce parthenogenetically is an ancestral feature present in many of the major groups of invertebrates that are components of ecosystems in more extreme environments, including the Diptera, Collembola, Acari, Nematoda, Rotifera and Tardigrada (Fig. 12.2c). In some cases, both parthenogenetic and sexual abilities are present within a single species (e.g. Nematoda – Andrássy, 1998; Tardigrada – Bertolani, 2001), but it is not clear whether the occurrence of both sexes, or the proportions found, are influenced by the conditions of polar or other extreme environments. Asexual reproduction brings the important practical benefits of potentially rapid increase in numbers, the retention of effective local adaptations, and removing the need to locate mates, all of which can be seen as advantageous in more extreme environments where physical stresses dominate over biological. Conversely, in circumstances where biotic interactions, particularly competition, become more important, the increased genetic variation generated by sexual reproduction is expected to lead to asexual strains being out-competed over evolutionary time.

Parthenogenetic species are thought to be disproportionately common amongst the faunas of extreme environments (Duckhouse, 1985), but they remain in the minority. In a slightly different context, the ability to reproduce parthenogenetically is clearly advantageous in facilitating the establishment of colonizing individuals, and it underlies the success of a number of inadvertent human-assisted introductions of non-native invertebrates into the sub- and maritime Antarctic environments (Delettre and Tréhen, 1977; Block *et al.*, 1984; Crafford, 1986; Frenot *et al.*, 2005).

Aphids (Hemiptera) are amongst the insects introduced to the sub-Antarctic, where they do not occur naturally. Their post establishment (and indeed global) success relies on the ability to reproduce both asexually (giving rapid population growth in summer) and sexually (giving overwinter survival and genetic variation for subsequent seasons). In contrast, aphids are part of the native fauna of the Arctic. *Acyrthosiphon svalbardicum*, a species endemic to the High Arctic Svalbard archipelago, shows a life-cycle feature that is unique within the aphids generally, with a proportion of individuals passing directly from the fundatrix to the sexual morphs, which produce the next generation of overwintering eggs without passing through intermediate asexual instars (Strathdee, *et al.* 1993a,b, 1995; see also Simon *et al.*, 2008). This guarantees survival over the subsequent winter, while also allowing the asexual individuals that are produced to take advantage of particularly mild summer conditions when they do occur to complete an additional generation, which increases the number of overwintering eggs by an order of magnitude. A second closely related and also endemic Svalbard species, *A. calvulus*, appears to have lost the ability to produce viviparae completely, as part of a life cycle that is closely synchronized with host plant phenology (Gillespie *et al.*, 2007).

12.3.5 Investment in physiological stress responses

The biochemical and physiological detail of stress responses in invertebrates is described elsewhere in this volume and is not central to this chapter. Rather, in this sub-section, we consider the implications of investment in the various stress-tolerance tactics to the overall life history. In the context of cold-tolerance, two broad strategies are widely recognized, termed "freezing-intolerance" (or "freeze-susceptibility") and "freezing-tolerance". In the former, any formation of ice within the animal's body is considered to be fatal, and this is avoided by a range of biochemical protection strategies. In the latter, extracellular freezing is tolerated (even encouraged by the use of ice-nucleating agents to seed ice-crystal formation at high sub-zero temperatures at which control of crystal growth is achievable), with the animal subsequently remaining viable and able to recover from exposure to (sometimes much) lower temperatures. A third strategy, termed "cryoprotective dehydration" has been described from a small number of Arctic springtails (Holmstrup and Sømme, 1998; Worland *et al.*, 1998) and the Antarctic midge, *Belgica antarctica* (Elnitsky *et al.*, 2008). Desiccation-stress is also a major challenge to invertebrates and, again, resistance and tolerance adaptations are well developed. At the level of cellular biochemistry, the details of responses to cold-stress and to desiccation-stress, both of which in essence are experienced as cellular dehydration, are very similar. This leads logically to the idea that tolerance of desiccation, which is likely to have been a relatively early evolutionary adaptation, may also then have provided a pre-adaptation in the terms of tolerance of exposure to cold (Ring and Danks, 1994). These ecophysiological strategies form the subject of a very large research literature, and many reviews (e.g. Block, 1990; Bale, 2002; Sinclair *et al.*, 2003a; Danks, 2005; Hennion *et al.*, 2006) provide good starting points to placing these strategies in broader ecological and also polar contexts.

An inherent feature of stress-tolerance strategies is that they share some form of biochemical cost, related to the synthesis and subsequent breakdown of protectant chemicals. Cryoprotectants such as glycerol or trehalose may contribute several percent of an animal's body mass while in the resistant state (Cannon and Block, 1988; Block, 1990; Convey, 1996b), while recent genomic studies illustrate that expression of elements of biochemical protection strategies, such as heat-shock proteins, may be permanently up-regulated in some life stages (e.g. Rinehart *et al.*, 2006; Lopez-Martinez *et al.*, 2008). However, the actual biochemical cost to the organism's energy budget does not appear to have been quantified. Furthermore, while the seasonal use of stress-tolerance strategies has long been recognized, more recently, the potential for rapid (and therefore repeated) triggering of stress responses has started to receive attention (e.g. Worland and Convey, 2001; Sinclair *et al.*, 2003b; Chown and Nicolson, 2004; Lee *et al.*, 2006; see also Chapter 2). In the absence of direct measurements or estimates of biochemical cost, circumstantial evidence is provided by life-history "trade-offs", such as described in the Antarctic oribatid mite *Alaskozetes antarcticus*, where populations from milder sub-Antarctic habitats appear to be able to invest greater resources in egg production than those from the more stressful maritime Antarctic (Convey, 1998). More generally, high-latitude arthropods typically have lower levels of reproductive output than their lower-latitude relatives (Danks, 1981; Convey, 1996b).

A particularly unusual form of reducing biochemical costs appears to have been developed by larvae of the Arctic lymantriid moth, *Gynaephora groenlandica* (Fig. 12.2b). This species is already recognized as being unusual in that its overwintering larvae are one of the largest insects known to use anhydrobiosis in order to survive the long winter period, as well as overwintering multiple times during their very extended life cycle (Kevan and Kukal, 1993). However, during the preparation phase before entry into anhydrobiosis, these larvae appear to reduce metabolic maintenance costs by dismantling approximately half of the mitochondria in cells (Kukal *et al.*, 1989; Bennett *et al.*, 2000; Levin *et al.*, 2003).

Further subtleties in strategy expression have also been identified (Fig. 12.2d,e). Bale *et al.* (2001), working on the sub-Antarctic perimylopid beetle *Hydromedion sparsutum*, showed that following repeated exposure to low temperatures in the field, two strategies became apparent, which can be seen as "bet-hedging" – individuals either retained a relatively high freezing point and were moderately freeze-tolerant and capable of surviving repeated freezing events, or they substantially depressed their freezing point, but were unable to survive freezing. Features such as this variability in life-history strategy expression remain topics urgently requiring research effort, not least as such flexibility in tactic use is likely to underlie responses to environmental change.

While research into stress-tolerance strategies has tended to concentrate on levels of stress sustainable before lethal effects take place, sub-lethal effects are also of considerable importance (Chen *et al.*, 1987; McDonald *et al.*, 1997, 2000). This remains a poorly investigated subject, in polar and alpine regions or elsewhere (Brown *et al.*, 2004, provide an exception). One study on sub-Antarctic Marion Island has examined the effect of repeated exposure to freezing events on larvae of the endemic moth, *Pringleophaga marioni*. While not finding any influence of repeated exposure to freezing conditions on survival, this experimental regime did result in decreased mass, gut mass and feeding activity (Sinclair and Chown, 2005a), indicating a real cost to repeated stress exposure. This cost was also prolonged, with larvae taking about one month to regain the lost mass and hence presumably generating a delay to the life history. Further autecological investigations demonstrated that larvae of this moth appear to avoid this potential environmentally imposed cost by selectively feeding on the material used in the nests of the wandering albatross (*Diomedea exulans*), which are maintained *c.* 5 °C higher than the surrounding environment (Sinclair and Chown, 2005b).

12.3.6 Competition, predation

A feature of organisms displaying "adversity-selected" life-history strategies is that they show very limited competitive abilities. In that selective pressures in polar terrestrial ecosystems are thought to be dominated by physical environmental stressors, it is already clear that biological stressors such as competition and trophic interactions (grazing, predation, parasitism) are likely to play a minor role in life-history strategies. As these activities are rare (or at least perceived to be so, see Hogg *et al.*, 2006), they have unfortunately received little research attention even though, for instance, predatory groups such as spiders and beetles are integral in both sub-Antarctic and Arctic terrestrial faunas (e.g. Gressitt, 1970; Convey, 2007; Coulson, 2007), along with predatory microarthropods and elements of the meiofauna. The simplest faunal communities yet described on the planet, lacking even the nematodes, include a predatory tardigrade (Convey and McInnes, 2005). Furthermore, in the context of current and rapid environmental changes affecting both polar regions, the importance and impacts of non-indigenous species are being recognized – many of these are predators or strong competitors, and they can have major impacts on indigenous species and communities (Frenot *et al.*, 2005; Convey *et al.*, 2006b) (Fig. 12.2f).

Direct evidence for the role of competition in native polar terrestrial invertebrate communities is remarkably limited. Davies (1987) and Chown (1992) inferred evidence from size distributions of sympatric beetle species on two sub-Antarctic islands. A study of the theoretical impacts of predation by mesostigmatid mites on their primary springtail prey in the maritime Antarctic suggested that, at current population levels, individual predator consumption rates could be supported by the lowest prey field densities (Lister *et al.*, 1988), in other words that predation *per se* was largely irrelevant in these microarthropod communities. However, a linked study also demonstrated evidence of prey-switching by the same mite species in the presence of a second predatory mite (Usher *et al.*, 1989), suggesting a role for competition. Set against these is the observation that predatory groups are amongst the first and most numerous of invertebrate groups to colonize recently exposed ground near retreating glaciers on Arctic Svalbard, even before the establishment of primary producers (Hodkinson *et al.*, 2001, 2002). There are thought to be few true herbivores in the ecosystems of cold environments, and the vast majority of carbon and energy flow passes through the detritivore cycle (Davis, 1981). Herbivory may not currently be a viable strategy through energetic constraints (Leather *et al.*, 1993). This has led to an assumption that the most dominant arthropods in these ecosystems will be generalist feeders, largely utilizing microbial groups and detritus. However, again, few rigorous autecological studies have addressed this assumption, and the few that have (e.g. Worland and Lukesová, 2000, on Antarctic springtails) have identified an unexpected degree of dietary selectivity. This raises the possibility that competition may yet be found to play a significant role in these ecosystems.

12.4 Summary and future directions

The single most striking element of the life-history strategies of insects and other invertebrates inhabiting the extreme environments of the world's polar and alpine regions is the degree of flexibility shown. This becomes apparent in many different elements of the animals' biology, not all of which are customarily included in descriptions of "life-history strategies". It is abundantly clear also that no one description is appropriate for all species in the context of any one specific feature. However, within this flexibility lies the key to understanding organism response both to natural environmental variability, and to systematic environmental change. Future work integrating biological processes across the full spectrum of elements contributing to the overall life-history strategy (i.e. from the gene, through genomic and metabolomic processes underlying ecophysiology, to reproductive strategies and the life cycle) must be central to furthering understanding of responses to change.

The features generally attributed to "adversity" selected life-history strategies (considerable physiological/biochemical investment in stress-tolerance tactics, reduced or no competitive or dispersal ability, low reproductive investment, extended life cycles) are all frequently encountered in organisms of these regions. However, whether or not these features have an evolutionary basis remains unclear and, again, modern genomic approaches applied to both polar invertebrates and their lower latitude relatives provide the potential for a rapid advance in understanding.

In the context of overall life-history strategy, one of the most important (and under-researched) consequences of investment in stress-tolerance tactics relates to the way in which it can be traded off against other life-history functions in the organism's energy budget. It is axiomatic that this trade-off must be achievable in order for survival to be possible in any given location, while subtle changes in the balance between different life-history elements will also form an important element of any responses to systematic environmental change. Thus, research effort is required in order to describe and quantify organism energy budgets, and better understand the various trade-offs possible within them, particularly in relation to imposed physical and biological environmental stresses.

Ecophysiological and reproductive strategies, and overall life-cycle duration, adopted by many species of these environments show considerable flexibility. It is currently unclear whether this is underlain by genotypic or phenotypic factors (or both), although evolutionary models of life-history strategy development, if they are to be applied, require an element of the former.

Elements of the life-history strategies commonly reported, including low competitive ability, low growth rates and low reproductive output are likely to mean that terrestrial species and communities have limited resistance to invasion by other species currently restricted to less extreme (lower altitude or latitude) environments, where competitive abilities or ruderal strategies are better developed. This factor is likely to be particularly important in the context of current scenarios of regional climate change, and central amongst predictions deriving from these are more rapid rates of change in the high latitude regions of the planet. Cold environment biota, amongst which the insects and other invertebrates feature highly, are therefore key to future work aimed at identifying biological responses to this change, placing us in a better position to understand, predict and even possibly mitigate analogous responses to be expected in lower-latitude ecosystems, where the majority of the world's human population lives and relies on resources.

References

Addo-Bediako, A., Chown, S. L. and Gaston, K. J. (2002). Metabolic cold adaptation in insects: a large-scale perspective. *Functional Ecology* **16**, 332–338.

Andrássy, I. (1998). Nematodes in the sixth continent. *Journal of Nematode Systematics and Morphology* **1**, 107–186.

Arnold, R. J. and Convey, P. (1998). The life history of the diving beetle, *Lancetes angusticollis* (Curtis) (Coleoptera: Dytiscidae), on sub-Antarctic South Georgia. *Polar Biology* **20**, 153–160.

Bale, J. S. (2002). Insects and low temperatures: from molecular biology to distributions and abundance. *Philosophical Transactions of the Royal Society of London, Series B* **357**, 849–862.

Bale, J. S., Hodkinson, I. D., Block, W., Webb, N. R., Coulson, S. J. and Strathdee, A. T. (1997). Life history strategies of Arctic terrestrial arthropods. In *The Ecology of Arctic Environments*, ed. Woodin, S. J. and Marquis, M. British Ecological Society Special Publication No. 13, Oxford: Blackwell, pp. 137–165.

Bale, J. S., Worland, M. R., and Block, W. (2001). Effects of summer frost exposures on the cold tolerance strategy of a sub-Antarctic beetle. *Journal of Insect Physiology* **47**, 1161–1167.

Barendse, J. and Chown, S. L. (2000). The biology of *Bothrometopus elongatus* (Coleoptera, Curculionidae) in a mid-altitude fellfield on sub-Antarctic Marion Island. *Polar Biology* **23**, 346–351.

Barendse, J. and Chown, S. L. (2001). Abundance and seasonality of mid-altitude fellfield arthropods from Marion Island. *Polar Biology* **24**, 73–82.

Bellido, A. and Cancela da Fonseca, J. P. (1988). Spatio-temporal organization of the oribatid mite community in a littoral turf of the Kerguelen archipelago. *Pedobiologia* **31**, 239–246.

Bennett, V. A., Kukal, O., and Lee, R. E. (2000). Seasonal metabolic depression and mitochondrial degradation in Arctic woollybear caterpillars, *Gynaephora groenlandica*. *American Zoologist* **40**, 942.

Bergstrom, D., Hodgson, D. A. and Convey, P. (2006). The physical setting of the Antarctic. In *Trends in Antarctic Terrestrial and Limnetic Ecosystems: Antarctica as a Global Indicator*, ed. Bergstrom, D. M, Convey, P. and Huiskes, A. H. L. Dordrecht: Springer, pp. 15–33.

Bertolani, R. (2001). Evolution of the reproductive mechanisms in tardigrades – a review. *Zoologischer Anzeiger* **240**, 247–252.

Block, W. (1990). Cold tolerance of insects and other arthropods. *Philosophical Transactions of the Royal Society of London, Series B* **326**, 613–633.

Block, W. (1996). Cold or drought – the lesser of two evils for terrestrial arthropods? *European Journal of Entomology* **93**, 325–339.

Block, W., Burn, A. J. and Richard, K. J. (1984). An insect introduction to the maritime Antarctic. *Biological Journal of the Linnean Society* **23**, 33–39.

Brown, C. L., Bale, J. S. and Walters, K. F.A. (2004). Freezing induces a loss of freeze tolerance in an overwintering insect. *Proceedings of the Royal Society, Series B* **271**, 1507–1511.

Caldwell, M. M., Björn, L. O., Bornman, J. F., Flint, S. D., Kulandaivelu, G., Teramura, A. H. and Tevini, M. (1998). Effects of increased solar radiation on terrestrial ecosystems. *Journal of Phytochemistry and Photobiology B: Biology* **46**, 40–52.

Callaghan, T. V. and Jonasson, S. (1995). Arctic terrestrial ecosystems and environmental change. *Philosophical Transactions of the Royal Society of London, Series A* **352**, 259–276.

Cannon, R. J.C. and Block, W. (1988). Cold tolerance of microarthropods. *Biological Reviews* **63**, 23–77.

Chen, C.-P., Denlinger, D. L. and Lee, R. E. (1987). Cold-shock injury and rapid cold hardening in the flesh fly *Sarcophaga crassipalpis*. *Physiological Zoology* **60**, 297–304.

Chown, S. L. (1992). A preliminary analysis of weevil assemblages in the sub-Antarctic: local and regional patterns. *Journal of Biogeography* **19**, 87–98.

Chown, S. L. and Convey, P. (2007). Spatial and temporal variability across life's hierarchies in the terrestrial Antarctic. *Philosophical Transactions of the Royal Society of London, Series B* **362**, 2307–2331.

Chown, S. L., Greenslade, P. and Marshall, D. J. (2006). Terrestrial invertebrates of Heard Island. In *Heard Island: Southern Ocean Sentinel*, ed. Green, K. and Woehler, E. J. Chipping Norton: Surrey & Beatty, pp. 91–104.

Chown, S. L. and Klok, C. J. (2003). Altitudinal body size clines: latitudinal effects associated with changing seasonality. *Ecography* **26**, 445–455.

Chown, S. L. and Nicolson, S. W. (2004). *Insect Physiological Ecology. Mechanisms and Patterns*. Oxford: Oxford University Press.

Chown, S. L. and Scholtz, C. H. (1989). Biology and ecology of the *Dusmoecetes* Jeannel (Col. Curculionidae) species complex on Marion Island. *Oecologia* **80**, 93–99.

Convey, P. (1994). Growth and survival strategy of the Antarctic mite *Alaskozetes antarcticus*. *Ecography* **17**, 97–107.

Convey, P. (1996a). Overwintering strategies of terrestrial invertebrates in Antarctica – the significance of flexibility in extremely seasonal environments. *European Journal of Entomology* **93**, 489–505.

Convey, P. (1996b). The influence of environmental characteristics on life history attributes of Antarctic terrestrial biota. *Biological Reviews* **71**, 191–225.

Convey, P. (1997). How are the life history strategies of Antarctic terrestrial invertebrates influenced by extreme environmental conditions? *Journal of Thermal Biology* **22**, 429–440.

Convey, P. (1998). Latitudinal variation in allocation to reproduction by the Antarctic oribatid mite, *Alaskozetes antarcticus*. *Applied Soil Ecology* **9**, 93–99.

Convey, P. (2000). How does cold constrain life cycles of terrestrial plants and animals? *Cryo-Letters* **21**, 73–82.

Convey, P. (2001a). Terrestrial ecosystem response to climate changes in the Antarctic. In *"Fingerprints " of Climate Change – Adapted Behaviour and Shifting Species Ranges,* ed. Walther, G.-R., Burga, C. A., and Edwards, P. J. New York: Kluwer, pp. 17–42.

Convey, P. (2001b). Antarctic ecosystems. In *Encyclopedia of Biodiversity*, ed. Levin, S. A. San Diego: Academic Press, vol. 1, pp. 171–184.

Convey, P. (2006). Antarctic climate change and its influences on terrestrial ecosystems. In *Trends in Antarctic Terrestrial and Limnetic Ecosystems: Antarctica as a Global Indicator*, ed. Bergstrom, D. M, Convey, P. and Huiskes, A. H.L. Dordrecht: Springer, pp. 253–272.

Convey, P. (2007). Antarctic ecosystems. In *Encyclopedia of Biodiversity*, 2nd (online) edition, ed. Levin, S. A. San Diego: Elsevier. doi:10.1016/B0-12-226865-2/00014-6.

Convey, P. and Block, W. (1996). Antarctic Diptera: ecology, physiology and distribution. *European Journal of Entomology* **93**, 1–13.

Convey. P., Chown, S. L., Wasley, J. and Bergstrom, D. M. (2006a). Life history traits. In *Trends in Antarctic Terrestrial and Limnetic Ecosystems: Antarctica as a Global Indicator*, ed.

Bergstrom, D. M, Convey, P. and Huiskes, A. H. L. Dordrecht: Springer, pp. 101–127.

Convey, P., Frenot, Y., Gremmen, N. and Bergstrom, D. M. (2006b). Biological invasions. In *Trends in Antarctic Terrestrial and Limnetic Ecosystems: Antarctica as a Global Indicator*, ed. Bergstrom, D. M, Convey, P. and Huiskes, A. H. L. Dordrecht: Springer, pp. 193–220.

Convey, P. and McInnes, S. J. (2005). Exceptional, tardigrade dominated, ecosystems from Ellsworth Land, Antarctica. *Ecology* **86**, 519–527.

Convey, P., Scott, D. and Fraser, W. R. (2003). Biophysical and habitat changes in response to climate alteration in the Arctic and Antarctic. In *Climate Change and Biodiversity: Synergistic Impacts*, ed. Lovejoy, T. E. and Hannah, L. Arlington, VA: Conservation International, Center for Applied Biodiversity Science. *Advances in Applied Biodioversity Science* **4**, 79–84.

Coulson, S. J. (2007). The terrestrial and freshwater invertebrate fauna of the High Arctic archipelago of Svalbard. *Zootaxa* **1448**, 41–58.

Crafford, J. E. (1986). A case study of an alien invertebrate (*Limnophyes pusillus*, Diptera, Chironomidae) introduced on Marion Island: selective advantages. *South African Journal of Antarctic Research* **16**, 115–117.

Crafford, J. E., Scholtz, C. H. and Chown, S. L. (1986). The insects of sub-Antarctic Marion and Prince Edward Islands; with a bibliography of entomology of the Kerguelen Biogeographical Province. *South African Journal of Antarctic Research* **16**, 41–84.

Danks, H. V. (1981). *Arctic Arthropods, a Review of Systematics and Ecology with Particular Reference to the North American Fauna*. Ottowa: Entomological Society of Canada.

Danks, H. V. (1992). Long life cycles in insects. *Canadian Entomologist* **124**, 167–187.

Danks, H. V. (1999). Life cycles in polar arthropods – flexible or programmed? *European Journal of Entomology* **96**, 83–102.

Danks, H. V. (2004). Seasonal adaptations in Arctic insects. *Integrative & Comparative Biology* **44**, 85–94.

Danks, H. V. (2005). Key themes in the study of seasonal adaptations in insects I. Patterns of cold hardiness. *Applied Entomology and Zoology* **40**, 199–211.

Davies, L. (1987). Long adult life, low reproduction and competition in two sub-Antarctic carabid beetles. *Ecological Entomology* **12**, 149–162.

Davis, R. C. (1981). Structure and function of two Antarctic moss communities. *Ecological Monographs* **5**, 125–143.

Delettre, Y. R. and Tréhen, P. (1977). Introduction à la dynamique des populations de Limnophyes pusillus Eaton dans les sols des Iles Australes Antarctiques Françaises. *Ecological Bulletins* **25**, 80–89.

Duckhouse, D. A. (1985). Psychodidae (Diptera, Nematocera) of the subantarctic islands with observations on the incidence of parthenogenesis. *International Journal of Entomology* **27**, 173–184.

Elnitsky, M. A., Hayward, S. A.L, Rinehart, J. P., Denlinger, D. L. and Lee, E. E. Jr. (2008). Cryoprotective dehydration and the resistance to inoculative freezing in the Antarctic midge, *Belgica antarctica*. *Journal of Experimental Biology* **211**, 524–530.

Farman, J. C., Gardiner, B. G. and Shanklin, J. D. (1985). Large losses of total ozone in Antarctica reveal seasonal ClO_x/No_x interaction. *Nature* **315**, 207–210.

Fogg, G., Thomas, D. N., Convey, P., Fritsen, C., Gilli, J.-M., Gradinger, R., Laybourne-Parry, J., Reid, K. and Walton, D. W.H. (2008). *The Biology of Polar Habitats.* Oxford: Oxford University Press.

Freckman, D. W. and Virginia, R. A. (1997). Low-diversity Antarctic soil nematode communities: distribution and response to disturbance. *Ecology* **78**, 363–369.

Frenot, Y., Chown, S. L., Whinam, J., Selkirk, P., Convey, P., Skotnicki, M. and Bergstrom, D. (2005). Biological invasions in the Antarctic: extent, impacts and implications. *Biological Reviews* **80**, 45–72.

Gaines, S. D. and Denny, M. W. (1993). The largest, smallest, highest, lowest, longest, and shortest: extremes in ecology. *Ecology* **74**, 1677–1692.

Gillespie, M., Hodkinson, I. D., Cooper, E. J., Bird, I.M and Jonsdottir, I. S. (2007). Life history and host-plant relationships of the rare endemic Arctic aphid *Acyrthosiphon calvulus* in a changing environment. *Entomologia Experimentalis et Applicata* **123**, 229–237.

Greenslade, P. J. M. (1983). Adversity selection and the habitat templet. *American Naturalist* **122**, 352–365.

Gressitt, J. L. (ed.). (1970). Subantarctic entomology, particularly of South Georgia and Heard Island. Pacific *Insects Monograph* **23**, 374.

Grime, J. P. (1988). The C-S-R model of primary plant strategies – origins, implications and tests. In *Population Dynamics*, ed. Anderson, R. M., Turner, B. D. and Taylor, L. R. Oxford: Blackwell, pp. 123–139.

Heal, O. W. and Ineson, P. (1984). Carbon and energy flow in terrestrial ecosystems: relevance to the microflora. In *Current Perspectives in Microbial Ecology*, ed. Klug, M. J. and Reddy, C. A. Washington DC: American Society for Microbiology, pp. 394–404.

Hennion, F., Huiskes, A., Robinson, S. and Convey, P. (2006). Physiological traits of organisms in a changing environment. In *Trends in Antarctic Terrestrial and Limnetic Ecosystems: Antarctica as a Global Indicator*, ed. Bergstrom, D. M, Convey, P. and Huiskes, A. H. L. Dordrecht: Springer, pp. 129–159.

Hodkinson, I. D. (2005a). Adaptations of invertebrates to terrestrial Arctic environments. *Det Konelige Norske Videnskabers Selskab, Skrifter*, **45** pp.

Hodkinson, I. D. (2005b). Terrestrial insects along elevation gradients: species and community responses to altitude. *Biological Reviews* **80**, 489–513.

Hodkinson, I. D, Coulson, S. J., Harrison, J. A. and Webb, N. R. (2001). What a wonderful web they weave: spiders, nutrient capture and early ecosystem development in the high Arctic – some counter intuitive ideas on community assembly. *Oikos* **95**, 349–352.

Hodkinson, I. D., Webb, N. R., Bale, J. S., Block, W., Coulson, S. J. and Strathdee, A. T. (1998). Global change and Arctic ecosystems: conclusions and predictions from experiments with terrestrial invertebrates on Spitsbergen. *Arctic and Alpine Research* **30**, 306–313.

Hodkinson, I. D., Webb, N. R., Bale, J. S. and Block, W. (1999). Hydrology, water availability and tundra ecosystem function in a changing climate: the need for a closer integration of ideas? *Global Change Biology* **5**, 359–369.

Hodkinson, I. D., Webb, N. R. and Coulson, S. J. (2002). Primary community assembly on land – the missing stages: why are the heterotrophic organisms always there first? *Journal of Ecology* **90**, 569–577.

Hodkinson, I. D. and Wookey, P. A. (1999). Functional ecology of soil organisms in tundra ecosystems: towards the future. *Applied Soil Ecology* **11**, 111–126.

Hogg, I. D., Cary, S. C., Convey, P., Newsham, K., O'Donnell, T., Adams, B. J., Aislabie, J., Frati, F. F., Stevens, M. I. and Wall, D. H. (2006). Biotic interactions in Antarctic terrestrial ecosystems: are they a factor? *Soil Biology and Biochemistry* **38**, 3035–3040.

Holmstrup, M. and Sømme, L. (1998). Dehydration and cold hardiness in the Arctic collembolan *Onychiurus arcticus* Tullberg 1876. *Journal of Comparative Physiology series B* **168**, 197–203.

Kennedy, A. D. (1993). Water as a limiting factor in the Antarctic terrestrial environment: a biogeographical synthesis. *Arctic and Alpine Research* **25**, 308–315.

Kevan, P. G. and Kukal, O. (1993). A balanced life table for *Gynaephora groenlandica* (Lepidoptera, Lymantriidae), a long-lived High Arctic insect, and implications for the stability of its populations. *Canadian Journal of Zoology* **71**, 1699–1701.

Kukal, O., Duman, J. G. and Serianni, S. (1989). Cold-induced mitochondrial degradation and cryoprotectant synthesis in freeze-tolerant arctic caterpillars. *Journal of Comparative Physiology B* **158**, 661–671.

Lawton, J. H. (1999). Are there general laws in ecology? *Oikos* **84**, 177–192.

Leather, S. R., Walters, K. F.A. and Bale, J. S. (1993). *The Ecology of Insect Overwintering*. Cambridge: Cambridge University Press.

Lee, R. E. Jr., Elnitsky, M. A., Rinehart, J. P., Hayward, S. A. L., Sandro, L. H. and Denlinger, D. L. (2006). Rapid cold-hardening increases the freezing tolerance of the Antarctic midge *Belgica antarctica*. *Journal of Experimental Biology* **209**, 399–406.

Levin, D. B., Danks, H. V. and Barber, S. A. (2003). Variations in mitochondrial DNA and gene transcription in freezing tolerant larvae of *Eurosta solidaginis* (Diptera: Tephritidae) and *Gynaephora groenlandica* (Lepidoptera: Lymantriidae). *Insect Molecular Biology* **12**, 281–289.

Lister, A. (1984). Studies on the Antarctic Terrestrial Mite *Gamasellus racovitzai*. PhD thesis, University of York.

Lister, A., Block, W. and Usher, M. B. (1988). Arthropod predation in an Antarctic terrestrial community. *Journal of Animal Ecology* **57**, 957–971.

Lopez-Martinez, G., Elnitsky, M. A., Benoit, J. B., Lee, R. E. Jr. and Denlinger, D. L. (2008). High resistance to oxidative damage in the Antarctic midge *Belgica antarctica*, and developmentally linked expression of genes encoding superoxidedismutase, catalase and heat shock proteins. *Insect Biochemistry and Molecular Biology* **38**, 796–804.

MacArthur, R. H. and Wilson, E. O. (1967). *The Theory of Island Biogeography*. Princeton: Princeton University Press.

McDonald, J. R., Bale, J. S. and Walters, K. F.A. (1997). Rapid cold hardening in the western flower thrips *Frankliniella occidentalis*. *Journal of Insect Physiology* **43**, 759–766.

McDonald, J. R., Head, J., Bale, J. S. and Walters, K. F.A. (2000). Cold tolerance, overwintering and establishment potential of *Thrips palmi*. *Physiological Entomology* **25**, 159–166.

Moorhead, D. L., Wall, D. H., Virginia, R. A. and Parsons, A. N. (2002). Distribution and life-cycle of *Scottnema lindsayae* (Nematoda) in Antarctic soils: a modeling analysis of temperature responses. *Polar Biology* **25**, 118–125.

Norton, R. A. (1994). Evolutionary aspects of oribatid mite life histories and consequences for the origin of the Astigmata. In *Ecological and Evolutionary Analyses of Life-History Patterns*, ed. Houck, M. New York: Chapman & Hall, pp. 99–135.

Panikov, N. S. (1995). *Microbial Growth Kinetics*. London: Chapman & Hall.

Peck, L. S., Convey, P. and Barnes, D. K. A. (2006). Environmental constraints on life histories in Antarctic ecosystems: tempos, timings and predictability. *Biological Reviews* **81**, 75–109.

Ricklefs, R. E. and Wikelski, M. (2002). The physiology/life-history nexus. *Trends in Ecology and Evolution* **17**, 462–468.

Rinehart, J. P., Hayward, S. A. L., Elnitsky, M. A., Sandro, L. H., Lee, R. E. Jr. and Denlinger, D. L. (2006). Continuous up-regulation of heat shock proteins in larvae, but not adults, of a polar insect. *Proceedings of the National Academy of Sciences, USA* **103**, 14227–14227.

Ring, R. A. and Danks, H. V. (1994). Desiccation and cryoprotection: overlapping adaptations. *Cryo-Letters* **15**, 181–190.

Rozema, J. (ed.) (1999). *Stratospheric Ozone Depletion, the Effects of Enhanced UV-B Radiation on Terrestrial Ecosystems*. Leiden: Backhuys.

Simon, J.-C., Bonhomme, J., Blackman, R. L. and Hulle, M. (2008). Winged morph of the high arctic aphid *Acyrthosiphon svalbardicum* (Hemiptera: Aphididae): abundance, reproductive status, and ecological significance. *Canadian Entomologist* **140**, 385–387.

Sinclair, B. J. (2001). Field ecology of freeze tolerance: interannual variation in cooling rates, freeze-thaw and thermal stress in the microhabitat of the alpine cockroach *Celatoblatta quinquemaculata*. *Oikos* **93**, 286–293.

Sinclair, B. J. and Chown, S. L. (2005a). Deleterious effects of repeated cold exposure in a freeze-tolerant sub-Antarctic caterpillar. *Journal of Experimental Biology* **208**, 869–879.

Sinclair, B. J. and Chown, S. L. (2005b). Caterpillars benefit from thermal ecosystem engineering by Wandering Albatrosses on sub-Antarctic Marion Island, *Biology Letters*, doi: 10.1098/rsbl.(2005).0384.

Sinclair, B. J., Vernon, P., Klok, C. J. and Chown, S. L. (2003a). Insects at low temperatures: an ecological perspective. *Trends in Ecology and Evolution* **18**, 257–262.

Sinclair, B. J., Klok, C. J., Scott, M. B., Terblanche, J. S. and Chown, S. L. (2003b). Diurnal variation in supercooling points of three species of Collembola from Cape Hallett, Antarctica. *Journal of Insect Physiology* **49**, 1049–1061.

Smith, R. I. L. (1984). Terrestrial plant biology of the sub-Antarctic and Antarctic. In *Antarctic Ecology*, ed. Laws, R. M. London: Academic Press, pp. 61–162.

Smith R. I. L. (1988). Recording bryophyte microclimate in remote and severe environments. In *Methods in Bryology*, ed. Glime, J. M. Nichnan: Hattori Botanical Laboratory.

Sømme, L. (1986). Ecology of *Cryptopygus sverdrupi* (Insecta: Collembola) from, Dronning Maud Land, Antarctica. *Polar Biology* **6**, 179–184.

Sømme, L. (1989). Adaptations of terrestrial arthropods to the alpine environment. *Biological Reviews* **64**, 367–407.

Sømme, L. (1995). *Invertebrates in Hot and Cold Arid Environments*. Berlin: Springer.

Southwood, T. R. E. (1977). Habitat, the templet for ecological strategies. *Journal of Animal Ecology* **46**, 337–365.

Southwood, T. R. E. (1988). Tactics, strategies and templets. *Oikos* **52**, 3–18.

Søvik, G. and Leinaas, H. P. (2003). Long life cycle and high adult survival in an arctic population of the mite *Ameronothrus lineatus* (Acari, Oribatida) from Svalbard. *Polar Biology* **26**, 500–508.

Søvik, G., Leinaas, H. P., Ims, R. A. and Solhoy, T. (2003). Population dynamics and life history of the oribatid mite *Ameronothrus lineatus* (Acari, Oribatida) on the High Arctic archipelago of Svalbard. *Pedobiologia* **47**, 257–271.

Strathdee, A. T., Bale, J. S., Block, W. C., Coulson, S. J., Hodkinson, I. D. and Webb, N. R. (1993a). Effects of temperature elevation on a field population of *Acyrthosiphon svalbardicum* (Hemiptera: Aphididae) on Spitsbergen. *Oecologia* **96**, 457–465.

Strathdee, A. T., Bale, J. S., Block, W. C., Webb, N. R., Hodkinson, I. D. and Coulson, S. J. (1993b). Extreme adaptive life cycle in a High Arctic aphid, *Acyrthosiphon svalbardicum*. *Ecological Entomology* **18**, 254–258.

Strathdee, A. T., Bale, J. S., Strathdee, F. C., Block, W., Coulson, S. J., Hodkinson, I. D. and Webb, N. R. (1995). Climatic severity and the response to warming of Arctic aphids. *Global Change Biology* **1**, 23–28.

Sugg, P., Edwards, J. S. and Baust, J. (1983). Phenology and life history of *Belgica antarctica*, an Antarctic midge (Diptera: Chironomidae). *Ecological Entomology* **8**, 105–113.

Usher, M. B., Block, W. and Jumeau, P. J. A. M. (1989). Predation by arthropods in an Antarctic terrestrial community. In *University Research in Antarctica (Antarctic Special Topic)*, ed. Heywood, R. B. Cambridge: British Antarctic Survey, pp. 123–129.

Vasseur, D. A. and Yodzis, P. (2004). The color of environmental noise. *Ecology* **85**, 1146–1152.

Walther, G.-R., Post, E., Convey, P., Parmesan, C., Menzel, M., Beebee, T. J.C., Fromentin, J.-M., Hoegh-Guldberg, O. and Bairlein, F. (2002). Ecological responses to recent climate change. *Nature* **416**, 389–395.

West, C. (1982). Life histories of three species of sub-Antarctic oribatid mite. *Pedobiologia* **23**, 59–67.

Worland, M. R. and Convey, P. (2001). Rapid cold hardening in Antarctic microarthropods. *Functional Ecology* **15**, 515–525.

Worland, M. R. and Convey, P. (2008). The significance of the moult cycle to cold tolerance in the Antarctic collembolan *Cryptopygus antarcticus*. *Journal of Insect Physiology* **54**, 1281–1285.

Worland, M. R., Grubor-Lajsic, G. and Montiel, P. (1998). Partial desiccation induced by sub-zero temperatures as a component of the survival strategy of the Arctic collembolan *Onychiurus arcticus* (Tullberg). *Journal of Insect Physiology* **44**, 211–219.

Worland, M. R. and Lukesová, A. (2000). The effect of feeding on specific soil algae on the cold-hardiness of two Antarctic micro-arthropods (*Alaskozetes antarcticus* and *Cryptopygus antarcticus*). *Polar Biology* **23**, 766–774.

PART III PRACTICAL APPLICATIONS

13

A template for insect cryopreservation

ROGER A. LEOPOLD AND JOSEPH P. RINEHART

13.1 Introduction

While there is a growing number of insect species that have been successfully cryopreserved during the almost two decades that have elapsed since publication of *Insects at Low Temperature*, practical use of the methodology has generally not been forthcoming. The benefits of having the capability to place insects into storage indefinitely are considerable, ranging from reducing rearing costs to preserving valuable genetic resources. However, traditional continuous rearing practices still prevail as the primary means to maintain multiple colonies of insects, even though cryopreservation protocols for some species are now available. It is understandable that scientists and insectary managers would be reluctant to make significant changes in rearing methods without adequate assurance that irreplaceable research stocks or production lines were not lost. With this in mind, we wish to use this communication to demonstrate how a cryopreservation protocol is developed and utilized for practical applications. Thus, we hope to help alleviate the concerns about the use of existing protocols and stimulate additional research on developing cryogenic technology for the long-term storage of insects important to agriculture and the scientific community.

The insect used as the example for this case study is the New World Screwworm, *Cochliomyia hominivorax*, an ectoparasitic calliphorid that causes myiasis in man and animals (Bishopp, 1915). It is the first insect to be eliminated through an area-wide release of mass-produced sterile insects and it no longer exists in North and Central America (Wyss, 2000). A barrier zone has been established by

Low Temperature Biology of Insects, ed. David L. Denlinger and Richard E. Lee Jr. Published by Cambridge University Press.

the continuous release of sterile screwworms in the Darien gap area of Panama (Galvin and Wyss, 1996) and, during the past few years it has been successful in blocking migration of screwworms northward from Colombia. The source of the sterile screwworms was the mass-production facility in Tuxtla Gutierrez, Mexico, but now a smaller Panamanian facility provides the flies for the barrier zone.

Successful cryopreservation of the house fly (Wang *et al.*, 2000) led to the development of a protocol for cryopreserving screwworms (Leopold *et al.*, 2001). This protocol was refined and used for preserving 12 screwworm strains, about 25 000 embryos (Suszkiw, 2005), prior to the moving of a USDA-ARS screwworm research facility from Lincoln, Nebraska to Pacora, Panama. Included in the 12 strains are insects currently used in mass production, mutants with visible genetic markers and a number of genetically transformed stocks possessing transgenes that express fluorescent protein markers (Allen *et al.*, 2004). In addition, it is planned that an expanded screwworm cryopreservation protocol will be used to amass a large number of embryos to serve as a back-up repository for the facilities in Mexico and Panama should there be a need to shut down production or replace the existing production strains. Consequently, not only has this method been used to preserve strains of flies valuable to the research community, it will aid an ongoing international control program for this important threat to human health and agriculture.

With this background, it is hopefully evident that insect cryopreservation technology has advanced to the status where it can be practically applied. While the following represents a case study for cryopreserving an insect, it should be mentioned that it is species specific. It has been our experience that each protocol needs to be tailored to fit the insect at hand and that this case study will serve as an adaptable template to be followed in development of a species-specific method for long-term storage of an insect.

13.2 Protocol development

This section is devoted to giving a brief explanation of the cryogenic foundation underlying the various steps taken when developing a protocol for cryopreserving insect embryos. A working knowledge of the various techniques and options available for processing organisms for cryopreservation is essential to make protocol adaptations to accommodate the individual traits of an insect species. For a more detailed analysis of the methods used for cryopreservation of non-mammalian organisms, see the review by Leopold (2007). At the end of this section, we have listed the procedure developed for cryopreservation of the screwworm (Table 13.1). The order of discussion of the following topics is not

Table 13.1 *The following is the method currently used for cryopreserving screwworm embryos and is presented here as an example of an insect cryopreservation protocol*

Screwworm Cryopreservation Protocol[1]

*Permeabilization (*Berkebile *et al., 2000)*

1. Egg collection (15–30 min).
2. Egg incubation (6.5 h – examine at 6 h)
3. Disperse egg masses by placing in 2.0% NaOH for 5 min and agitate.
4. Rinse in H_20 and then agitate in 5.0% $NaClO_3$ for 2 min.
5. Rinse in H_20 for 3 min.
6. Agitate in isopropyl alcohol for 30 s.
7. Air dry in moistened stream of air for 3 min.
8. Agitate in hexane for 30 s.
9. Air dry 1 min.
10. Agitate and rinse in *Drosophila* Ringer's 2 min or, if clumped, break up clumps with spray bottle.

1,2-Propanediol (EG) Loading & Dehydration

1. Embryos placed in 1.8 M EG in Schneider's insect cell culture media (Schneider, 1964) 20 min at ambient.
2. Embryos transferred to vitrification solution of 37 wt % EG + 6 wt % polyethylene glycol + 8 wt % trehalose for 8 min on ice.

Vitrification and Storage in Liquid Nitrogen

1. Transfer <100 embryos to 25 mm polycarbonate membrane.
2. Blot on tissue to remove excess solution.
3. Hold membrane with embryos in liquid nitrogen vapor 1 min.
4. Plunge membrane with embryos into liquid nitrogen for storage.

Recovery from Liquid Nitrogen Storage

1. Remove from liquid nitrogen and hold in vapor for 1 min.
2. Quickly plunge membrane with embryos into 0.5 M trehalose and 10% fetal bovine serum in Schneider's media for 2 min.
3. Replace with Schneider's media and wash 3 times at intervals of 10 min.
4. Remove membrane and leave embryos in fresh Schneider's media overnight for hatching.
5. Place hatched larvae on larval diet for growth to adulthood.

[1] Revised from Leopold *et al.* (2001).

presented in a manner that corresponds directly, step by step, to the screwworm cryopreservation procedure. Instead, the topics with the more general concepts are listed first and followed by issues relating to specific aspects of the method used to cryopreserve the screwworm and other insects.

13.2.1 Intracellular ice

The successful preparation and recovery of cells and organisms to be cryopreserved in liquid nitrogen demands that certain requirements be met. Leading the list of these requirements is the necessity to avoid intracellular ice formation (IIF) (Mazur, 2004). Avoidance of IIF can be accomplished by osmotic elimination of water by slowly cooling the cells to subzero temperatures. When cooling conditions are favorable, cells supercool and lose water through the osmotic imbalance created between the extracellular-frozen and intracellular-unfrozen fractions. Whether cells or organisms can escape the lethal formation of intracellular ice in this manner is dependent on the supercooling capacity, cooling rate, membrane permeability and composition of the surrounding milieu. Insect eggs are relatively small and, depending upon the species, can supercool to temperatures between −25 and −40 °C before freezing (Sømme, 1964). While having the capacity to cool to low subzero temperatures before freezing would appear to be an attribute favorable for losing water through formation of an osmotic imbalance, most of the insects tested for possible embryo cryopreservation thus far do not survive a slow supercooling regimen (Heacox *et al.*, 1985; Mazur *et al.*, 1992; Miles and Bale, 1995). One notable exception is the diapausing drosophilid, *Chymomyza costata*, which survives to the pupal stage after cooling the embryos to −40 °C at 0.1 °C min^{-1} prior to placing in liquid nitrogen (Moon *et al.*, 1996). In this case, osmotic dehydration was apparently initiated by nucleating the surrounding medium at −2 °C before cooling to −40 °C and it was confirmed by cryomicroscopic examination that ice-crystal formation remained in the extracellular spaces of the organism.

13.2.2 Developmental stage tolerance

Treating the embryos at the correct stage of development is another crucial aspect of successful insect cryopreservation. Some of the steps leading up to storage in liquid nitrogen are variously toxic to early-stage embryos (Wang *et al.*, 2000) and the most tolerant stage for treatment found thus far has been the stage just prior to development of the embryonic exoskeleton (Mazur *et al.*, 1992; Steponkus and Caldwell, 1993; Wang *et al.*, 2000; Leopold *et al.*, 2001; Rajamohan *et al.*, 2003; Rajamohan and Leopold, 2007). At this stage for dipterans, head involution has begun, dorsal closure is complete, segmentation is evident and a minimum of yolk is contained in a coiled midgut (Fig. 13.1). Circumstantial evidence suggests that physical or chemical damage to the yolk reserves by treatment of a younger embryo causes a cessation in development (Leopold, 1991). Furthermore, treatment following formation of an embryonic cuticle during the final stages of development is also ineffective, indicating that removal of water and loading of the embryo with a chemical cryoprotectant were structurally blocked.

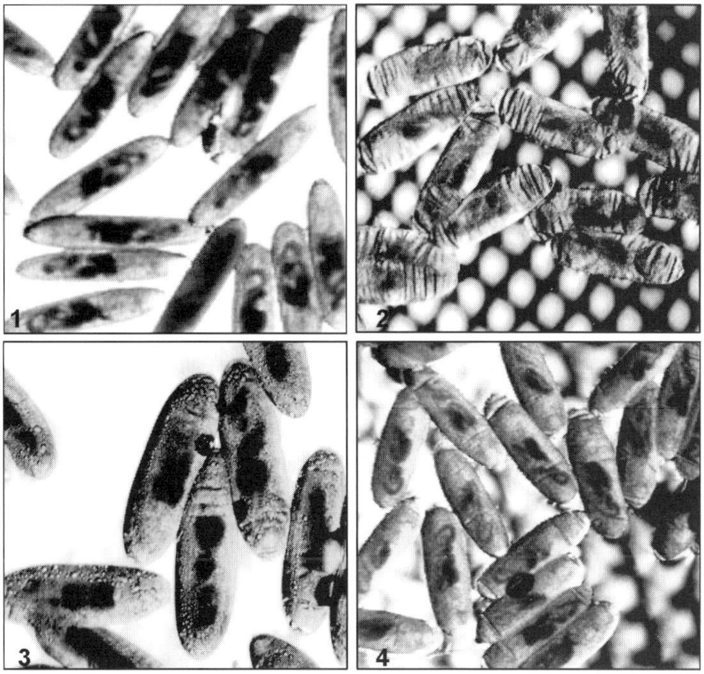

Figure 13.1 Screwworm embryos at different and treatment stages in the cryopreservation protocol. Panel 1 shows permeablized embryos at the beginning of treatment. Note that the embryos have coiled midguts and are at the correct stage to begin treatment. Panel 2 shows extremely flattened, dehydrated embryos in vitrification solution. Panel 3 shows embryos that are in the process of being unloaded of the cryoprotectant. Note the small, clear bubbles on the membrane surfaces. Panel 4 shows embryos in the process of being loaded with cryoprotectant. Some shrinkage is evidenced by the wrinkles occurring at the ends of the embryos.

13.2.3 Embryo permeabilization

Key to the preparation of embryos prior to exposure to the rapid cooling step required for vitrification is that the membranes of the cells/organisms must be permeable to allow the exit of water into the extracellular medium and entry of the cryoprotective agent (CPA). Gaining permeability for insect eggs, such as those of the screwworm, requires application of a series of chemical treatments to remove the inherent barriers to water loss. With screwworms, these barriers include a proteinaceous material that binds the egg mass together, an egg shell or chorion and a lipid layer that coats the vitelline membrane surrounding the embryo (Berkebile *et al.*, 2000). To date, alkanes have been used as solvents to remove the lipid layer associated with the membranes of embryos. The lipid layer shows considerable variation in hydrocarbon composition between species (Nelson and Leopold, 2003) which may explain why there is variation in the effectiveness of 5- to 8-carbon alkanes as solvents for the lipids associated with the vitelline membrane.

13.2.4 Slow versus rapid cooling

The slow cooling method endeavors to maintain an osmotic equilibrium between the exit of water and the entry of the CPA at a 1–2 M concentration (Mazur, 1990). Alternatively, as mentioned above, IIF can be avoided by exposing cells/organisms to multimolar concentrations of a CPA followed by rapid cooling. This rapid non-equilibrium cooling results in vitrification, whereby the cellular liquid fraction becomes a glassy, extraordinarily viscous matrix, as opposed to the formation of crystalline ice. Rall (1987) and Rall and Fahy (1985) were among the first to develop this technique for the cryopreservation of mouse embryos. Since then, the vitrification method has been used to successfully cryopreserve a variety of plant and animal cells, tissues, germ cells and animal embryos, including insects (Takahashi *et al.*, 1986; Jutte *et al.*. 1987; Van Wagen-De Leeuw *et al.*, 1997; Nishizawa *et al.*, 1993; Leopold, 2007).

The first vitrification technique for insects was simultaneously developed for *Drosophila melanogaster* by Mazur *et al.* (1992) and Steponkus and Caldwell (1993). The *Drosophila* technique developed by these two groups is a variation of that of Rall and Fahy (1985), whereby permeabilized embryos are loaded with a low concentration of a CPA (~2 M) followed by a short-term placement in a higher concentration (~8 M) that concentrates the amount of intracellular CPA within the embryos by removing additional cell water.

13.2.5 Cryoprotective agents

Circumventing lethality in insects like screwworms, which are intolerant to slow cooling in the embryonic stage, has been accomplished by altering the method of introducing the cryoprotectant and by increasing cooling to a rapid rate. CPAs, such as 1,2-ethanediol or dimethylsulfoxide, are ordinarily used in cryopreservation protocols to combat the potentially harmful effects of cell-solute concentration upon dehydration and to depress the occurrence of IIF. Cryoprotectants are generally classified as penetrating or non-penetrating chemicals. Dimethylsulfoxide and 1,2-ethanediol are penetrating CPAs, while chemicals like polyethylene glycol, hydroxyethyl starch and sucrose are grouped with the non-penetrating CPAs. The benefits of using non-penetrating chemicals as the singular CPA or in a CPA mixture are not entirely clear, but depending upon the chemical characteristics, the values put forth include membrane protection, vitrification enhancement, toxicity reduction and removal of intracellular water (Meryman, 1971, 2007).

To date, the cryoprotectant of choice for insects has been 1,2-ethanediol (Mazur *et al.*, 1992; Steponkus and Caldwell, 1993; Wang *et al.*, 2000; Leopold *et al.*, 2001; Nunamaker and Lockwood, 2001; Rajamohan *et al.*, 2003; Rajamohan and Leopold, 2007). Polyethylene glycol and trehalose were added to 1,2-ethanediol to create a solution used for vitrification of house fly, screwworm and Mexican fruit fly embryos (Wang *et al.*, 2000; Leopold *et al.*, 2001; Rajamohan and Leopold, 2007). Addition of the polyethylene glycol reduces fracturing of the vitrified solution surrounding the Mexican fruit fly embryos upon rapid cooling (Rajamohan and Leopold, 2007).

The cryopreservation method developed for the screwworm and the other dipterans was patterned after that of Rall (1987) and Rall and Fahy (1985) and requires a CPA loading step followed by a dehydration step to raise the intracellular concentration of CPA to a level where damaging intracellular ice is not formed upon rapid cooling. The secondary dehydration step typically employs a multimolar concentration of primarily the same CPA (~8 M) to remove additional cell water that was not eliminated during the loading step and to raise the intracellular level of the CPA. Cryoprotectant toxicity varies widely depending upon the cell type, organism or species, especially at high concentrations. For this reason, the secondary dehydration step for dipteran embryos is tightly time-regulated and is conducted at 0 °C to reduce toxicity of CPA that occurs at higher temperatures. The time required for dehydration/CPA concentration varies with the particular size of the insect embryo and the relative permeability of the membranes surrounding the embryo. It is somewhat astonishing that the dehydrated, flattened embryos (Fig. 13.1, panel 2) can survive a potentially lethal situation that is nevertheless critical to obtaining the vitrified condition.

13.2.6 *Pre- and post-storage treatment*

As mentioned above, species variation, such as length of embryogenesis, permeability, chilling-tolerance, or CPA sensitivity, can dictate the design of the individual steps in the cryopreservation process. Dealing with this variation is explained in detail in the optimization section of this chapter. The pre- and post-storage treatment steps are areas where the need to adapt to the individuality of a particular species is especially evident. For example, the cooling regime for the *Drosophila* and *Culicoides* techniques (Mazur *et al.*, 1992; Steponkus and Cauldwell, 1993; Nunamaker and Lockwood, 2001) employed prior to storage in liquid nitrogen (LN) differs from that used for the house fly. Wang *et al.* (2000) found increased survival for the house fly was obtained by cooling the embryos in liquid nitrogen vapor until vitrification occurred, as opposed to the more rapid cooling by direct exposure to liquid nitrogen slush or liquid propane. Subsequent studies with the screwworm and two tephritids showed similar results (Leopold *et al.*,

2001; Rajamohan *et al.*, 2003; Rajamohan and Leopold, 2007). Holding the embryos in LN vapor for up to one minute initiates vitrification and apparently mitigates the abrupt phase change, along with the possibility of fracturing the vitreous mass (Rajamohan and Leopold, 2007).

Recovery of the vitrified embryos from LN storage also requires exacting stepwise treatment. Like the cooling rate, the warming of the embryos is conducted at a rapid rate. Following removal from LN into LN vapor for >30 seconds, it is essential that immersion into a room temperature solution be instantaneous to prevent devitrification and resulting damage by ice-crystal formation. Equally important is the prompt removal of the CPA from the embryos, by placement in a solution containing a non-penetrating solute such as sucrose or trehalose. This procedure sets up an imbalance in the osmotic equilibrium between the extra- and intracellular water, thus reversing the CPA loading and dehydration steps used earlier in the protocol (Fig. 13.1, panel 3).

Because the vitelline membrane remains permeable and the embryonic cuticle has yet to form at this stage in the protocol, it is recommended that an insect cell culture medium be used as the diluent for the non-penetrating solute to guard against potential loss of essential salts and minerals by the embryo during CPA removal. An insect cell culture medium can also be beneficial in aiding embryo survival during the latter stages of development and also after hatching, until they are placed on larval medium. It should be noted, however, that the insect culture medium is easily contaminated and will cause larval mortality if it is tainted.

13.3 Optimization

13.3.1 Pre-treatment incubation

Although selection of the correct embryonic stage of development for treatment is of primary importance when establishing a cryopreservation protocol, subtle aspects of the pre-treatment incubation period can have a marked effect on post-storage survival. For example, Mazur *et al.* (1993) doubled the hatching rate of *Drosophila melanogaster* embryos following vitrification by switching from 12- to 14-hour-old embryos. Likewise, with the screwworm, a threefold variation in hatching rates after recovery from storage was obtained solely by making slight variations in the amount of time spent at different pre-treatment incubation temperatures (Leopold *et al.*, 2001). In a related study, Rajamohan and Leopold (2007) increased the hatching rate of previously vitrified Mexican fruit fly embryos from 40 to nearly 70% by physically selecting embryos deemed to be the correct treatment stage from a sample collected from ovipositing females over only a 30 min period. Further, it has been our experience when developing protocols for five dipteran species, comprising three different taxonomic groups, that there is natural variation in the rate of embryonic development. Thus, collecting adequate numbers of eggs from ovipositing females in the shortest time possible reduces the number of embryos that are not at the optimum stage for treatment.

13.3.2 Osmotic equilibrium

After the chorion and lipid layers are removed from the insect embryo, osmotic stress becomes an immediate concern. At the cellular and molecular levels, exposure to extreme cold and the process of vitrification does not harm the organism, but problems associated with water imbalance and the resulting concentration of cellular solutes can be fatal. Minimizing extraordinary fluctuations in the osmotic equilibrium throughout the cryopreservation and recovery processes can dramatically improve the yield of viable insects after cryopreservation. For example, the screwworm protocol uses *Drosophila* Ringer's solution (Steponkus and Caldwell, 1993) as an equilibrating/transfer liquid immediately after hexane is used to remove the lipid layer from the vitelline membrane (Leopold *et al.*, 2001). In most species tested so far, replacing Ringer's solution with distilled water or other solutions that do not confer osmotic equilibrium greatly reduces survival after cryopreservation (personal observations). While *Drosophila* Ringer's is currently the solution of choice for this step, it is likely that adjustments will need to be made as protocols are developed for insects other than dipterans.

Accommodations for maintaining osmotic equilibrium within the embryos must be made for the steps leading up to vitrification. Both the loading and vitrification solutions used for preserving screwworms are prepared with insect cell culture media (Schneider's Insect Cell Culture Medium, Sigma Chemical Co.) to ameliorate the effects of osmotic stress during these steps. During loading with the CPA, water leaves the egg, causing it to shrink visibly (Fig. 13.1, panel 4), followed by re-swelling as an equilibrium is re-established between the extra- and intracellular areas. Using a diluent for the CPA that emulates the internal milieu of the embryo allows the exchange of water for the CPA without loss of essential intracellular components.

The two areas in an insect cryopreservation protocol, where osmotic stress is of greatest concern, are dehydration of the embryos in vitrification solution and the subsequent removal of the CPA during recovery from LN storage. A combination of three actions is taken to reduce detrimental effects of severe dehydration of the embryos prior to vitrification. Specifically, these actions are loading the embryos with a CPA, conducting the dehydration step at 0 °C and using an insect cell culture medium as the diluent for the solutions. All of these activities are intended to lessen the probability of damage caused by solute concentration within the cells of the embryos.

The purpose of the recovery phase of the protocol is to remove the CPA and to rehydrate the embryo. These steps also require the use of an insect cell culture medium or the equivalent to re-establish osmotic equilibrium in the embryo as it is rehydrated. Considerable attention has been given to the process by which rehydration is achieved. Some mammalian cryopreservation protocols employ a stepwise rehydration, using a series of solutions to gradually re-establish osmotic normalcy (Isachenko *et al.*, 1997, 2004; Wessel and Ball, 2004), the theory being that immediately returning the organism directly to equilibrium with the external medium causes osmotic shock and/or mechanical damage upon rapid rehydration. Although the rate of rehydration of insect embryos is relatively rapid when examined through a microscope, it is important to note that none of the current insect cryopreservation protocols have been improved by including a stepwise rehydration regime.

13.3.3 Strain variation

With continuing advancements in the area of genetic transformation, new strains of insects are being developed rapidly. This is, of course, in addition to the accumulation of strains that have been developed via conventional means down through the years. Clearly, one of the paramount forces for the development of cryopreservation techniques is to alleviate the increasing pressure that these additional strains put upon limited resources. Hence, a fundamental question emerges; does strain variation affect the development of cryopreservation protocols? The aforementioned effort to cryopreserve multiple strains of screwworms gives an insight into this issue. Success rates for cryopreservation varied widely for strains of the screwworm maintained at the USDA-ARS facility in Lincoln, Nebraska. The average hatch rates after cryopreservation varied from 17% to over 55% and 2 of 14 strains processed could not be successfully cryopreserved (Table 13.2). Thus, strain variation will likely need constant attention as the technology is developed for most insects. It is expected that for many strains, slight adaptations in the protocol will be sufficient to achieve acceptable survival rates. For example, if a particular strain differs from the wild-type with respect to the rate of embryonic development, then adjusting the pre-incubation period of the protocol for the strain in question may be sufficient to achieve successful cryopreservation. Other differences between strains, including chorion thickness or lipid layer composition, could be accommodated by protocol modification as well. Of course, as this technology is applied to more strains, some will likely be intractable, regardless of protocol modification. This may become especially evident in genetically transformed lines, where genes that are essential to cryopreservation survival are suppressed. Strain characteristics such as low fecundity, low survival, or heightened reactivity to certain environmental stressors can

Table 13.2 *Mean hatching rates of cryopreserved screwworm strains. Wild-type P95 strain was developed from collections in Panama in 1995. Mutant strains designated as: CECH (Cherry Eye), Limon (Lemon Eye), LH (Yellow Eye) and PA34 (Orange Eye) are autosomal mutations obtained from collections made in Mexico in 1984–1985. Mutant Strain C9 (white eye) is an uncharacterized recessive mutation obtained from collections in Panama in 1995. Transgenic strains are designated as previously reported by Allen* et al.*, 2004*

Strain Designation	Hatching after Cryopreservation (%)
Wild-type	
P95	39
Mutant	
CECH	33
LIMON	21
LH	30
PA34	0
C9	28
Transgenic	
CLAY	22
CLIX	34
CLOX	17
COTY	0
FOLY	2
GARY	30
GIZA	55
SUEZ	18

render a strain unsuitable for cryopreservation. For example, when developing a protocol for cryopreservation of a temperature-sensitive mutant strain of the Mediterranean fruit fly, it was discovered that using a pre-incubation temperature that approached 30 °C yielded a preponderance of male flies. Results such as these would not favor reconstitution of the strain following cryopreservation. However, slight modification in the pre-incubation temperature remedied the situation.

13.3.4 Diet and environment

Survival during recovery from cryopreservation can be greatly enhanced by adjusting the larval diet and environment in which the insects develop.

Optimization can be made at two stages during post-recovery development of the insects. The first stage is the environment in which the embryo finishes development and into which the larva emerges. Like the earlier steps in the protocol, removing the chorion and lipid layer also renders the developing embryo exceptionally susceptible to osmotic stress. Hence, optimal results are usually obtained by incubating the embryo in the same cell culture medium used as the diluent for the CPA. For the screwworm, survival was further enhanced by the addition of 10% fetal bovine serum to the culture medium (Leopold *et al.*, 2001).

Proper incubation temperature is, of course, also essential to survival, and usually closely follows the optimal temperature for the insect under normal conditions. For example, embryos and emerging larvae of cryopreserved screwworms survive best when held at 37 °C while the Mexican fruit flies have better survival at < 28 °C. Further, the newly hatched larvae should be removed from the culture media promptly to minimize the negative effects of microbial contamination. The addition of antibiotics (Leopold *et al.*, 2001) and the use of proper sterile tissue-culture techniques can greatly reduce this problem. An alternate technique to circumvent osmotic stress to the developing embryo is to cover the embryos in mineral oil until hatching occurs, thereby preventing dehydration and possible contamination. This method was successfully implemented for *D. melanogaster* following recovery from cryopreservation (Mazur *et al.*, 1992; Steponkus and Caldwell, 1993).

Once hatched, larvae can be transferred to a larval diet and allowed to develop normally. In the case of screwworms, the larvae develop on the standard larval media used for rearing this insect in the laboratory (Berkebile *et al.*, 2000). However, utilization of a standard laboratory diet may not be optimal for all cryopreserved species. For example, with the Mediterranean fruit fly, pupation rates varied from 0% to 80%, depending on what bulking agent was used in the post-cryopreservation larval diet (Rajamohan *et al.*, 2003). Thus, if one experiences good hatching rates after cryopreservation, but poor larval survival, careful consideration of all the components of the larval diet may prove productive in improving survival to adulthood.

Finally, all aspects of the larval environment should be considered when optimizing the post-cryopreservation procedure. An intriguing problem with screwworm post-cryopreservation incubation involves aggregation behavior of the feeding larvae. Screwworm larvae perform best when they are part of a feeding aggregation within the larval media (personal observations). Since the number of larvae processed through the limited scale of laboratory cryopreservation is usually less than the number required for an effective feeding mass, larval survival can be adversely affected. This issue can be corrected by either increasing the scale of the cryopreservation procedure, or by using a mutant marker strain as a "surrogate mass" to augment the number of feeding screwworm larvae that can be separated from the cryopreserved individuals after adult emergence.

13.3.5 Fitness testing

Maintaining quality in insects submitted to cryopreservation is a necessity, especially when cryopreserved insects or their progeny are to be released as part of a control program. There are a number of laboratory tests that can be applied beyond the basic reproduction tests such as fertility, fecundity, hatching and mating propensity. Additional tests assaying flight ability and duration, and field reproduction have been used with several tephritid species following cryopreservation (Leopold *et al.*, 2010). No significant differences were found between controls and the cryopreserved insects in these studies.

Some deleterious-appearing effects of the cryopreservation procedure may be transient in expression. In the case of the screwworm, a significantly lower pupal weight was obtained following recovery of the embryos from cryopreservation (Leopold *et al.*, 2001). However, this effect was not evident in the progeny of the next generation. Thus, depending on the assay and the ultimate use of the insects, negative results obtained from certain fitness tests may not be an important issue when viewed over several generations. Further, a direct connection between reduced fitness to a step or series of steps in the protocol is not always clearly obvious. For example, a reduction in adult fertility following cryopreservation does not point to any one step in the procedure, while a reduced size or mass of the pupae probably relates to the larval diet or the larval rearing environment.

Recent attention has been given to the molecular genetic effects of cryoprotectants and cryopreservation on plant and animal cells (Yildiz *et al.*, 2007; Ogawa *et al.*, 2008), but it remains to be seen how many of these observations will carry over into insect species. Preliminary studies using suppressive subtractive hybridization to compare gene expression in Mexican fruit flies that were previously cryopreserved with their untreated counterparts showed that specific genes were affected by the cryopreservation process (Rajamohan *et al.*, 2005). However, these results were tentative and connections to specific steps in the cryopreservation protocol were not readily evident. Hopefully, as cryopreservation becomes more widely practiced in model insect species (e.g. *D. melanogaster*), the molecular characterization of insect cryopreservation and the development of molecular biomarkers for fitness testing will become readily available.

13.4 Conclusions and future directions

Information obtained from past studies concerned with the development of cryopreservation technology for insects indicates that certain obligatory elements must be satisfied when developing new protocols. The organism must be, or rendered, permeable to the loss of water and the influx of a cryoprotectant. It must be in a stage of development that tolerates chilling, osmotic manipulations and chemicals at concentrations that promote vitrification upon rapid cooling. As a consequence of these conditions, one is generally directed towards working with the insect embryo because of size limitations and ease of permeabilization.

While there are cryobiological limitations that must be followed when developing an insect cryopreservation protocol, there was one aspect not discussed in this chapter and that is: how much of the protocol is devoted to science and how much is "art" or technique? The science versus art issue cannot be ignored because it should be apparent that maintaining life through the rigorous manipulations of cryopreservation leaves little room for handling errors or carelessness. For most of the procedure, timing is critical, because of the required exposure of a vulnerable embryonic stage to potentially toxic chemicals. As with all relatively new technologies, continued use generates new ideas for ease of handling and modification. Cryopreservation procedures for mammalian embryos improved dramatically over years, and it is our hope that methods for insect cryopreservation can be simplified and also benefit from new molecular-based cryogenic approaches.

Along with creation of long-term storage technology for insects there can be underlying rewards for this type of endeavor. It is clearly evident that when a method is available to alleviate the tedious and expensive continuous rearing of numerous strains of insects, the researcher using those insects gains an instant recompense. However, the benefits would, no doubt, be expanded considerably should a national insect germplasm repository be constructed that would allow storage and dissemination of insects important to research and industry. With the development of each new protocol, especially for those insects that have undergone genomic sequencing, the need for such a system becomes critical for the entire research community, particularly as resources diminish. Such a system could become an entity essential to entomological pursuits, providing dependable access to a wide variety of insect germplasm, while ensuring the safe-keeping of an extremely valuable commodity for future generations of researchers.

References

Allen, M. L., Handler, A. M., Berkebile, D. R. and Skoda, S. R. (2004). piggyBac transformation of the New World screwworm *Cochliomyia hominivorax*, produces multiple distinct mutant strains. *Journal of Medical and Veterinary Entomology* **18**, 1–9.

Berkebile, D. R., Chirico, R. J. and Leopold, R. A. (2000). Permeabilization of *Cochliomyia hominivorax* (Diptera: Calliphoridae) embryos. *Journal of Medical Entomology* **37**, 968–972.

Bishopp, F. C. (1915). Flies which cause myiasis in man and animals – some aspects of the problem. *Journal of Economic Entomology* **8**, 317–329.

Galvin, T. J. and Wyss, J. H. (1996). Screwworm eradication program in Central America. *Annals of the New York Academy of Sciences* **791**, 233–240.

Heacox, A. E., Leopold, R. A. and Brammer, J. D. (1985). Survival of house fly embryos cooled in the presence of dimethyl sulfoxide. *CryoLetters* **6**, 305–312.

Jutte, N. H., Heyse, P., Jansen, H. G., Bruining, G. J. and Zeilmaker, G. H. (1987). Vitrification of human islets of Langerhans. *Cryobiology* **24**, 403–411.

Isachenko, V. V., Isachenko, E. F., Ostashko, F. I. and Grishchenko, V. I. (1997). Ultrarapid freezing of rat embryos with rapid dilution of permeable cryoprotectants. *Cryobiology* **34**, 157–164.

Isachenko, V., Montage, M., Isachenko, E., Nawroth, F., Dessole, S. and Van Der Ven, H. (2004). Developmental rate and ultrastructure of vitrified human pronuclear oocytes after step-wise versus direct rehydration. *Human Reproduction* **19**, 660–665.

Leopold, R. A. (1991). Cryopreservation of insect germplasm: cells, tissues and organisms. In *Insects at Low Temperature*, ed. R. E. Lee and D. L. Denlinger. New York, NY: Chapman and Hall, pp. 379–407.

Leopold, R. A. (2007). Cryopreservation of nonmammalian metazoan systems. In *Advances in Biopreservation*, ed. J. G. Baust and J. M. Baust. Boca Raton, FL: CRC Taylor and Francis Group, pp. 271–298.

Leopold, R. A., Rajamohan, A., Shelly, T. E. and Handler, A. M. (2010). Quality testing of three species of tephritid fruit flies after embryo cryopreservation. *Annals of Entomological Society of America* (in press).

Leopold, R. A., Wang, W. B., Berkebile, D. R. and Freeman, T. P. (2001) Cryopreservation of embryos of the New World Screwworm, *Cochliomyia hominivorax* (Diptera: Calliphoridae). *Annals of Entomological Society of America* **94**, 695–701.

Mazur, P. (1990). Equilibrium, quasi-equilibrium and nonequilibrium freezing of mammalian embryos. *Cell Biophysics* **17**, 53–92.

Mazur, P. (2004). Principles of cryobiology. In *Life in the Frozen State*, ed. B. J. Fuller, N. Lane and E. E. Benson. Boca Raton, FL: CRC Taylor Francis Group, pp. 3–65.

Mazur, P., Cole, K. W., Hall, J. W., Schreuders, P. D. and Mahowald, A. P. (1992). Cryobiological preservation of *Drosophila* embryos. *Science* **258**, 1932–1935.

Mazur, P., Cole, K. W., Schreuders, P. D. and Mahowald, A. P. (1993) Contributions of cooling and warming rate and developmental stage to the survival of *Drosophila* embryos cooled to −205 °C. *Cryobiology* **30**, 45–73.

Meryman, H. T. (1971). Cryoprotective agents. *Cryobiology* **8**, 173–183.

Meryman, H. T. (2007). Cryopreservation of living cells: principles and practice. *Transfusion* **47**, 935–945.

Miles, J. E. and Bale, J. S. (1995). Analysis of chilling injury in *Aphidoletes aphidimyza*. *Cryobiology* **32**, 45–73.

Moon, I., Fujikawa, S. and Horie, Y. (1996). Cryopreservation of *Chymomyza costata* larvae (Diptera: Drosophilidae) at −196 °C with extracellular freezing. *CryoLetters* **17**, 105–110.

Nelson, D. R. and Leopold, R. A. (2003). Composition of the surface hydrocarbons from the vitelline membrane of dipteran embryos. *Comparative Biochemistry and Physiology* **136**, 210–308.

Nishizawa, S., Sakai, A. Amano, Y. and Matsuzawa, T. (1993). Cryopreservation of asparagus (*Asparagus officinalis* L.): embryogenic suspension, cells and subsequent plant regeneration by vitrification. *Plant Science* **91**, 67–73.

Nunamaker, R. A. and Lockwood, J. A. (2001). Cryopreservation of embryos of *Culicoides sonorensis* (Diptera: Ceratopogonidae). *Journal of Medical Entomology* **38**, 55–58.

Ogawa, Y. Suzuki, H., Sakurai, N. Aoki, K., Saito, K.and Shibata, D. (2008) Cryopreservation and metabolic profiling analysis of *Arabidopsis* T87 suspension-cultured cells. *CryoLetters* **29**, 427–436.

Rajamohan, A. and Leopold, R. A. (2007). Cryopreservation of Mexican fruit flies by vitrification: stage selection and avoidance of thermal stress. *Cryobiology* **54**, 44–54.

Rajamohan, A., Leopold, R. A., Wang, W. B., Harris, M., McCombs, S. D., Peabody, N. C. and Fisher, K. (2003). Cryopreservation of Mediterranean fruit fly embryos. *CryoLetters* **24**, 125–132.

Rajamohan, A., Yocum, G. D. and Leopold, R. A. (2005). Differential gene expression in Mexican fruit flies after cryopreservation. *Cryobiology* **51**, 406.

Rall, W. F. (1987). Factors affecting the survival of mouse embryos cryopreserved by vitrification. *Cryobiology* **24**, 387–402.

Rall, W. F. and Fahy, G. M. (1985) Ice-free cryopreservation of mouse embryos at −196 °C by vitrification. *Nature* **313**, 573–575.

Schneider, I. (1964). Differentiation of larval *Drosophila* eye-antennal discs *in vitro*. *Journal of Experimental Zoology* **156**, 91–104.

Sømme, L. (1964). Effects of glycerol on cold hardiness in insects. *Canadian Journal of Zoology* **42**, 87–101.

Steponkus, P. L. and Caldwell, S. (1993). An optimized procedure for the cryopreservation of *Drosophila melanogaster* embryos. *CryoLetters* **14**, 377–380.

Suszkiw, J. (2005) Frozen flies safeguard research, screwworm eradication efforts. *Agricultural Research* **53**, No. 2, 14–15.

Takahashi, T., Hirsh, A., Erbe, E. F., Bross, J. B., Steere, R. L. and Williams, R. J. (1986). Vitrification of human monocytes. *Cryobiology* **23**, 103–115.

Van Wagtendonk-De Leeuw, A. M., Den Daas, J. H. G., Kruip, Th. A. M. and Rall, W. F. (1997). Comparison of the efficacy of conventional slow freezing and rapid cryopreservation methods for bovine embryos. *Cryobiology* **32**, 157–167.

Wang, W. B., Leopold, R. A., Nelson, D. R. and Freeman, T. P. (2000). Cryopreservation of *Musca domestica* (Diptera: Muscidae) embryos. *Cryobiology* **41**, 153–166.

Wessel, M. T. and Ball, B. A. (2004). Step-wise dilution for removal of glycerol from fresh and cryopreserved equine spermatozoa. *Animal Reproduction Science* **84**, 147–156.

Wusteman, M. C., Simmonds, J., Vaughan, D. and Pegg, D. E. (2008). Vitrification of rabbit tissues with propylene glycol and trehalose. *Cryobiology* **56**, 62–71.

Wyss, J. H. (2000). Screwworm eradication in the Americas. *Annals of the New York Academy of Sciences* **916**, 186–193.

Yildiz, C., Ottaviani, P., Law, N., Ayearst, R., Liu, L. and McKerlie, C. (2007). Effects of cryopreservation on sperm quality, nuclear DNA integrity, *in vitro* fertilization, and *in vitro* embryo development in the mouse. *Reproduction* **133**, 585–595.

14

Implications of cold-tolerance for pest management

JEFFREY S. BALE

14.1 Introduction

In a global context "pest management" usually refers to the control of invertebrate organisms, the great majority of which are insects, with some important examples among the mites and nematodes. Also, whilst these pests are primarily associated with agricultural and horticultural crops and forestry, the impact of arthropod disease vectors on human and animal populations can be devastating.

As a proportion of the Class Insecta as a whole (>1 million species so far described), there are relatively few pest species, mainly because most insects have a range of natural enemies (including diseases, other insects and higher animals) that collectively suppress their numbers below the level at which they would become categorized as pests. The situations under which previously innocuous species or minor pests achieve a much higher or widespread pest status are numerous, but include, for example: (a) global cultivation of a crop, such as potato, which is attacked by the Colorado beetle, *Leptinotarsa decemlineata*, across different continents; (b) transfer of pest species to new environments without their natural enemies e.g. cottony-cushion scale, *Iceryae purchasi*, from Australia to citrus crops in the US; (c) introduction of new crop species into agroecosystems e.g. increased cultivation of oil seed rape (canola) in Europe over the last 40 years; (d) extensive annual monocultures that favor pests rather than natural enemies e.g. maize and other cereals; (e) removal of refugia and sources of alternative prey e.g. widescale destruction of hedgerows in the UK; and (f) application of a pesticide against a

Low Temperature Biology of Insects, ed. David L. Denlinger and Richard E. Lee Jr. Published by Cambridge University Press.

"problem pest" that kills the natural enemies of a "minor pest" that then escalates in importance.

Insects are poikilothermic animals with limited ability to regulate their body temperature. It might be expected therefore that there would be some large-scale macrophysiological determination of insect distributions at the level of biomes and distinct climatic regions, and in general terms this is true. Thus there are insects that are predominantly associated with polar or tropical climates, and many species that have a more widespread distribution, from the Mediterranean to subpolar Scandinavia in Europe, and from the southern USA to Canada in North America. The implications of cold-hardiness and more broadly, thermal biology, for the management of pest insects (and related arthropods such as mites) are diverse, but primarily include their distribution and various inter-related rate-based processes (development, voltinism, annual abundance), and in the context of biological control, their establishment potential after release; here it is important to note that long-term establishment is essential for classical biocontrol of pests and weeds, but may be an undesirable consequence when non-native agents are released into glasshouses and escape into the wider environment. However, these processes do not operate in isolation, and this complex of interactions is subject to the dynamic impacts of a changing climate.

There is a rapidly expanding literature on the thermal biology of insects, covering areas such as cold-tolerance, winter survival, distributions, abundance and the impacts of a warmer climate on range margins. Against this background, it is not possible to present a comprehensive review of all of these areas, not least because many studies have focused on so-called indicator species or those of conservation rather than agricultural importance (e.g. Parmesan *et al.*, 1999; Thomas *et al.*, 2001). The aim of this chapter, therefore, is to exemplify the scope of thermal effects on pests and their natural enemies, focusing on both well-studied and more recent examples. The first part of the chapter describes the effects of low temperatures on the abundance of insect pests, and then highlights examples of how temperature determines the distribution of such species. It is evident that many of the most recent studies have considered how the abundance and distribution of important pests have changed over recent decades in response to climate warming and may continue to change into the foreseeable future; the next section of the chapter therefore considers examples where data on the cold-tolerance of insects, climate records and models of predicted climate scenarios have been integrated to provide the most up to date "forward look" at pest problems. The chapter then briefly reviews the use of temperature as part of quarantine control for invasive pests, and concludes with a section on the role of temperature in determining the establishment potential of non-native biocontrol agents.

14.2 Abundance of pest species

The likelihood and extent of damage caused to crops by insects is determined by the growth stage of the plant at the time of colonization and the density of the pest, and this principle applies to both long-term ecosystems, such as forestry, as well as annual monocultures, and to insects with different types of life cycles and levels of cold-hardiness.

The autumnal moth, *Epirrita autumnata*, overwinters as eggs on birch (*Betula pubscens* spp. *Tortuosa*) in parts of northern Scandinavia. The eggs hatch in spring and the emerging caterpillars feed on the new leaf growth. Severe outbreaks have occurred at irregular intervals, sometimes resulting in the complete defoliation and death of trees, lowering the forest limit (tree line) over large tracts of land. However, even during these outbreak years there are areas of the forest in which there is much less damage. Climate records revealed that in these "damage-free" areas the minimum winter temperatures fell below the mean freezing temperature (supercooling point or SCP) of the overwintering eggs (around −35 to −36 °C) resulting in high levels of mortality; in other areas, although the minimum temperature was low (−30 °C), it was not sufficiently low to kill the eggs, which therefore survived to produce caterpillars in the spring (Nilssen and Tenow, 1990; Tenow and Nilssen, 1990). There is therefore a direct correlation between the cold-hardiness of the overwintering stage and both the extent and local occurrence of feeding damage in the following spring and summer. It is also interesting to note that *E. autumnata* is one of the few species so far studied in which there seems to be little or no pre-freeze mortality above the measured SCP.

The aphid, *Myzus persicae*, is one of the most serious pests of temperate agriculture transmitting virus diseases to potato (potato virus Y and potato leaf roll virus) and sugar beet (beet yellows virus). As with many aphids of economic importance, *M. persicae* overwinters in one of two distinct life cycles. In the holocyclic life cycle, sexual morphs are induced by the decreasing photoperiod in the autumn and then produce the overwintering eggs. Aphid eggs are very cold-hardy with SCPs below −35 °C (Strathdee *et al.*, 1995) and unlikely to be threatened by even the most severe winters. By contrast, anholocyclic clones reproduce asexually throughout the year, but are much less cold-hardy than the eggs of holocyclic clones (Bale *et al.*, 1988; Clough *et al.*, 1990).

Analysis of 40 years of annual flight activity and aerial density of *M. persicae* and other aphids from a suction trap in southern England (part of a network of traps coordinated by Rothamsted Research, Harpenden, UK) has found a strong correlation between the January–February mean temperature and the date of first record for an alate *M. persicae* in the suction trap (Harrington *et al.*, 1990); aphid migratory

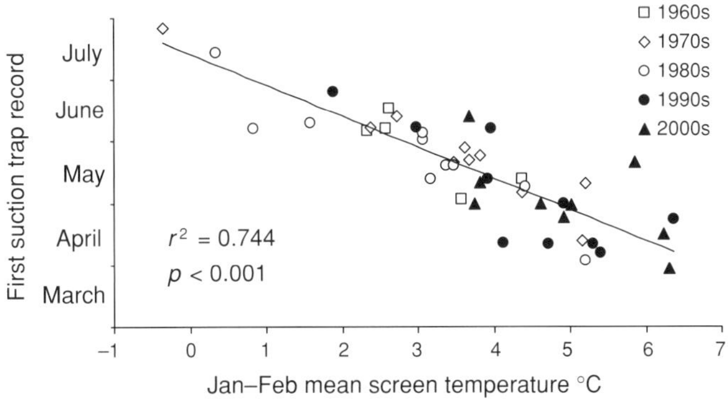

Figure 14.1 Date of first record of the aphid *Myzus persicae* in the Rothamsted Research suction trap from 1965 to 2008.

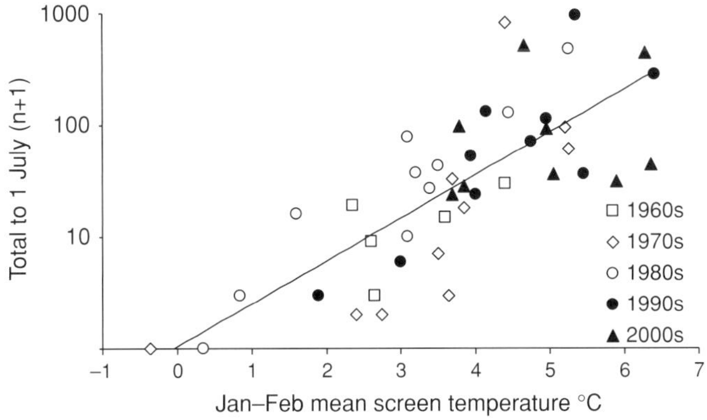

Figure 14.2 Total migrant population (to July 1) of the aphid *Myzus persicae* in the Rothamsted Research suction trap from 1965 to 2008.

flight occurs earliest (early April) after very mild winters (January–February mean temperature of around 6 °C) and latest (mid-July) after the coldest winters (January–February mean of around 0 °C). Additionally, after milder winters, the general abundance of alate aphids from spring to early summer is also correspondingly higher (Harrington *et al.*, 2007). When the data for date of first record and total migrant population are analysed at decade intervals since 1965, there is a clear trend toward earlier migrations and higher migratory populations since 2000, which have been some of the mildest winters in the UK over the last 40 years (Figs. 14.1 and 14.2). Because the predicted migratory (crop colonization) date in any year is known by the end of February, growers can receive this advice before the sowing dates of potato and sugar beet, and take appropriate crop-protection measures, i.e. insecticides might be used in years with an expected early migration and high pest density, but chemical control can be avoided in years when the migration will be late, aphid density low and thus minimal crop damage (see Harrington *et al.*, 1989; Werker *et al.*, 1998 and Qi *et al.*, 2004 for a description of recent developments in the forecasting of aphid transmission of sugar beet viruses).

The precision of this forecasting system can be explained by studies that have investigated the cold-tolerance of the anholocyclic clones of *M. persicae* and species with similar life cycles. Adults and nymphs of *M. persicae* have SCPs below −20 °C, but 50% of populations die after exposures of less than 1 h to temperatures around −10 to −14 °C, depending on the age group and level of acclimation (Bale *et al.*, 1988; Clough *et al.*, 1990). Thus after a severe winter there will be few survivors and population build up to the critical densities that induce the alate generations that migrate to crops will not occur until mid-summer; after very mild winters, survival will be high, population increase rapid and alates will be produced as much as 12 weeks earlier than after the coldest winters (Harrington *et al.*, 1990).

Myzus persicae was one of the first well-documented insect pests to develop resistance to insecticides, now involving a number of distinct biochemical mechanisms that confer resistance to different chemical groups of insecticides (see Foster *et al.*, 2007 for a review). It appears, however, that there is a trade-off between resistance and cold-tolerance, in which the most insecticide resistant clones of *M. persicae* are the least cold-tolerant with a lower overwintering success (Foster *et al.*, 1996, 1997, 2002).

Although winter climate has a major influence on the population dynamics and economic status of aphids such as *M. persicae*, the overwintering biology of anholocyclic aphids is different from that of many other insects. Whereas most insects overwinter in one stage of the life cycle, which survives (or not) from autumn through to spring, aphids continue to develop and reproduce throughout winter such that individuals found on plants in early spring are likely to be the granddaughters or great-granddaughters of the aphids that entered the winter approximately 4–6 months earlier. This in turn sheds further light on some of the atypical features that underpin aphid overwintering, as recently discussed by Powell and Bale (2008). Firstly, there is a limit to which aphids might be expected to invest in increasing the cold-hardiness of a single life-cycle stage of a generation or even a particular generation, as no individuals of the early winter generations will survive until the end of winter; however, some acclimatory ability is desirable, at least to the level that allows immature stages to become adult and reproduce, and aphids do show this within-stage acclimation (Clough *et al.*, 1990; Powell and Bale, 2008). Secondly, as aphids are relatively short-lived and may experience diurnal fluctuations in temperature, an ability to rapidly cold-harden would aid survival; recent studies have shown that both "summer" populations of *Sitobion avenae* (reared at 20 °C) and an acclimated culture (maintained at 10 °C) are both able to rapidly cold-harden by about 3 °C below their respective discriminating temperatures (temperature at which 80–90% die in a "direct plunge" exposure) of around −8 and −11 °C respectively (Powell and Bale, 2004, 2005). However, the so-called "telescoping of generations" provides a further cold-hardening mechanism that does not occur in most other insects. During the summer months and for anholocyclic clones throughout the year, aphid reproduction is both parthenogenetic and viviparous, i.e. progeny are born as nymphs without mating of the female parent. Thus, when an aphid (nymph or adult) becomes more cold-hardy in response to lower temperatures, the developing nymphs within their body will experience the same decreasing temperatures as will the embryos within the developing nymphs. So, when aphids are maintained at 10 °C for three generations (or presumably when low winter temperatures persist for similar periods of time), there is an increase in cold-tolerance with each generation, resulting from a combination of pre-natal and post-natal acclimation. In *S. avenae* this decrease in the LT_{50} is from −8.0 °C in the control (20 °C population) to −14.8 °C after three generations at 10 °C; first-born nymphs were also more cold-hardy than those born later in the birth sequence, with a decrease in LT_{50} down to −16.6 °C after three generations at 10 °C for the earliest-born progeny (Powell and Bale, 2008). Given that minimum winter temperatures in the UK are usually between −5 (mild winter) and −15 °C (severe winter), these acclimatory responses may be sufficient to increase survival, at least in acute exposures.

14.3 Distribution of pest species

There is no doubt that climate, though not exclusively winter low temperatures, exerts a major influence on the distribution of insects, and the relative abundance and voltinism of species within the occupied range. As an example, the Large Yellow Underwing moth *Noctua pronuba* (family Noctuidae) is common in the UK, overwintering as larvae (commonly known as cutworms) that feed on the leaves, stems and roots of a wide range of plants, including beet, potato and carrot; feeding occurs on warm days in winter, suggesting that there is no winter diapause. Adults of the related Dark Sword-grass moth *Agrotis ipsilon* are often found in the UK in large numbers in late summer (August to October), but these migrants seem unable to establish permanent populations. The larvae (black cutworms) of *A. ipsilon* are major pests on crops of world importance (maize, cotton, soybean). When the cold-tolerance of the "overwintering" larvae of these species was compared, the SCP of the larvae were similar (around −12 °C), decreasing to −15.4 °C after acclimation in *N. pronuba*, but with no response in *A. ipsilon*

Table 14.1 *Comparative thermal data for larvae of the noctuid moths,* Noctua pronuba *and* Agrotis ipsilon

Thermal index	*Noctua pronuba*	*Agrotis ipsilon*
SCP (non-acclimated) ± SE (°C)	−12.1 ± 0.3	−12.4 ± 0.5
SCP (acclimated) ± SE (°C)	−15.4 ± 0.5	−12.6 ± 0.5
$LTemp_{50}$ acclimated ± 95% fiducial limits (°C)	−15.8	−8.9
−15.0 to −16.5	−7.4 to −10.1	
$LTime_{50}$ acclimated at −5 °C ± 95%	9.0	5.8
Fiducial limits (days)	6.2 to 11.2	5.2 to 6.2
Field survival	>6 months	6 weeks

(Table 14.1). The $LTemp_{50}$ of acclimated larvae was much lower in *N. pronuba* (−15.8 compared with −8.9 °C) indicating some mortality above the SCP in *A. ipsilon*, and the $LTime_{50}$ at −5 °C was also longer in *N. pronuba* (9 and 5.8 days respectively). The most striking difference was observed in winter field survival, where populations of *A. ipsilon* provided with food died out after 6 weeks, whereas 40% of *N. pronuba* larvae survived for over 6 months until spring. The lack of larval cold-hardiness seems the most likely explanation for the failure of *A. ipsilon* to establish in the UK (Bale and Walters, 2001).

Similar temperature constraints appear to restrict the "resident" distributions of the Diamond-back moth, *Plutella xylostella* and Silver-Y moth, *Autographa gamma. Plutella xylostella* is a widely distributed pest of cruciferous crops. In tropical and subtropical areas the moth breeds continuously, but in the more northerly and southerly parts of its distribution it appears that there are few, if any, permanent populations and the summer occurrence of larvae and outbreaks are related directly to a late-spring–early-summer immigration of adult moths (Talekar and Shelton, 1993). Outbreaks of *P. xylostella* occurred in the UK in 1958, 1966 and 1978, based on a large-scale immigration of moths, apparently travelling over 2000 km from Scandinavia (Chapman *et al.*, 2002). A novel radar tracking system was devised to deduce the origin of the 2000 migration of *P. xylostella* into the UK (see Chapman *et al.*, 2002 for details). A detailed analysis of the radar tracks of *P. xylostella* on specific nights in May 2000 combined with aerial and light traps indicated that the immediate source of the immigrant population was from the Netherlands and directly neighbouring countries, with an estimated nightly influx of around 3000 moths per km along a "migration front" extending over several hundred km. Chapman *et al.* (2002) speculate that the original source of the 2000 *P. xylostella* infestation was from the Mediterranean area with moths flying 400–500 km per night over several consecutive nights. In the same year, large numbers of *P. xylostella* were found on Svalbard, approximately 800 km to the north of mainland Norway (Coulson, 2000). At the present time it seems that *P. xylostella* is unable to overwinter in the UK (and similar cool temperate areas) and that early season migrations originate from southern Europe, whereas the later season influxes such as those observed in 1958, 1966 and 1978 are from the Baltic region (Chapman *et al.*, 2002).

The radar tracking technique has also been applied to another migratory moth, the Silver Y, *Autographa gamma*. As recognized by Chapman *et al.* (2008), the migratory period of nocturnal migrant moths is limited to a few days between adult emergence and sexual maturity (Hill and Gatehouse, 1993). In the case of *A. gamma*, this moth migrates to the UK in spring and early summer, reproduces for a few generations, and then undertakes a return migration to overwinter in the Mediterranean region. The larvae feed on the leaves of sugar beet and other crops. Whilst it has hitherto been assumed that high-altitude, long-distance nocturnal migrations of insects were largely "uncontrollable" by the organisms, effectively being transported and dispersed by prevailing wind direction, Chapman *et al.* (2008) have shown that in the return migration of *A. gamma* leaving the UK in August, moths are able to select fast-moving airstreams at particular altitudes, utilize some "inherited compass" that identifies the required migratory direction, and compensate for cross-winds that would otherwise displace the insects from their preferred southerly route. Given that in the 2003 return migration an estimated 200 million moths migrated south along a 150 km "migration front" across the UK, these results shed new light on the orientation abilities of such migratory pests.

Agrotis ipsilon, P. xylostella and *A. gamma* are thus all capable of long-distance migrations, often arriving in northern Europe in large numbers in summer. In taking a forward look at the likely impacts of climate warming on insect distributions, Lawton (1995) contrasted the rapid range expansion capabilities of insects with the much slower movement of any required host plant. With these lepidopteran species it is clear that suitable hosts are available in many of the areas to which the moths regularly migrate and establish transient summer-only populations. The failure to become permanent residents thus seems attributable, at least in part, to the absence of a diapause state and a lack of adequate winter cold-hardiness, although diapause is not an essential requirement for overwintering success.

The ability to make accurate predictions on changes in winter survival linked to different scenarios of global climate change is difficult because there is often an absence of data on the cold-tolerance of pest species, and even where field experiments have reported "no survival", the nature of the cold stress (acute or chronic) causing the mortality may not be readily decipherable. This is an important consideration because whilst increases in mean temperature may lead to higher survival, the greater frequency of abnormally cold (or warm) "extreme events" may be more detrimental (or favorable) in the longer term. These interactions are considered in the next section.

14.4 Abundance, distributions and climate change

There is much current interest in the possible impacts of global climate change on both the abundance and distribution of pest species. All insects have a developmental threshold below which there is no development, and above this threshold, a "thermal budget", the number of day degrees required to complete a generation. By way of generalization, it might be expected therefore that under a warmer climate, insects would develop more rapidly within existing distributions (increasing abundance and possibly annual voltinism) and also expand their range margins to higher latitudes (and altitudes) – though there are other factors at play that may limit a "favorable" impact of climate warming on insect pests. For example, higher temperatures may favor natural enemy species rather than pests, or lead to aestivation in summer; and in those species that overwinter in diapause, whilst a warmer climate would not affect photoperiod, which is the dominant inducing cue for facultative diapause, higher temperatures may interact with critical daylengths to modify the diapause response (see also Chapter 10, Bradshaw and Holzapfel).

The interactions between cold-tolerance, climate warming and range expansion have been investigated for the southern green stinkbug, *Nezara viridula*, which occurs at the northern margin of its Asian distribution in Japan (see Musolin, 2007 for review). In the 1960s, the northern limit of *N. viridula* was at 34.1°N delineated by the 5 °C mean temperature isothermal line for the coldest month of the winter (January). By 2000, *N. viridula* was found 70 km further north (34.7°N), progressively displacing the closely related *N. antennata*. The northern limit of *N. viridula* seems to be determined by winter cold rather than availability of host plants or the failure of summer temperatures to support development and reproduction. Analysis of climate records indicates that the lowest and mean winter temperatures increased by 1–2 °C between 1950–1960 and 2000 (Musolin, 2007). The study on *N. viridula* also highlighted how the critical photoperiod for diapause induction can act to constrain range expansion. *Nezara viridula* has 5–6 generations a year in the study area and 50% of individuals enter diapause when the daylength decreases to between 13 to 12.5 h of light at 20 to 25 °C (Musolin and Numata, 2003a). Experiments have shown that only diapausing adults are able to overwinter; eggs and nymphs always die in winter (Musolin and Numata, 2003b). Adults that emerged before September 1 mated and reproduced, but unless their progeny became adult before winter, they were all predestined to die; adults emerging in October were all in diapause. *Nezara viridula* was found to have the shortest critical photoperiod of eight Heteropteran species, resulting in a relatively "late" induction of diapause under natural conditions. Musolin (2007) concluded that *N. viridula* is not well adapted to conditions experienced at its northern margin in Japan because the timing of diapause induction means that over the two-month period from mid-September to early November, diapause is not induced in all individuals in the population, but for those adults that reproduce, decreasing temperatures prevent their offspring from completing development to become diapausing adults. Whilst the high mortality associated with the late induction of diapause may select for longer critical daylengths (and correspondingly earlier diapause), the migratory nature of *N. viridula* will maintain a genetic intermixing for this trait at the range margin. Thus the longer-term completion (and continuation) of the northwards colonization and invasive process will depend on re-establishing phenological synchrony with both the physical and biological environments (Musolin, 2007).

Studies on the tropical root weevil, *Diaprepes abbreviatus*, show how winter climate can limit the distribution of pest species within a relatively small area (Lapointe *et al.*, 2007). This weevil is a pest of citrus and ornamental plants and was first recorded in an area of Florida, USA in 1964. Since its introduction it has spread to the south, but not to the north of the state. Larvae have a developmental threshold around 12 °C, oviposition does not occur below 15 °C, and 95% of eggs are killed in about 4 days at 12 °C. Analysis of climate records found that the northern limit of *D. abbreviatus* in Florida correlated with areas where there were 15–20 days a year with a mean temperature at or below 12 °C. Furthermore, establishment of egg parasitoids of the weevil in Florida occurred in regions with less than 25 days a year with a temperature below 15 °C, indicating that the parasitoid would be unable to survive in California, Texas and Arizona, though the weevil could establish in some parts of these states (Lapointe *et al.*, 2007).

The winter pine processionary moth, *Thaumetopoea pilyocampa*, is a defoliating pest of pine trees of Mediterranean origin that has expanded its range both altitudinally and latitudinally in Europe, apparently in response to higher winter temperatures associated with climate warming, which increases larval survival (Battisti *et al.*, 2006; Buffo *et al.*, 2007); over the last 30 years, the altitudinal range in the Italian Alps has increased by up to 230 m in a gradual pattern (Battisti *et al.*, 2005). The larvae of *T. pilyocampa* are highly gregarious, spinning nests made from silk on branches of host trees, usually in November; depending on temperature, larvae leave the nest at night to feed on pine needles. The number of hours per day that larvae are able to feed has been quantified, based on thresholds of daytime nest temperature (>9 °C) and night air temperature (0 °C), termed the "realized feeding threshold" or RTF (Battisti *et al.*, 2005; Buffo *et al.*, 2007). In a transplant experiment, colonies of larvae were transferred to sites at different altitudes in the Italian Alps (as a proxy for increases in temperature arising from climate change) and onto known host tree species as well as "new" hosts. These sites were selected to represent the "historical" distribution of *T. pilyocampa*, recently colonized areas, and locations outside of the known range. The dominant factor affecting post-winter survival was the number of feeding hours per day during the "cold period" from December to February (Battisti *et al.*, 2005; Buffo *et al.*, 2007). A recent study has examined the cold-tolerance of *T. pilyocampa* by a range of indices, as well as the ability of overwintering larvae to withstand starvation, to further understand the rapid range expansion of this insect (Hoch *et al.*, 2009). The mean SCP of larvae in winter was around −7 °C with a range from −1 to −13 °C; approximately 40% of larvae were alive immediately after freezing, skewed toward those with the highest SCPs. When exposed to −17 °C for 1 h inside nests, there was 74% survival of starved larvae and 67% of larvae provided with food, and depending on the temperature treatment, up to 90% of larvae were alive after 8 weeks of starvation (9 °C day, −5 °C night). Based on these data it appears that *T. pilyocampa* may have a dual overwintering strategy, in which larvae with a high SCP are freeze-tolerant (to around 10 °C below the SCP), whereas those with a lower SCP survive by supercooling, but die if they freeze. However, although the freeze-tolerant state provides an explanation for the recently observed range of expansion and survival at high altitude (Battisti *et al.*, 2005, 2006; Buffo *et al.*, 2007), there are combinations of day and night temperatures (negative thermal sum) that prove lethal to the overwintering larvae, which Hoch *et al.* (2009) speculate to be low daytime temperatures that limit or prevent the repair of cryo-injuries experienced during subzero exposure.

Whilst global climate change is often characterized by increases in minimum or mean temperatures, therefore favoring winter survival, and abundance and range expansion respectively, there are also predictions of increasing frequency of "extreme events" that may modify patterns predicted or actually occurring under current climate. Thus, as Battisti *et al.* (2006) have recognized, shifts in range margins may be temporary and revert to previous distributions as either variation in prevailing climate or an extreme event kills off "new colonies" at the range margin. However, *T. pilyocampa* provides an example of an extreme event promoting a rapid range expansion that may become permanent (Battisti *et al.*, 2006). During the summer of 2003, the moth colonized high-altitude sites in the Italian Alps, promoted by very warm nights during the June to August flight period, when temperatures above the flight take-off threshold (around 14 °C) were much more frequent than normal, including in areas around the existing range margin. This "single year" expansion of the *T. pilyocampa* range was unprecedented in the previous 30 years. Colonies at these elevated sites survived through the 2003–2004 winter and produced offspring in the following year. Whilst an abnormally cold winter could reverse this expansion, the overwintering larvae appear to possess adequate cold-hardiness to survive in the newly colonized areas, and in combination with the unusual trait of a pupal diapause extending over several years, may allow this abnormally rapid extension to the range margin to become permanent (Battisti *et al.*, 2006).

Studies on the cabbage root fly (*Delia radicum*) illustrate the complexity of some of the interactions between diapause, voltinism and climate warming. *Delia radicum* is a widely distributed pest of cultivated brassica crops in northern Europe and the USA, damage being caused by root-feeding larvae; *D. radicum* overwinters in the soil as diapausing pupae. Diapause is induced in the last larval instar by decreasing photoperiod and temperatures below 15 °C (Hughes, 1960); the period of diapause development varies from 12 to 22 weeks, with temperatures between 0 and 10 °C, terminating most quickly at lower temperatures within this range (Collier and Finch, 1983a,b). The developmental threshold is around 6 °C (with some interstage variation) with a thermal budget of 580 day degrees for the summer non-diapause generation(s), and an additional 100 day degrees post-winter required for the overwintered generation, reflecting the time taken for post-diapause pupae to develop into flies (Collier and Finch, 1988). In the UK, there are two generations a year in Scotland and northern England, and three generations from the Midlands southwards (Collier *et al.*, 1991). Interestingly, the overwintering pupal population consists of genetically distinct "early"- and "late"-emerging flies; early emerging flies commence post-diapause development above 4.3 °C (Collier and Finch, 1986), whereas for the late-emerging flies the post-diapause threshold is around 7 °C (Collier *et al.*, 1989). There is also intraspecific variation in summer pupal aestivation, with an increasing proportion aestivating as the temperature increased from 20 to 27 °C; normal development resumed as soon as the temperature decreased (Finch and Collier, 1985). Application of a simulation model (Phelps *et al.*, 1993) based on an increase in temperature of 3 °C indicated that whilst spring emergence would be about a month earlier, it would be less synchronized than at present, and there would still be three generations a year, even in the south of England; an increase of 5 °C or higher would be required for four generations, with aestivation then likely to disrupt egg-laying (Collier *et al.*, 1991). In this example, increases of temperature at the extreme end of current predictions would be required to modify annual voltinism.

In a review of the impacts of global warming and land-use change on the pest status of rice and fruit bugs (Heteroptera) in Japan, Kiritani (2007) reports evidence of an increase in importance of both groups of pests over the last 30–40 years, as a result of changes in agronomic practices and plant diversity. The reduction in the area used for rice cultivation has resulted in approximately 10% of former paddy fields lying fallow, providing a resource for rice bugs, which feed and reproduce on forage crops and grass weeds. Similarly, the introduction of coniferous plantations from the 1950s onwards that produced the first cones about 20 years later is linked to increased damage by "fruit bugs" (*Plautia crossota stali, Glaucias subpunctatus* and *Halyomorpha halys*) on crops including pear, peach, apple and mandarin orange. These bugs are associated with conifers, such as Japanese cedar (*Cryptomeria japonica*) and cypress (*Chamaecyparis obtusa*); the annual abundance of the bugs is dependent on the masting (cone production) by the host trees which occurs at approximately 3-year intervals, preceded by a hot, dry summer the year before. The bugs cannot reproduce on their fruit-tree hosts, but when abundant after a masting year on their coniferous hosts, cause damage by feeding punctures on developing fruits.

The analysis of heteropteran pest problems in Japan also identified changes in distribution since 1970 and the importance of winter temperatures in determining subsequent pest abundance (Kiritani, 2007). Following a general increase in mean temperature in Japan of about 1 °C over the past 40 years, there has been a northward shift of the mirid, *Stenotus rubrovittatus*, and the increase in temperature has been sufficient to allow an extra generation each year (Kiritani, 2006). Comparison of levels of survival in areas or winters of different severity indicate that for each 1 °C increase in mean temperature above 4 °C in January (for *N. viridula*) and in January–February (for *H. halys*) there is a 16.4 and 13.5% reduction, respectively, in winter mortality. Thus in general, higher temperatures have the potential to increase winter survival, annual abundance and voltinism accompanied by shifts in range margins.

The outbreak areas of the autumnal moth, *E. autumnata* (see earlier section on abundance of pest species), have also changed in recent years. Records of birch forest damage in northern Scandinavia dating back over 100 years have revealed differences in the expansion of outbreak areas attributable to two geometrid moths, *E. autumnata* and the winter moth, *Operophtera brumata* (Jepsen *et al.*, 2008). *Epirrita autumnata* has a lower egg (overwintering stage) SCP than *O. brumata*, which has been regarded as one of the explanations for the historical differences in their distributions and outbreak areas. Over the last 15–20 years there has been an increase in both the minimum winter and mean annual temperatures in northern Scandinavia; over the same period, *O. brumata* has shifted its range north-eastwards, becoming dominant in areas where outbreaks were previously caused solely by *E. autumnata*, whereas the autumnal moth has expanded its range into even colder continental parts of its distribution (Jepsen *et al.*, 2008), where, in some winters, there are now no days below −35 °C, the mean SCP of the overwintering eggs (Tenow and Nilssen, 1990).

A series of studies on pine beetles in the USA and Canada have combined direct measurement of winter cold-hardiness with climate records to parameterize models that explain recent expansion in range margins and identify areas of forest that are at risk in the event of continuing climate change. The southern pine beetle, *Dendroctonus frontalis*, is distributed in the south-eastern states of the USA from Texas and Florida northwards to Pennsylvania and New Jersey, an area with a warming trend of around 3 °C over the past 40 years (Tran *et al.*, 2007). The beetle kills mature pine trees and outbreaks have become more common at the northern range margin in recent years (Tran *et al.*, 2007). By measuring the SCPs of various life cycle stages it was concluded that a minimum winter temperature of −16 °C or lower would kill >90% of adults and virtually all larvae and pupae; furthermore, the northern distribution of *D. frontalis* coincided with the −16 °C isoline, where there was a 90% annual probability of this temperature occurring (Ungerer *et al.*, 1999). By modeling the expected minimum temperatures in different areas of the distribution with allowance for the thermal buffering provided within host tissues (up to 4 °C) and measuring the SCP of adults and pre-pupae (last larval instars that move from the phloem to the outer bark) Tran *et al.* (2007) found that pre-pupae had the lowest mean SCP (−14.6 °C), with some individuals freezing as low as −19.9 °C, and this was the only life cycle stage found alive in relatively large numbers at the northern range margin in New Jersey. The winter population of *D. frontalis* at the current northern limits of its distribution is therefore biased toward the most cold-hardy stage of the life cycle, a phenomenon described as "adaptive seasonality" by Logan and Bentz (1999).

The thermal ecology of the mountain pine beetle, *Dendroctonus ponderosae*, has also been extensively studied and modeled. The success of this species is dependent on cohort synchrony and voltinism that, in the absence of diapause, result from the interaction of annual temperature regimes and developmental thresholds of all life stages (Bentz *et al.*, 1991; Logan and Bentz, 1999; Powell *et al.*, 2000). The univoltine life cycle ensures a single overwintering period per generation; stage-specific developmental thresholds result in overwintering in the most cold-hardy life-cycle stages, leading to a synchronized adult emergence that can produce a mass attack capable of overwhelming the chemical defenses of the host pine trees. In addition, because tree defenses are greatly affected by drought stress, precipitation is an important determinant of attack success (Safranyik, 1978). However, the most important weather factor affecting survival is low temperature (Cole, 1981). The cold-tolerance of *M. ponderosae* is highly dynamic, and responds rapidly to daily temperature fluctuations (Bentz and Mullins, 1999; Régnière and Bentz, 2007); larvae are the most cold-hardy stage, whilst eggs and pupae are most susceptible to low temperature. Models combining voltinism, drought stress on host trees and winter mortality have been developed and used to predict the likely geographical distribution of *M. ponderosae* risk in North America (personal communication, B. J. Bentz, USDA Forest Service, Logan, Utah, United States of America and J. Régnière, Canadian Forest Service, Quebec, Quebec, Canada). As climate changes, the distribution of *M. ponderosae* is expected to shift rather than expand. In the short-term, population success will increase at high elevations and northern latitudes. Indeed, it is likely that recent winter warming in western Canada has contributed greatly to the development of unprecedented outbreaks of the mountain pine beetle (Carroll *et al.*, 2006). In 2006, a massive outbreak of *M. ponderosae* in British Columbia spread into north-western Alberta, on the eastern side of the Rocky Mountains, which had previously constituted a geographical barrier. However, model predictions suggest that over the next 70 years, winter temperature regimes in the Canadian boreal forest east of the Rocky Mountains from Alberta to Ontario are likely to remain too cold for *M. ponderosae* to thrive, despite generally higher temperatures.

The pink bollworm, *Pectinophora gossypiella*, is a cotton pest of worldwide importance. The bollworm invaded the USA over 40 years ago and currently occurs in parts of Arizona and California, with concerns that it may extend its range to the main cotton-growing area in the Central Valley of California (Gutierrez *et al.*, 2006). The current distribution is thought to be limited to frost-free areas with <10% of diapausing larvae surviving through winter in Arizona. Using a modeling approach that integrates the agronomy of cotton growth with estimates of bollworm mortality below the developmental threshold of 10 °C and climate records for different localities in Arizona and California, Gutierrez *et al.* (2006) predicted that within the two states between 1996 and 2003, winter survival of larvae would vary from 0 to 100%, with very high mortality over much of the Central Valley; importantly, areas where permanent establishment was predicted were those where *P. gossypiella* is currently a serious pest. When increases in mean temperature of between 0.5 and 2.5 °C were incorporated into this model, there was no marked change in the current distribution until a warming effect of 1.5 °C was applied; as the temperature increased to 2.5 °C, the winter survival of larvae increased, and *P. gossypiella* was then predicted to become permanently established in the Central Valley (Gutierrez *et al.*, 2006). In an extension of this modeling approach, Gutierrez *et al.* (2008) predict that within California, there will be changes to the distribution of the olive fly, *Bactrocera oleae* and phenology of olive trees, modification to food-web structure in alfalfa (*Medicago sativa*), and detrimental effects on the biological control of vine mealy bug (*Planococcus ficus*) on grape (*Vitis vinifera*) by its natural enemies *Cryptolaemus montrouzieri, Anagyrus pseudococci* and *Leptomastidae abnormis.*

14.5 Thermal treatment of quarantine pests

The application of high and low temperatures provides a chemical-free alternative to the use of fumigants to "disinfect" fruits and related crop products imported into countries such as the USA and Canada. All insects have upper and lower "instantaneously" lethal thermal limits, but these temperatures are also damaging to some products, especially in the subzero range. Low-temperature treatments are usually applied over periods of days or weeks to impose a chronic non-freezing lethality on target pests, whereas heat treatments are effective after exposures of only minutes or hours; both methods can be applied pre-shipment, in transit, or on arrival at an import location (see Heather and Hallman, 2008 for a review of this area).

Many fruits are tolerant of low temperatures of around 0 to 2 °C over extended periods of time, including those of tropical origin (though for not more than 10–40 days) and in some cases, the cold regime used to store fruits (e.g. apples) also functions as a phytosanitary treatment. A selected treatment must be effective against the most cold-tolerant stage of the life cycle, which can be identified by preliminary screening of different life-cycle stages across a range of temperatures and exposure periods. A further complication is that for a pest such as the Mediterranean fruit fly (*Ceratitis capitata*), the minimum exposure time required to achieve 100% mortality of the most cold-tolerant stage varies depending on the host fruit (Heather and Hallman, 2008); in the USA, the presently recommended treatments for *C. capitata* are 14, 16 and 18 days at 1.1, 1.6 and 2.2 °C respectively. Whilst pests of tropical and Mediterranean origin can be effectively killed by relatively short exposures, temperate pests may require much longer treatment to ensure quarantine security, especially if the species is able to diapause: for example, Heather and Hallman (2008) cite exposure periods at 2.2 °C of 10 days for the non-diapausing oriental fruit fly, *Bactrocera dorsalis*, and 42 days for the diapausing apple maggot, *Rhagoletis pomonella*. Cold treatments can also be combined with methyl bromide fumigation and this reduces the required period of exposure e.g. at 2.2 °C from 18 to 4 days for *C. capitata*, which enables products to be treated "in transit".

Quarantine pests can also be controlled by heat treatments, though high temperatures are potentially damaging to fresh fruits and vegetables (also reviewed by Heather and Hallman, 2008), and their effectiveness is more variable than with cold and fumigation. Heat treatments are "carried" in air or water; both of these treatments have increased in importance following the loss of ethylene dibromide (a carcinogen) and concerns over the environmental safety of methyl bromide. For a range of tephritid fruit flies, heated-air treatments were effective between 43.5 and 47 °C with exposures of 10 min to 5 h (combined time of "heating up" and "holding" temperature), depending on the fruit product and pest species. For a different group of fruits and ornamental plants, and a wider range of pests, hot-water treatments have been applied at 46 to 52 °C for exposures of 5 min to 1.5 h (Heather and Hallman, 2008).

14.6 Biological control

The term "biological control" has a range of definitions, but generally refers to the use of predators (insects and mites), parasitoids (insects) and microbes (bacteria, fungi and viruses) to "suppress" pest populations, usually comprising insects and mites (Bale *et al.*, 2008). However, this definition of biological control also includes the use of entomopathogenic nematodes, semiochemicals and botanical "pesticides". The concept of "suppression" relates primarily to classical biocontrol, where the aim is to establish a long-term self-sustaining control in which eradication of the pest would be undesirable as it would be followed inevitably by the loss of the control agent through starvation, and thus any re-invasion of the pest may lead to its rapid expansion in the absence of any natural enemies.

Biological control pre-dates the modern pesticide era in which the control of the scale insect, *Iceryae purchasi*, is widely regarded as the first well-documented example. *Iceryae purchasi* was introduced accidentally from Australia to citrus crops in California and rapidly became a major pest. Two natural enemies of the scale were found in Australia, a predatory ladybird, *Rodolia cardinalis*, and the tachynid parasitoid, *Cryptochaetum iceryae*, which gave effective control soon after release in California, and this management program was sustained throughout the twentieth century. Whilst "establishment" is an essential pre-requisite for success of a classical biological control agent, this is either not the aim, or may be impossible to achieve with other types of biocontrol. For example, augmentation refers to all forms of biocontrol in which natural enemies are periodically released, and in most cases such agents are produced commercially by companies. Inundation is a form of augmentative control in which large numbers of the control agent are released, such as the *Trichogramma* egg parasitoids for control of lepidopteran pests, including the sugar cane borer, *Diatraea saccharalis* and the European corn borer, *Ostrinia nubilalis* (Bigler, 1986; van Lenteren and Bueno, 2003). In inundative control the major objective is to create an overwhelming ratio in favour of the natural enemy, thus producing a rapid control, but there is little or no reproduction via the target hosts to produce even transient establishment; and as far as is known, the parasitoids do not disperse far from the area of release and do not persist through winter, at least not in large numbers (Babendreier *et al.*, 2005).

In cool temperate climates such as in northern Europe, many horticultural crops and ornamental plants are produced in glasshouses, as the outdoor conditions are too severe, especially in winter, for such plants to survive. These crops are frequently attacked by a complex of pests (whiteflies, aphids, thrips, leafminers, mites), for which biological control is now the preferred, and sometimes the only, management option. Large numbers of commercially produced agents are released into the glasshouses in a form of inundative (augmentative) control, described as "seasonal inoculative control" by van Lenteren and Woets (1988). This technique is particularly suitable for short-term crops, where many generations of the pest can occur within a single growing season, enabling the control agents to increase in numbers as long as hosts or prey are available.

Conservation biocontrol refers to use of indigenous predators and parasitoids, usually against native pests. As the concept of this technique implies, most native insects have a range of native natural enemies, but the level of "control" is normally insufficient to prevent economic damage. Various measures are taken to "tip the balance" in favor of the natural enemy species, including for instance, creation of "beetle banks" in the middle of fields to act as overwintering refugia, or increasing the diversity of flowers in field headlands to provide essential nutrients for adult parasitoids and aphidophagous hover flies (Gurr *et al.*, 2000; Wäckers, 2003).

When viewed at a global scale and across all types of biological control, temperature has one dominant influence: determining, at least in the part, the natural distribution and range margins of the control agents (or of species with the potential to become biocontrol agents), and then a series of modifying effects on important processes within the occupied range, such as rate of development, voltinism and mobility, all of the latter being measurable, and therefore comparable to that of the target pest species. Information on this climate–distribution relationship then forms the basis of predictions of climatic "eco-regions", within which an introduced species might be expected to be able to establish long-term viable populations. The critical aspect of this climatic analysis, often overlooked, is that depending on the type of biocontrol program being undertaken, permanent establishment of an introduced natural enemy may be essential for success, as with the classical control of arthropod pests or weed biocontrol, or highly undesirable, as with non-native agents that escape from glasshouses and "unexpectedly" establish outdoors, potentially attacking non-target hosts and prey. Thus, in addition to climate allowing or preventing survival of a non-native biocontrol in a new environment, establishment is also dependent on the ability of the agent to utilize wild prey, either as an aid to survival of a classical agent during periods when the target pest occurs at low density, or as an essential requirement for species that escape from "protected cultivation" (glasshouses and poly-tunnels). Establishment is therefore usually the first factor to be evaluated in a potential novel control agent, as summarized in the hierarchical screening process (Figure 14.3) described by van Lenteren *et al.* (2006). From this analytical approach it is immediately apparent that if a species is intended for release as a classical biocontrol agent and it cannot establish in the new climate, then it is certain to fail and no further investigation would be worthwhile; although it is apparent that some of the early attempts to introduce non-native agents were conducted on

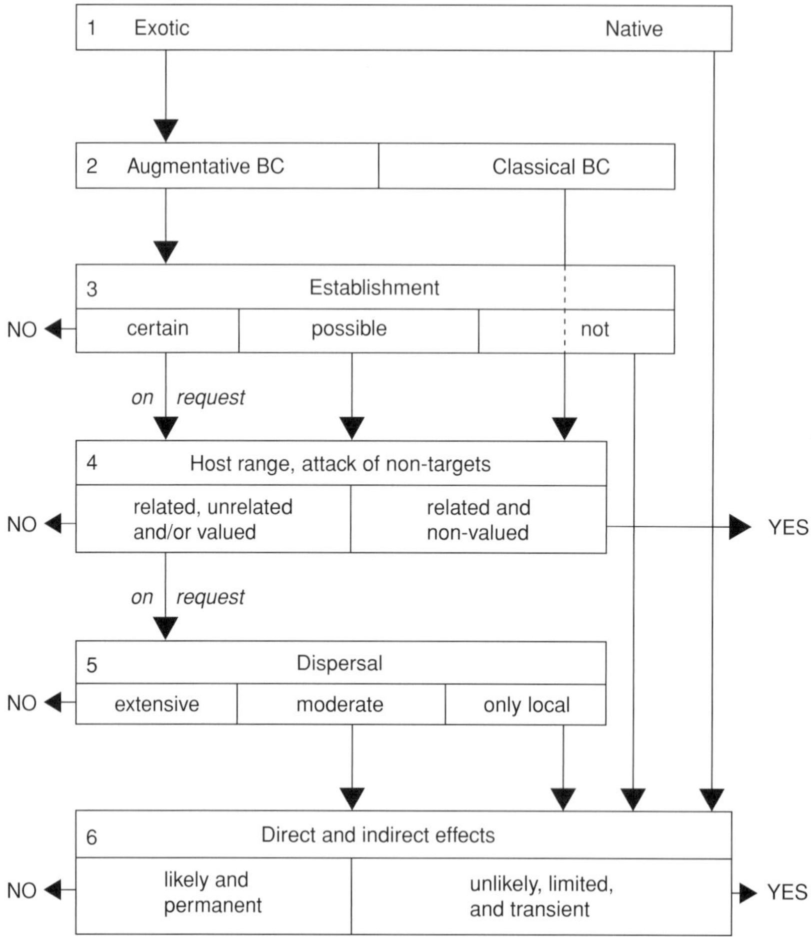

Figure 14.3 Flow chart summarizing a hierarchical environmental risk assessment scheme for arthropod biocontrol agents (modified after van Lenteren *et al.*, 2006).

a "trial and error" basis with little or no pre-release assessment of cold-tolerance or overwintering ability. As an example, the parasitoid, *Aphelinus mali*, released as a control for the woolly apple aphid, *Eriosoma lanigerum*, failed to establish in UK apple orchards apparently because of a lack of winter cold-hardiness, even though the parasitoid had been successfully introduced into other countries (Greathead, 1976). However, by the careful selection of agents, biocontrol can be successfully implemented in areas with seasonal climates and severe winters, such as the introduction from Europe of the parasitoids, *Cyzenis albicans* and *Agrypon flaveolatum*, to control the winter moth, *Operopthera brumata*, in Canada (DeBach and Rosen, 1991).

In cool temperate and colder climates, a significant proportion of the annual agricultural and horticultural production is carried out under protected cultivation, where temperature and humidity are both controlled and generally favorable for pest development and reproduction. The vegetable crops and ornamental plants grown in glasshouses are "high-value" commodities, where market value is determined largely by appearance. This has led to the frequent application of pesticides, and the consequent problems of pest resistance and chemical residues on produce, sometimes exceeding toxicologically determined "safe limits." Against this background there has been a marked increase in glasshouse biocontrol in northern Europe and elsewhere over the past 20–30 years. In addition to the need to find alternative control options because of the problems associated with pesticides, the reputation of glasshouse biocontrol was enhanced by the success of two widely used agents – the parasitoid, *Encarsia formosa*, against the whitefly, *Trialeurodes vaporariorum*, and the mite, *Phytoseiulus persimilis*, as a predator of the spider mite, *Tetranychus urticae* – neither of which had any reported negative environmental effects. An increasing number of commercially produced agents are now available and whilst there is no evidence that these introduced species pose any threat to human health (apart from some allergic reactions in production personnel) or crop plants (except in a few zoophytophagous species such as *Macrolophus caliginosus*), it was recognized that if they became established in new countries, there might be undesirable effects on non-target species. Various systems of environmental risk assessment (ERA) for non-native biocontrol agents are now operational in the United States, Canada, Australia and New Zealand (see Hunt *et al.*, 2008 for a review of the regulatory framework in different countries).

In Europe, there is no coordinated system of regulation, and those countries that require an ERA prior to granting a license to release a non-native species often have different information requirements, which in some cases are now known to lack sufficient scientific rigour. As an example, the UK regulatory system that operated in the 1990s required an assessment of the "likelihood of outdoor establishment" of species released exclusively into glasshouses. There was no requirement for companies to carry out direct assessments of cold-tolerance or overwintering ability of the candidate agents, and this information was rarely available in the published literature; instead, "climate matching" was used as a proxy for likely winter survival – if a species was of tropical, subtropical or Mediterranean origin, it would not be expected to survive through UK winters. The long-standing annual releases of *E. formosa* and *P. persimilis* without any establishment bolstered the apparent reliability of this index. However, the licensed release of the predatory mite, *Neoseiulus californicus*, in 1991 and the mirid, *M. caliginosus*, in 1995 highlighted the inadequacies of the climate-matching approach, linked in the case of

N. californicus to another overlooked attribute – the existence of strains with the ability to diapause. In the case of *N. californicus*, it seems that the original source population used for commercial production was non-diapausing, but was refreshed with new material, containing individuals with the ability to diapause. Once released into and escaping from UK glasshouses, the seasonal decrease in photoperiod and temperature selected for the diapause trait, leading to the establishment of outdoor populations (Jolly, 2000). However, a detailed study by Hart *et al.* (2002a) found that non-diapausing *N. californicus* could survive throughout a UK winter, and that some oviposition occurred during this time. There has been no study on the possible non-target effects arising from establishment of *N. californicus* in the UK. With *M. caliginosus*, individuals have been reported outside of glasshouses in winter, though it is not yet clear whether "full" establishment has occurred (Hart *et al.*, 2002b).

The studies on *N. californicus* and *M. caliginosus* highlighted the problem of assessing the likelihood of establishment by the system of climate matching and encouraged the development of methods based on the direct assessment of cold-hardiness in the laboratory and of winter survival in the field. This approach has now been applied to seven species of control agents, including *N. californicus* and *M. caliginosus*, revealing a strong correlation between the duration of survival in the laboratory at 5 °C and the maximum period of survival in winter (Figure 14.4; Bale, 2005; Hatherly *et al.*, 2005). This system enables a direct comparison of different physiological states within a species (e.g. adults and immature stages, acclimated and non-acclimated, and, where appropriate, diapause and non-diapause populations), and between species. As is evident from Fig. 14.4, species can be effectively categorized into three groups: (a) species with little or no cold-tolerance that typically die out after 2–4 weeks of winter field exposure, irrespective of the severity of the winter; (b) species with a greater level of cold-hardiness, but that would be normally insufficient to allow survival throughout a winter; (c) species with cold-tolerance attributes (possibly including diapause) comparable to native species. If the intention was to identify agents that would be "safe" to release into glasshouses such that any escaping individuals would rapidly succumb at temperatures of 5 °C and lower, then the selection could be made with confidence from category (a) and any species in category (c) would be unsuitable and "risky." However, if the purpose was to find a classical control agent capable of surviving in the climate of northern Europe, then the species would have to be in category (c), otherwise there would be no prospect of permanent establishment.

The environmental risk assessment summarized in Fig. 14.3 is intentionally hierarchical. The biocontrol companies supplying parasitoids, and predatory insects and mites in Europe are small- to medium-size enterprises with limited

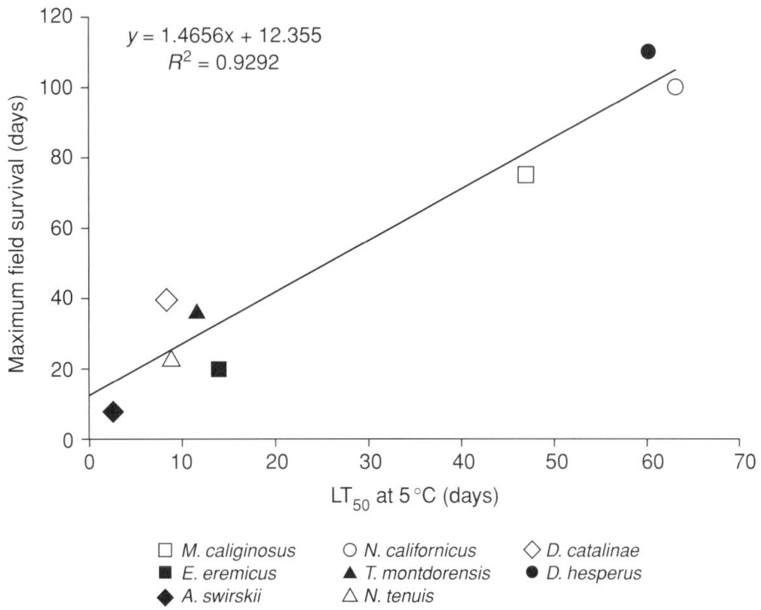

Figure 14.4 Relationship between $LTime_{50}$ at 5 °C (laboratory) and winter field survival of eight non-native glasshouse biocontrol agents (*Macrolophus caliginosus*, *Neoseiulus californicus*, *Delphastus catalinae*, *Eretmocerus eremicus*, *Typhlodromips montdorensis*, *Dicyphus hesperus*, *Amblyseius swirskii* and *Nesidiocoris tenuis*).

resources to invest in R&D activities. There is therefore an understanding between regulators, industry and environmental protection agencies on the need to develop a balanced regulatory system that minimizes costs for industry without compromising environmental safety. For example, if a candidate agent is intended for glasshouse use and has no potential to establish outdoors in the area of release, in those countries with a licensing system, the organism could be granted a permit; it could also be agreed with a regulatory authority that it would be unnecessary for the company to carry out any host range tests (see Fig. 14.3), as the failure to establish would prevent any non-target effects. However, if a species could establish outdoors after escaping from glasshouses (see Fig. 14.4), or if it was intended for "open-field" release in areas where climate would not impact on survival (e.g. in southern Europe), then an assessment of host range and possible non-target effects would be essential; and in some cases, some knowledge or study of dispersal ability may also become important.

There has been considerable recent interest in Europe in the environmental risk assessment of non-native biocontrol agents, involving a consolidation of guidelines produced by different organizations (FAO, OECD, EPPO), as summarized by Bigler *et al.* (2005) and refinement of research methods for ERA (Bigler *et al.*, 2006; van Lenteren *et al.*, 2006). This interest has been stimulated by a number of driving forces: firstly, as more pesticides have been withdrawn from the market and there is continued public resistance to the cultivation of GM crops, there has been a renaissance of interest in biocontrol; secondly, the lack of any uniform regulatory policy across Europe means that whilst some countries have strict regulatory policies on the import and release of non-native species (e.g. Switzerland), directly neighboring countries (e.g. France and Italy) have no restrictions; and thirdly, this lack of consistency has enabled some species to be released in countries without regulation, that have subsequently spread across large areas of Europe, of which the Harlequin ladybird, *Harmonia axyridis*, is the most infamous example.

Harmonia axyridis was first released in western Europe in field trials in France in 1990 as a predator of aphids, and then marketed commercially in 1995 (see Brown *et al.*, 2008 for review). For several years after release and commercialization there were no reports of any increase in numbers, spread or impacts on non-target species. In 2000 and 2001, the ladybird spread from France and Belgium across parts of northern Europe, and in 2003 or 2004 across the English Channel into the UK, from where it has spread westwards and northwards over the next four years, as well as into Scandinavia (Brown *et al.*, 2008). Interestingly, if the ERA presented in Fig. 14.3 had been applied to *H. axyridis* in 1990, and all EU countries had the same regulatory policy, it seems likely that no release licence would have been issued as the species "fails" all of the normally applied ERA tests: the ladybird can overwinter throughout Europe, it is polyphagous and can disperse over long distances.

Assessment of cold-hardiness therefore provides a clear guide to establishment potential, noting that in addition to overwintering ability, outdoor establishment of species released for glasshouse biocontrol cannot occur unless the agent can find suitable alternative prey (Hatherly *et al.*, 2009). Temperature also has a dominant influence on other processes that directly or indirectly can affect the success of introduced species as biocontrol agents. Many temperate insects and mites have developmental thresholds of around 5 °C or sometimes lower; even if an introduced species can survive in a "cooler climate" it is unlikely to be effective if its development threshold is markedly higher than the target pest – and it would be advantageous for the natural enemy to have a lower threshold and faster generation time than its prey or host. The developmental threshold of biocontrol agents of tropical or Mediterranean origin is often in the region of 10 °C; for example, 9.2 and 10.3 °C, respectively, for *Eretmocerus eremicus* and *Typhlodromips montdorensis* (Hatherly *et al.*, 2005). However, as many glasshouse pests are also of "warm climate origin," the developmental thresholds of the target and the control agent may well be similar. Developmental thresholds are estimated by weighted linear regression (Draper and Smith, 1981) of the rate of development at different temperatures, from which the day degree requirement per generation can be derived as the reciprocal of the slope (see Hart *et al.*, 2002a,b for details of *A. californicus* and *M. caliginosus*). Once the day degree requirement is known, this value can be compared to annual temperature records for any country or eco-region (taking the mean over a period of 10 years) as an indication of the possible voltinism of an introduced or escaped organism. If the temperature data is separated into nominal 6-month periods comprising autumn–winter and spring–summer, it is also possible to determine if the species can develop or complete a generation over the winter period or if dormancy or diapause is necessary for winter survival (see Tullett *et al.*, 2004). Whilst this type of analysis is important in the screening of classical and weed biocontrol agents (where establishment and maintenance of relatively high populations are vital for success), it is also a relevant part of the risk assessment of glasshouse agents, as it is impossible to prevent the escape of such species.

Temperature has a direct effect on one other critical process that affects the success of biocontrol – the relative mobility of the control agent compared with that of its prey or host. For all insects and mites there is a "critical minimum temperature" (CT_{min}) at which the organism becomes immobile and then at a lower temperature, ceases all "twitching" of body appendages (legs, antenna); there is considerable intraspecific variation in the temperature at which the first and last individuals become immobile (which may be related to a sensing of the decreasing temperature rather than a cold-induced immobility per se), but a much more consistent pattern in the temperatures at which the "final" movement of an appendage occurs, e.g. typically between 1 and −1 °C for *M. persicae*. Above the CT_{min}, activity increases with temperature up to the CT_{max}, when heat stupor or coma induces immobility. There is much less variation in the CT_{max} than CT_{min} (<1 °C for *M. persicae*), and the CT_{max} is usually within 1–2 °C of the upper lethal temperature (Hazell *et al.*, 2008). The temperature at which a predator is active relative to its prey, or at which a parasitoid can fly and disperse within a glasshouse or across a field will have an important influence on the number of prey that are encountered and killed, or hosts that are found. There are a range of techniques by which flight and activity thresholds can be assessed by direct observation of organisms placed within an environment which can be cooled at different rates (Powell and Bale, 2006), though the video-capture technique described by Hazell *et al.* (2008) has the advantages of not requiring the constant visual monitoring of behavior during each experiment combined with a permanent record for retrospective analysis.

14.7 Conclusions and future directions

It is clear that winter low temperatures can determine the subsequent spring and summer abundance of some major pests, such as the aphid *M. persicae* and the tree-defoliating moth, *E. autumnata*. Likewise, the distribution of some pests of world importance (e.g. *P. xylostella*) appear to be constrained by the inadequate cold-tolerance of overwintering stages such that damage in some regions is caused by transient populations derived from mass migration. With biological control, both the essential establishment of classical control agents and the undesirable establishment of species that escape from glasshouses, there is again strong evidence for a dominant influence of low temperature. In an era of global climate change, this raises the interesting question of "how much warmer" does it need to become to allow year-round survival of such pests and natural enemies beyond their current ranges. It seems reasonable to conclude that for some species, survival is currently "marginal" and relatively small increases in temperature might allow localized establishment and an expansion of current range margins; for other species, the difference between their cold-hardiness and the stress imposed by winter conditions may be too great for climate warming to modify this relationship, at least within the currently predicted increases in temperature.

On the basis of the species and case studies that have been highlighted in this chapter, there is evidence of considerable variation in the responses of pest insects to climate warming: an increase in mean annual temperature of 1 °C would allow an additional generation per year of the mirid, *S. rubrovittatus*, in Japan, but an increase of 5 °C would be required for an extra generation of the cabbage root fly, *D. radicum*, in the UK. Likewise, whereas increases in temperature over the last 40 years in the USA have promoted a range expansion of the southern pine beetle, *D. frontalis*, a similar response has not been observed in the cotton bollworm, *P. gossypiella*. The reasons for these differences are in some cases explainable by differences in the ecophysiology of the subject species: *D. frontalis* is more cold-hardy than *P. gossypiella* and therefore more able to exploit relatively small changes toward a more favorable climate.

Beyond the scientific exploration of the impacts of temperature and a changing climate on pests and their distributions and abundance, there are major socio-economic implications arising from these interactions, not least the potential threat to world food supplies. For this reason, various governments and international agencies have engaged with the scientific community to seek "reliable predictions" of future patterns. Even at its most basic level, such studies have identified some of the key requirements that must be in place for any prediction system to be robust – and these features have been clearly exemplified in this chapter: it is very difficult to predict the impacts of future climate scenarios without a detailed knowledge of a species' ecophysiology (cold-tolerance) and performance (development, reproduction) under prevailing climate – and for many species, including some major pests, this information is still lacking; and the rate of change in features such as timing of migrations and changes in distribution are only readily tractable in species for which there is a long history of monitoring, which again, are relatively few.

Overall, these examples confirm the views expressed by Cannon (1998) and Bale *et al.* (2002) that species-specific differences in factors such as life cycles (voltinism, diapause), host plants, feeding guild and natural-enemy responses make it difficult to reach a general consensus on future distributions and abundance in a changing climate, or to construct realistic models of these interactions in a wider context. The trends that are decipherable from the species, and studies considered in this chapter suggest the pest species will be generally favored by a warmer climate, increasing in abundance within the current range and extending into new areas; and natural control will not be sufficient to constrain such changes. With fewer pesticide options and concerns about the environmental consequences of continued GM cultivation, alternative pest management technologies are urgently required.

References

Babendreier D., Bigler F. and Kuhlmann U. (2005). Methods used to assess non-target effects of invertebrate biological control agents of arthropod pests. *BioControl* **50**, 821–870.

Bale, J. S., Harrington, R. and Clough, M. S. (1988). Low temperature mortality of the peach potato aphid *Myzus persicae*. *Ecological Entomology* **13**, 121–129.

Bale, J. S. (2005). Effects of temperature on the establishment of non-native biocontrol agents: the predictive power of laboratory data. *Second International Symposium on Biological Control of Arthropods (IBSCA)* Vol. II, 593–602.

Bale, J. S. and Walters K. F. A. (2001). Overwintering biology as a guide to the establishment potential of non-native arthropods in the UK. In *Environment and Animal Development: Genes, Life Histories and Plasticity*, ed. D. A. Atkinson and M. Thorndyke. Oxford, UK: Bios, pp. 343–354.

Bale, J. S., Masters, G. J, Hodkinson, I. D., Awmack, C., Bezemer, T. M., Brown, V. K., Butterfield, J. E. L., Buse, A., Coulson, J. C., Farrar, J., Good, J. E. G., Harrington, R., Hartley, S., Jones, T. H., Lindroth, R. L., Press, M. C., Symrnioudis, I., Watt, A. and Whittaker, J. B. (2002). Herbivory in global climate change research: direct effects of rising temperatures on insect herbivores. *Global Change Biology* **8**, 1–16.

Bale, J. S., van Lenteren, J. C. and Bigler, F. (2008) Biological control. In *Sustainable Agriculture, special issue of Philosophical Transactions of the Royal Society* **363**, 761–776.

Battisti, A., Stastny, M., Netherer, S., Robinet, C., Schopf, A., Roques, A. and Larsson, S. (2005). Expansion of geographic range in the pine processionary moth caused by increased winter temperatures. *Ecological Applications* **15**, 2084–2096.

Battisti, A., Stastny, M., Buffo, E. and Larsson, S. (2006). A rapid altitudinal range expansion in the pine processionary moth produced by the 2003 climatic anomaly. *Global Change Biology* **12**, 662–671.

Bentz, B. J. and Mullins, D. E. (1999). Ecology of mountain pine beetle (Coleoptera: Scolytidae) cold hardening in the intermountain West. *Environmental Entomology* **28**, 577–587.

Bentz, B. J., Logan, J. A. and Amman, G. D. (1991). Temperature-dependent development of the mountain pine beetle (Coleoptera: Scolytidae) and simulation of its phenology. *Canadian Entomologist* **123**, 1083–1094.

Bigler, F. (1986). Mass production of *Trichogramma maidis* Pint. Et Voeg. and its field application against *Ostrinia nubilalis* Hbn in Switzerland. *Journal of Applied Entomology* **101**, 23–29.

Bigler, F., Bale, J. S., Cock, M. J. W., Dreyer, H., Greatrex, R., Kuhlmann, U., Loomans, A. J. M. and van Lenteren, J. C. (2005). Guidelines on information requirements for import and release of invertebrate biological control agents in European countries. *Biocontrol News and Information* **26**, 115–123.

Bigler, F., Babendreier, D. and Kuhlmann, U. (eds.) (2006). *Environmental Impact of Invertebrates for Biological Control of Arthropods: Methods and Risk Assessment*, Wallingford, UK: CABI.

Brown, P. M. J., Adriaens, T., Bathon, H., Cuppen, J., Goldarazena, A., Hagg, T., Kenis, M., Klausnitzer, B. E. M., Kovar, I., Loomans, A. J. M., Majerus, M. E. N., Nedved, O., Pedersen, J., Rabitsch, W., Roy, H. E., Ternois, V., Zacharov, I. A. and Roy, D. B. (2008). *Harmonia axyridis* in Europe: spread and distribution of a non-native coccinellid. *BioControl* **53**, 5–21.

Buffo, E., Battisti, A., Stastny, M. and Larsson, S. (2007). Temperature as a predictor of survival of the pine processionary moth in the Italian Alps. *Agricultural and Forest Entomology* **9**, 65–72.

Cannon, R. J. C. (1998). The implications of predicted climate change for insect pests in the UK, with emphasis on non-indigenous species. *Global Change Biology* **4**, 785–796.

Carroll, A. L., Régnière, J., Logan, J. A., Taylor, S. W., Bentz, B. J. and Powell, J. A. (2006). Impacts of climate change on range expansion by mountain pine beetle. *Mountain Pine Beetle Initiative Working Paper* 2006–14 ISBN 0-662-44349-7 Cat. No. Fo143–3/2006–14E.

Chapman, J. W., Reynolds, D. R., Smith, A. D., Riley, J. R., Pedgley, D. E. and Woiwod, I. P. (2002). High-altitude migration of the diamondback moth *Plutella xylostella* to the UK: a study using radar, aerial netting and ground trapping. *Ecological Entomology* **27**, 641–650.

Chapman, J. W., Reynolds, D. R., Mouritsen, H., Hill, J. K., Riley, J. R., Sivell, D., Smith, A. D. and Woiwod, I. P. (2008). Wind selection and drift compensation optimize migratory pathways in a high-flying moth. *Current Biology* **18**, 514–518.

Clough, M. S., Bale, J. S. and Harrington, R. (1990). Differential cold hardiness in adults and nymphs of the peach-potato aphid *Myzus persicae*. *Annals of Applied Biology* **116**, 1–9.

Cole, W. A. (1981). Some risks and causes of mortality in mountain pine beetle populations: a long-term analysis. *Researches in Populations Ecology* **23**, 116–144.

Collier, R. H. and Finch, S. (1983a). Completion of diapause in field populations of the cabbage root fly (*Delia radicum*). *Entomologia Experimentalis et Applicata* **34**, 186–192.

Collier, R. H. and Finch, S. (1983b). Effects of intensity and duration of low temperatures in regulating diapause development of the cabbage root fly (*Delia radicum*). *Entomologia Experimentalis et Applicata* **34**, 193–200.

Collier, R. H. and Finch, S. (1986). Accumulated temperatures for predicting cabbage root fly, *Delia radicum* (L.), (Diptera: Anthomyiidae) emergence in the spring. *Bulletin of Entomological Research* **75**, 395–404.

Collier, R. H. and Finch, S. (1988). Thermal requirements for cabbage root fly, *Delia radicum*, development. In *Progress on Pest Management in Field Vegetables*, ed. R. Cavalloro, C. Pelerents and P. P. Rotondo – D. G. XIII – Luxembourg No. EUR 10514. Rotterdam: Balkema, pp. 21–26.

Collier, R. H., Finch, S. and Anderson, M. (1989). Laboratory studies on late emergence in the cabbage root fly (*Delia radicum*). *Entomologia Experimentalis et Applicata* **50**, 233–240.

Collier, R. H., Finch, S., Phelps, K. and Thompson, A. R. (1991). Possible impact of global warming on cabbage root fly (*Delia radicum*) activity in the UK. *Annals of Applied Biology* **118**, 261–271.

Coulson, S. J. (2000). A review of the terrestrial and freshwater invertebrate fauna of the High Arctic archipelago of Svalbard. *Norwegian Journal of Entomology* **47**, 41–63.

DeBach, P. and Rosen, D. (1991). *Biological Control by Natural Enemies*, 2nd edn., Cambridge: Cambridge University Press.

Draper, N. R. and Smith, H. (1981). *Applied Regression Analysis*. 2nd edn., New York: J. Wiley & Sons, Inc.

Finch, S. and Collier, R. H. (1985). Laboratory studies on aestivation in the cabbage root fly (*Delia radicum*). *Entomologia Experimentalis et Applicata* **38**, 137–143.

Foster, S. P., Harrington, R., Devonshire, A. L., Denholm, I., Devine, G. J., Kenward, M. G. and Bale, J. S. (1996). Overwintering success of insecticide-susceptible and resistant peach-potato aphids, *Myzus persicae* (Sulzer) (Hemiptera: Aphididae) *Bulletin of Entomological Research* **86**, 17–27.

Foster, S. P., Harrington, R., Devonshire, A. L., Denholm, I., Clark, S. J. and Mugglestone, M. A. (1997) Evidence for a possible fitness trade-off between insecticide resistance and low temperature movement that is essential for survival of UK populations of *Myzus persicae* (Hemiptera:Aphididae). *Bulletin of Entomological Research* **87**, 573–579.

Foster, S. P., Harrington, R., Dewar, A. M., Denholm, I. and Devonshire, A. L. (2002). Temporal and spatial dynamics of insecticide resistance in *Myzus persicae* (Sulzer). *Pest Management Science* **58**, 895–907.

Foster, S. P., Devine, G. J. and Devonshire, A. L. (2007). Insecticide resistance. In *Aphids as Crop Pests*, ed. H. F. van Emden and R. Harrington. Wallingford, UK: CABI Publishing.

Greathead, D. J. (1976). *A Review of Biological Control in Western and Southern Europe.* Technical Communication No. 7. Commonwealth Institute of Biological Control, Farnham Royal, Slough, UK, 182 pp.

Gurr, G. M., Wratten, S. D. and Barbosa, P. (2000) Success in conservation biological control of arthropods. In *Measures of Success in Biological Control*, ed. G. Gurr and S. Wratten. Dordrecht: Kluwer Academic Publishers, pp. 105–132.

Gutierrez, A. P., D'Oultremont, T., Ellis, C. K. and Ponti, L. (2006). Climatic limits of pink bollworm in Arizona and California: effects of climate warming. *Acta Oecologica* **30**, 353–364.

Gutierrez, A. P., Ponti, L., D'Oultremont, T. and Ellis, C. K. (2008). Climate change effects on poikilothermic tritrophic interactions. *Climatic Change* **87**, 167–192.

Harrington, R., Tatchell, G. M. and Bale, J. S. (1990). Weather, life cycle strategy and spring populations of aphids. *Acta Phytopathologica & Entomologica Hungarica* **25**, 423–432.

Harrington, R., Clark, S. J., Welham, S. J., Verrier, S. J., Denholm, C. H., Hullé, M., Maurice, D., Rounsevell, M. D. A., Cocu, N. and EU EXAMINE Consortium (2007). Environmental change and the phenology of European aphids. *Global Change Biology* **13**, 1550–1564.

Harrington, R., Dewar, A. M., and George, B. (1989). Forecasting the incidence of virus yellows in sugar beet in England. *Annals of Applied Biology* **114**, 459–469.

Hart, A. J., Bale, J. S., Tullett, A. G., Worland, M. R. and Walters, K. F. A. (2002a). Effects of temperature on the establishment potential of the predatory mite *Amblyseius californicus* McGregor (Acari: Phytoseiidae) in the UK. *Journal of Insect Physiology* **48**, 593–600.

Hart, A. J., Tullett, A. G. Bale, J. S. and Walters, K. F. A. (2002b). Effects of temperature on the establishment potential in the UK of the non-native glasshouse biocontrol agent *Macrolophus caliginosus*. *Physiological Entomology* **27**, 112–123.

Hatherly, I. S., Hart, A. J., Tullett, A. G. T. and Bale, J. S. (2005). Use of thermal data as a screen for the establishment potential of non-native biocontrol agents in the UK. *BioControl* **50**, 687–698.

Hatherly, I. S., Pedersen, B. P. and Bale, J. S. (2009). Effect of host plant, prey species and intergenerational changes on the prey preferences of the predatory mirid *Macrolophus caliginosus*. *BioControl* **54**, 35–45.

Hazell, S. P., Pedersen, B. P., Worland, M. R., Blackburn, T. M. and Bale, J. S. (2008). A method for the rapid measurement of thermal tolerance traits in studies of small insects. *Physiological Entomology* **33**, 389–394.

Heather, N. and Hallman, G. J. (2008). *Pest Management and Phytosanitary Trade Barriers.* Wallingford, UK: CABI.

Hill, J. K. and Gatehouse, A. G. (1993). Phenotypic plasticity and geographical variation in the pre-reproductive period of *Autographa gamma* (Lepidoptera: Noctuidae) and its implications for migration in this species. *Ecological Entomology* **18**, 39–46.

Hoch, G., Toffolo, E. P., Netherer, S., Battisti, A. and Schopf, A. (2009). Survival at low temperature of larvae of the pine processionary moth, *Thaumetopoea pityocampa* from an area of range expansion. *Agricultural and Forest Entomology* **11**, 313–320.

Hughes, R. D. (1960). Induction of diapause in *Erioischia brassicae* (Bouché) (Dipt., Anthomyiidae). *Journal of Experimental Biology* **37**, 218–223.

Hunt, E., Kuhlmann, U., Sheppard, A., Qin, T. K., Barratt, B. I. P., Harrison, L., Mason, P. G., Parker, D., Flanders, R. V. and Goolsby, J. (2008). Review of invertebrate biological control regulation in Australia, New Zealand, Canada and the USA: recommendations for a harmonized European system. *Journal of Applied Entomology* **132**, 89–123.

Jepsen, J. U., Hagen, S. B., Ims, R. A. and Yoccoz, N. G. (2008). Climate change and outbreaks of the geometrids *Operophtera brumata* and *Epirrita autumnata* in subarctic birch forest: evidence of a recent outbreak range expansion. *Journal of Animal Ecology* **77**, 257–264.

Jolly, N. (2000). The predatory mite *Neoseiulus californicus*: its potential as a biological control agent for the fruit tree red spider mite, *Panonychus ulmi*. *BCPC conference at Brighton, Pests and Diseases*, 487–490.

Kiritani, K. (2006). Predicting impacts of global warming on population dynamics and distribution of arthropods in Japan. *Population Ecology* **48**, 5–12.

Kiritani, K. (2007). The impact of global warming and land-use change on the pest status of rice and fruit bugs (Heteroptera) in Japan. *Global Change Biology* **13**, 1586–1595.

Lapointe, S. L., Borchert, D. M. and Hall, D. G. (2007). Effect of low temperatures on mortality and oviposition in conjunction with climate mapping to predict spread of the root weevil *Diaprepes abreviatus* and introduced natural enemies. *Environmental Entomology* **36**, 73–82.

Lawton, J. H. (1995). The response of insects to environmental change. In *Insects in a Changing Environment*, ed. R. Harrington and N. E. Stork. New York: Academic Press, pp. 3–26.

Lenteren, J. C. van and Bueno, V. H. B. P. (2003). Augmentative biological control of arthropods in Latin America. *BioControl* **48**, 123–139.

Lenteren, J. C. van and Woets, J. (1988). Biological and integrated pest control in greenhouses. *Annual Review of Entomology* **33**, 239–269.

Lenteren, J. C. van, Bale J. S., Bigler F., Hokkanen H. M. T., and Loomans A. J. M. (2006). Assessing risks of releasing exotic biological control agents of arthropod pests. *Annual Review of Entomology* **51**, 609–634.

Logan, J. A. and Bentz, B. J. (1999). Model analysis of mountain pine beetle (Coleoptera: Scolytidae) seasonality. *Environmental Entomology* **28**, 924–934.

Musolin, D. H. (2007). Insects in a warmer world: ecological, physiological and life history responses of true bugs (Heteroptera) to climate change. *Global Change Biology* **13**, 1565–1585.

Musolin, D. H. and Numata, H. (2003a). Photoperiodic and temperature control of diapause induction and colour change in the southern green stink bug *Nezara viridula*. *Physiological Entomology* **28**, 65–74.

Musolin, D. H. and Numata, H. (2003b). Timing of diapause induction and its life history consequences in *Nezara viridula*: is it costly to expand the distribution range? *Ecological Entomology* **28**, 694–703.

Nilssen, A. and Tenow, O. (1990). Diapause, embryo growth and supercooling capacity of *Epirrita autumnata* eggs from Northern Fennoscandia. *Entomologia Experimentalis et Applicata* **57**, 39–55.

Parmesan, C., Ryrholm, N., Stefanescu, C., Hill, J. K., Thomas, C. D., Descimon, H., Huntley, B., Kaila, L., Kullberg, J., Tammaru, T., Tennent, W. J., Thomas, J. A. and Warren, M. (1999). Polewards shifts in geographical ranges of butterfly species associated with regional warming. *Nature* **399**, 579–583.

Phelps, K., Collier, R. H., Reader, R. J. and Finch, S. (1993). Monte Carlo simulation method for forecasting the timing of pest insect attacks. *Crop Protection* **12**, 335–342.

Powell, J. A., Jenkins, J. L., Logan, J. A. and Bentz, B. J. (2000). Seasonal temperature alone can synchronize life cycles. *Bulletin of Mathematical Biology* **62**, 977–998.

Powell, S. J. and Bale, J. S. (2004). Cold shock injury and ecological costs of rapid cold hardening in the grain aphid *Sitobion avenae* (Hemiptera: Aphididae). *Journal of Insect Physiology* **50**, 277–284.

Powell, S. J and Bale, J. S. (2005) Low temperature acclimated populations of the grain aphid *Sitobion avenae* retain ability to rapidly cold harden with enhanced fitness. *Journal of Experimental Biology* **208**, 2615–2620.

Powell, S. J and Bale, J. S. (2006) Effect of long term and rapid cold hardening on the cold torpor temperature of an aphid. *Physiological Entomology* **31**, 348–352.

Powell, S. J. and Bale, J. S. (2008). Intergenerational acclimation in aphid overwintering. *Ecological Entomology* **33**, 95–100.

Qi, A., Dewar, A. M. and Harrington, R. (2004). Decision making in controlling virus yellows of sugar beet in the UK. *Pesticide Management Science* **60**, 727–732.

Régnière, J. and Bentz, B. J. (2007). Modeling cold tolerance in the mountain pine beetle, *Dendroctonus ponderosae*. *Journal of Insect Physiology* **53**, 559–572.

Safranyik, L. (1978). Effect of climate and weather on mountain pine beetle populations. In *Theory and Practice of Mountain Pine Beetle Management in Lodgepole Pine Forests*, ed. D. L. Kibbee, A. A. Berryman, G. D. Amman, and R. W. Stark. Conference held at Washington State University, Pullman WA in 1978. Forest, Wildlife and Range Experiment Station, University of Idaho, Moscow ID. 224 pp.

Strathdee, A. T., Howling, G. G. and Bale, J. S. (1995). Cold hardiness of aphid eggs. *Journal of Insect Physiology* **41**, 653–657.

Talekar, N. S. and Shelton, A. M. (1993). Biology, ecology and management of the diamondback moth. *Annual Review of Entomology* **38**, 275–301.

Tenow, O. and Nilssen, A. (1990). Egg cold hardiness and topographical limitations to outbreaks of *Epirrita autumnata* in Northern Fennoscandia. *Journal of Applied Ecology* **27**, 723–734.

Thomas, C. D., Bodsworth, E. J., Wilson, R. J., Simmons, A. D., Davies, Z. G., Musche, M. and Conradt, L. (2001). Ecological and evolutionary processes at expanding range margins. *Nature* **411**, 577–581.

Tran, J. K., Ylioja, T., Billings, R. F., Régnière, J. and Ayres, M. P. (2007). Impact of minimum winter temperatures on the population dynamics of *Dendroctonus frontalis*. *Ecological Applications* **17**, 882–899.

Tullett, A. G. T., Hart, A. J., Worland, M. R. and Bale, J. S. (2004). Assessing the effects of low temperature on the establishment potential in Britain of the non-native biological control agent *Eretmocerus eremicus*. *Physiological Entomology* **29**, 363–371.

Ungerer, M. J., Ayres, M. P. and Lombardero, M. J. (1999). Climate and the northern distribution limits of *Dendroctonus frontalis* Zimmermann (Coleoptera: Scolytidae). *Journal of Biogeography* **26**, 1133–1145.

Wäckers F. L. (2003). The parasitoids' need for sweets: sugars in mass rearing and biological control. In *Quality Control and Production of Biological Control Agents: Theory and Testing Procedures*, ed. J. C. van Lenteren. Wallingford, UK: CABI, pp. 59–72.

Werker, A. R., Dewar, A. M., and Harrington, R. (1998). Modelling the incidence of virus yellows in sugar beet in the UK in relation to numbers of migrating *Myzus persicae*. *Journal of Applied Ecology* **35**, 811–818.

Index